BIOCHEMISTRY RESEARCH TRENDS

INDOLEAMINES

SOURCES, ROLE IN BIOLOGICAL PROCESSES AND HEALTH EFFECTS

BIOCHEMISTRY RESEARCH TRENDS

Additional books in this series can be found on Nova's website
under the Series tab.

Additional e-books in this series can be found on Nova's website
under the e-book tab.

BIOCHEMISTRY RESEARCH TRENDS

INDOLEAMINES

SOURCES, ROLE IN BIOLOGICAL PROCESSES AND HEALTH EFFECTS

ÁNGEL CATALÁ
EDITOR

nova publishers
New York

Library of Congress Cataloging-in-Publication Data

ISBN: 978-1-63482-097-4

Library of Congress Control Number: 2015931169

Published by Nova Science Publishers, Inc. † New York

CONTENTS

PREFACE

This book presents an overview of the role in biological processes and health effects produced by indoleamines, which are ubiquitous and versatile molecules with important physiological functions. The topics analyzed in this book cover a broad spectrum of functions played by indoleamines including the role of indoleamines in reducing free radical damage and oxidative stress; melatonin; endogenous sources and role in the regulation of fish reproduction; the neuroimmune roles of indoleamine pathways in cancer; tryptophan damage signal and indoleamine 2, 3-dioxygenase; the health effects of indoleamines on cognition, in the prevention of brain aging and age-related diseases; melatonin as an ocular anti-inflammatory; the relations between water-soluble B vitamins and indoleamines; melatonin in neurodegenerative, psychiatric and systemic inflammatory disorders; tryptophan, indolamines and sleep; from human milk to tryptophan-enriched diets; function of indoleamines in biological processes with emphasis on lipid peroxidation and indoleamines in edible plants; and the role in human health effects.

Chapter 1 – This chapter presents an overview of the chemical processes involved in the protection against oxidative stress exerted by indoleamines, which are ubiquitous and versatile molecules with important physiological functions. Eight representative indoleamines were chosen to illustrate the various aspects of their protecting effects. They are melatonin, 6-hydroxymelatonin, tryptophan, 5-hydroxytriptophan, serotonin, N-acetylserotonin, tryptamine, and 5-methoxytryptamine. The possible reaction mechanisms for the free radical scavenging activity of these compounds are analyzed, as well the kinetic data gathered so far. Both experimental and theoretical information are taken into account. In addition, the role of free radicals in oxidative damage, as well as the chemical reactivity of different free radicals of biological importance, is discussed. The features desirable for an ideal antioxidant are presented, paying particular attention to the characteristics of indoleamines related to this behaviour. It is concluded that indoleamines are capable of reducing oxidative stress by several pathways including, at least, free radical scavenging and metal chelating processes. This family of compounds fulfills most of the features necessary to be considered ideal protectors against oxidative stress. The literature has been thoroughly reviewed, and the gathered data is expected to contribute to a deeper understanding on the important protective role exerted by indoleamine against oxidative processes in living organisms.

Chapter 2 – Melatonin (*N*-acetyl-5-methoxytryptamine), since first isolation, purification and chemical characterization from the extracts of bovine pineal gland, has been detected in several extra-pineal organs, more specifically in retina, Harderian gland, gut, and ovary in vertebrates as well as in a wide variety of organisms, ranging from invertebrates to various plants and even in microorganisms. This tiny, but versatile, indoleamine hormone is synthesized from amino acid tryptophan following four enzymatic reactions. The regulatory mechanisms of melatonin synthesis in the pineal and retina are synchronized with the environmental light-dark cycle, with a nocturnal peak and nadir at midday, irrespective of the habit and habitat of studied animals. Conversely, so far no significant influences of environmental light-dark conditions have been evident in the regulation of melatonin synthesis in any extra-pineal or extra-retinal tissues. Hence, the daily and seasonally changing melatonin rhythm patterns in serum are depicted as the result of varying melatonin synthesis in the pineal and/or retina, which are involved in transmission of information on 'time of day' and 'time of year' and thereby serve as a biological clock and/or a biological calendar. Past few decades have witnessed an enormous progress in understanding the sources and mechanism of melatonin actions in the regulation of diverse physiological functions especially seasonal reproduction in a wide range of vertebrates including fish. However, relative contribution of different melatonin producing tissues to circulating profiles of melatonin and their temporal relationship with reproductive events in an annual cycle in any vertebrates remained speculative. Accordingly, findings on a carp *Catla catla* appeared important, as this is the single fish species that has been used for collection of data on the daily and seasonal profiles of melatonin in the pineal, retina, gut and serum as well as the density of arylalkylamine *N*-acetyltransferase (AANAT) protein, the key regulator of melatoninergic pathway, in the pineal, retina and gut under natural and diverse experimental conditions as well. Notably, daily concentrations of melatonin and the band intensity of a ~23kDa AANAT protein in the pineal and serum titers of melatonin showed parallel changes in a daily cycle with a peak in late dark-phase (during the preparatory phase) or at midnight (in the remaining parts of annual reproductive cycle). On the other hand, levels of melatonin and AANAT protein in the gut, irrespective of sampling months, exhibited a peak at midday, and thereby depicted a negative correlation between the melatoninergic system in gut and pineal in the same species. Likewise, the influences of altered photoperiods on the activity of both the testis and the ovary as well as the levels of melatonin and the density of AANAT protein in the pineal varied in relation to the reproductive seasons, though daily rhythmic profiles of gut melatonin remained identical irrespective of the reproductive status of the fish and the light-dark regimens to which the carp were held. The obvious outcome of such study is that both the pineal and gut appeared as major sites of melatonin synthesis, but their environmental as well as physiological synchronizer(s) might not be identical. Nonetheless, convincing data gathered from a number of carefully controlled studies employing exogenous administration of melatonin to several fish species, especially carp, implicated melatonin as a potent candidate in the regulation of seasonal reproduction in fish. Remarkably, an identical regimen of exogenous melatonin treatment led to stimulation or inhibition or no influences at all on the functions of gonads depending on the reproductive status of concerned fish. Cross-talk between the ovarian steroids and melatonin has also been demonstrated by an *in vitro* study, in which melatonin accelerated the action of 17α, 20β-dihydroxy-4-pregnen-3-one or maturation inducing hormone (MIH) on meiotic cell cycle resumption in carp oocytes by formation of maturation promoting factor (MPF) - a complex of two proteins, cyclin B and cyclin dependant kinase (Cdk1). While several lines of evidence suggest melatonin actions on the hypothalamo-hypophyseal-gonadal axis, localization and dynamics of a 37-kDa melatonin receptor protein in carp oocytes argued in favor of extra-hypothalamic action of melatonin on fish reproduction. The carp study also indicated that response of ovary to photoperiods might be related to varied profiles of serum melatonin, but not to any rhythm parameters of MT1 or MT2 receptors. Very recently, melatonin has been detected and measured in the carp ovary. Though it remained unclear whether measured melatonin is synthesized within the ovary itself or is derived from the circulation, ovarian melatonin concentrations displayed significant seasonal variations with a peak at spawning and nadir during post-spawning, and thereby underlined its proximity to the course of ovarian development in carp. Moreover, search for a correlation between melatonin concentrations, oxidative status and oocyte dynamics in the carp ovary led to emergence of an idea that ovarian melatonin may be associated with an enhanced activity/level of antioxidative agents for reduction in oxidative stress to augment ovarian functions during spawning. This article summarizing available information on the endogenous sources of melatonin in fish deals with external factor(s) that may act as synchronizer of rhythmic synthesis of melatonin in different organs and finally discusses the modalities by which this indoleamine hormone impacts on physiological regulations of fish reproduction mainly focusing the recent findings on carp.

Chapter 3 – A Latin aphorism "*Mens sana in corpore sano*" implies that the correlation between mental and physical health has been known for thousands of years. This phenomenon has been observed in several diseases, such as cardiovascular diseases, autoimmune diseases, and cancer. Currently, cancer kills more than six million people each year and has become a heavily economic burden caused by the treatments. Miovic and Block reported that 5-26% of advanced cancer patients have major depression, and are prescribed with antidepressants. A new medical field named "psycho-oncology" is launched based on these clinical observations. In the field of psyco-oncology discusses both the effects of cancer on a person's mental health as well as the behavioral factors that may affect the progress of cancer.

Serotonin is a major neurotransmitter contributing to the feeling of happiness, and there is a strong correlation between the serotonin level and depression. It has been shown that the serotonin level of cancer patients is significantly lower than that of healthy subjects. Serotonin is synthesized from tryptophan which is one of the essential amino acids for mammals. Dysregulation of tryptophan metabolic pathways could be responsible for developing depression symptoms in cancer patients. Tryptophan is metabolized through the kynurenine pathway by up-regulated indoleamine 2,3-dioxygenase (IDO) in cancers instead of the serotonin pathway used in neurons, resulted in the decrease of serotonin production. In addition, the intermediates and final product of the kynurenine pathway, such as 3-hydroxykynurenine (OHK) and quinolinic acid (QUIN), are potent neurotoxins. Increased levels of these neurotoxins may contribute to the apoptosis of astrocytes and other neurons, affect various networks in the brain, and result in depression symptoms or other mental disorders. The up-regulated kynurenine pathway in cancer patients not only strikes the survival of neurons and regulations of several networks in the brain, but impacts the antitumor immunity of T cells. The shortage of tryptophan inhibits the proliferations of CD8+ T cells and CD4+ T-helper cells. In addition, kynurenine could cause these T cells to apoptosis. It also has been reported that numbers and activities of T cells are highly correlated with the prognosis and survival of patients. Other than decreasing the numbers and functions of T cells, kynurenine also turns tumor microenvironment toward more immunosuppressive.

Tryptophan is converted into 5-hydroxyl-L-tryptophan (5-HTP) by tryptophan hydroxylase (TPH) and its cofactor tetrahydrobiopterin (BH4) under the normal physical condition. However, the BH4 is also the cofactor of nitric oxide synthase (NOS) which could be induced by proinflammatory cytokines, such as interleukin-6 (IL-6), in the tumor microenvironment. Inducible nitric oxide synthase (iNOS) competes BH4 with TPH and causes the lower production of serotonin and other neurotransmitters, both worsen the depression symptoms in cancer patients. On the other hand, BH4 helps iNOS to convert arginine to nitric oxide (NO) that is further metabolized into reactive nitrogen species (RNS), such as peroxynitrite. These RNS cause post-translational modifications of proteins, and the RNS-modified cytokines would affect the recruitments of immune cells and antitumor immunity.

In overall, the metabolism of tryptophan significantly influences the synthesis of serotonin and generation of kynurenine, and is closely correlated to the cancer-induced depression and immunosuppression. Hence, the roles of indoleamines, such as tryptophan and serotonin, and their related pathways will be discussed in the following. Moreover, how these molecules modulate the immunity and mood in cancer patients will also be covered.

Chapter 4 – Tryptophan (TRP) has been linked to the immune response mainly through the immunosuppressive activities of indoleamine 2, 3-dioxygenase (IDO). Here the authors show evidence that TRP acts as a biochemical signal from damaged cells that attracts and enables immune responses. TRP levels above 2-3μmol/L are essential for T-Lymphocyte proliferation, but it has never been previously established whether normal physiological levels of TRP are sufficiently low to limit T-Lymphocyte proliferation. Although blood plasma TRP in humans varies from 30 to 65μmol/L, most of this TRP is bound to protein so free-TRP levels are much lower. Interstitial fluid levels of drugs and metabolites are generally regarded as being similar to plasma free-levels. Human cerebrospinal fluid (CSF) TRP levels vary from 0.9 to 3.3μmol/L in non-inflammatory conditions, but albumin is also present at levels of 1.5 to 5.2μmol/L and able to bind TRP in a 1 : 1 ratio. Using ultrafiltration, the authors measured free TRP levels in CSF from 10 non-inflammatory patients and found these were 22% lower than total TRP in these same patients. This indicated a normal reference interval of 0.5 to 2.9μmol/L for CSF free-TRP which is either below or bordering upon the threshold for T-Lymphocyte proliferation previously determined in tissue culture medium. The authors also measured free TRP in 4 patients with central nervous system inflammatory disease and found free-TRP levels of 0.5+/-0.3 which were well below the threshold.

Activated T-lymphocytes in inflammatory conditions release γ-interferon, which induces IDO in a wide range of different cells including epithelial cells, fibroblasts, dendritic cells, neurones, astrocytes, microglia, trophoblasts and macrophages. IDO is able to lower TRP levels below the threshold for T-lymphocyte proliferation, thus acting as a TRP noise suppression system and feedback control mechanism. Here the authors show that neutrophils exhibit strong TRP chemotaxis, migrating towards higher TRP levels, which raises the possibility that other myeloid and lymphoid cells might also migrate in a TRP gradient. They observed that freeze-thaw damaged liver tissue slices or fresh slices incubated with *Staphylococcus aureus* or *Escherichia coli* produce very large quantities of TRP along with other amino acids. The authors propose that the TRP signal arises from damaged tissue, raising TRP levels well above the threshold for T-lymphocyte proliferation, that IDO dampens signal noise by removing excess TRP, highlighting inflammatory foci, maintaining a TRP gradient for inward chemotaxis of neutrophils and possibly other immune cells.

Chapter 5 – New treatments for several diseases of young and middle age have increased life expectancy significantly, but as a consequence a variety of diseases are emerging, especially disorders associated with brain dysfunction. Even in the absence of brain disorders, aging is associated with cognitive deficits, such as memory and learning impairment as well as motor disabilities that compromise the quality of life. It is generally accepted that damage resulting from reactive oxygen species, or oxidative stress, plays a role in the aging process, and that the brain is especially vulnerable to its effects. In this context, a main objective of investigation is the search for treatments or strategies that delay or reduce the effects of brain aging, i.e., the neuroprotective effects of antioxidants.

The indoleamines serotonin and melatonin are biosynthesized from the essential amino acid tryptophan. The rate-limiting enzyme in the synthesis of serotonin, tryptophan hydroxylase (TPH) determines the effective concentration of serotonin *in vivo*. At night, serotonin is used as a substrate for melatonin synthesis. Thus, there is a balance between serotonin's use as a neurotransmitter and its availability as precursor for melatonin synthesis, which is deeply dependent on the environmental light. Serotonin is one of the major neurotransmitters influencing hippocampal neurotransmission, which is involved in many functions throughout the brain such as learning and memory; while melatonin, besides its known role in the circadian control, has an important role in the aging process since it acts as a powerful and direct toxic free radicals and reactive intermediates scavenger. The synthesis of serotonin and melatonin decreases with age, partly as a consequence of reduced plasma tryptophan by its degradation to kynurenic derivatives, and this may explain some physiological alterations during brain aging. In addition, TPH has been shown to be susceptible to oxidative damage by reactive oxygen species, and consequently its reduced activity could explain the decline of indoleamines in aged brain. Some studies suggest that treatments aimed at modulating the activity of TPH enzyme, which could be accompanied by restorative effect of the cognitive functions, were worsened in aging. In this manner, the synthesis of serotonin, and consequently that of melatonin can be modulated with potent antioxidants, and thus the age-related decrease of serotonin and melatonin could be prevented, and perhaps some complaints of aging could also be delayed. In summary, dietary intake of antioxidants could be used as a therapy to mitigate the cognitive decline associate with aging and it might aid to delay some aging-related diseases emerging as the greatest health threats of the Twenty First century.

Chapter 6– Uveitis is a prevalent intraocular inflammatory disease, and one of the most damaging ocular conditions. Currently available therapies are relatively effective in the treatment of uveitis, but are often associated with severe side effects. In that context, uveitis remains a challenging field to ophthalmologists and represents a significant public health concern. Oxidative stress-induced inflammation is a major contributor to uveitis. In addition, altered nitric oxide (NO) metabolism, and increased prostaglandins (PGs) and tumor necrosis factorα (TNFα) levels also play a significant role in the pathogenesis of uveitis. Therefore, an antioxidant and anti-nitridergic therapy, as well as reducing PG and TNFα levels may be useful strategies to diminish tissue damage and ocular dysfunction induced by uveitis. Several lines of evidence strongly support that melatonin is a potent antioxidant molecule which is effective in scavenging free radicals that are generated in ocular tissues, and that it is able to reduce the retinal nitridergic pathway activity along with PG and TNFα levels. In view of the fact that melatonin lacks significant adverse collateral effects even at high doses, the application of melatonin could potentially protect ocular tissues by effectively scavenging free radicals, and reducing excessive amounts of NO, PGs, and TNFα generated in the uveitic eye. This chapter summarizes recent investigations showing that melatonin not only prevents clinical, biochemical, histological, ultrastructural, and functional consequences of experimental uveitis, but it is also capable of suppressing the actively ongoing ocular inflammatory response. These results strongly support the role of melatonin as a therapeutic strategy for uveitis treatment.

Chapter 7 – B vitamin concentrations affect 5-hydroxytryptamine (5HT). 5HT neurons degenerate in the Wernicke-Korsakoff syndrome marked by thiamine deficiency. Likewise, pyrithiamine-induced thiamine-deficiency in mice reduced 5HT-immunoreactive cell numbers and tryptophan hydroxylase activity in medial and dorsal raphe nuclei along with their target structures. 5HT is also affected by folate. Indeed, a folate-deficient diet decreased brain 5HT levels in rats, probably mediated by S-adenosyl-methionine (SAM). Folate deficiency is prevalent in patients with major depression, an effect likely involving 5HT and SAM. As with folate, vitamin B_{12} deficiency was associated with low CSF 5HIAA levels, reflecting brain 5HT, an effect of possible relevance in major depression as well. Vitamin B_6-deficient rats had lower brain 5HT concentrations, probably because pyridoxal 5-phosphate acts as a cofactor for the 5HT-synthesizing enzyme, amino acid decarboxylase. The vitamin may be used in treating amino acid decarboxylase deficiency, an inborn error of metabolism leading to 5HT deficiency.

Chapter 8 – The release of melatonin from the pineal gland has a long-standing association with the regulation of the circadian rhythm and, via this, with a plethora of medical conditions. Alterations in the melatonergic pathways are associated with a) neurodegenerative conditions, including: Alzheimer's disease, multiple sclerosis and Parkinson's disease; b) psychiatric conditions, including: bipolar disorder, depression and schizophrenia; and c) a wide array of other medical conditions, including obesity, cancers and migraines. Recent work has suggested that melatonin effects are not simply mediated via circadian regulation, but are also driven by variations in local mitochondrial melatonin synthesis by a wide array of cells, including central glia and systemic immune cells. In glia and immune cells, melatonin decreases reactivity and increases phagocytosis, suggesting that melatonin has significant impacts on such a wide array of medical conditions via the modulation of systemic and central immune system responses. As such, the melatonergic pathways are important in the etiology, course and treatment of a host of medical conditions. In this chapter, the authors look at the role of the melatonergic pathways in neurodegenerative, psychiatric and wider medical conditions, emphasizing its interactions with oxidative and nitrosative stress, immune inflammation and the tryptophan catabolite pathway.

Chapter 9 – Chrononutrition is the field of Chronobiology that establishes the principle of consuming foodstuffs at times of the day when they are more useful for health, improving, therefore, biological rhythms and physical performance. Tryptophan is an essential amino acid which has to be taken from diet and its ingestion has been widely studied by the authors' research group. Previous studies carried out in laboratory animals reported that after tryptophan ingestion, serotonin and melatonin levels increase in plasma. The next step was to study tryptophan oscillations in human milk and its possible effects on sleep/wake circadian rhythm, urinary indolamine metabolites in newborns and its possible role as sleep/wake synchronizer. Taking into account tryptophan circadian rhythm in breastfeeding milk and its association with sleep conditions, the authors' research was focused in creating dissociated formula diets and to enhance both indolamines levels and sleep conditions in other ages (children with sleep problems and elderly people) and children with neurological disorders. These different studies reported similar conclusions: enhancing indolamine levels through tryptophan supplements or tryptophan-enriched diets sleep conditions are able to improve both in animals and humans.

Chapter 10 – The three main actors involved in the panorama described are: indoleamines, lipid peroxidation and retina. First, a brief introduction about the history of these players is presented.

Chapter 11 – Melatonin (MEL) and Serotonin (SER), also known as [5-Hydroxytryptamine (5-HT)] are reported to be in a range of plant types which are edible. Their occurrence in plants species appears to be ubiquitous. Their presence in high quantities in plants assumes significance owing to their physiological effects upon consumption by human beings. MEL is a well-know animal hormone mainly released by the pineal gland known to influence circadian rhythm, sleep, apart from immune enhancement. Similarly, SER is a well-known neurotransmitter that regulates mood, sleep and anxiety in mammals. It is implicated in memory, behavioral changes, scavenging reactive oxygen species, antipsychotic, etc. Similarly Role of SER and MEL in plant morphogenesis, and various physiological processes through intense research is beginning to unfold. These molecules are in common foods viz banana, pineapple, plum, nuts, milk, grape wine. N- Feruloyl serotonin and p-coumaroyl serotonin found in certain seeds are found to possess antioxidant, anti-inflammatory, antitumor, antibacterial and anti-stress potential apart from reducing depression and anxiety. MEL is found in Mediterranean diets, nuts, cherries, tomato berries, and olive products. Consumption of foods rich in MEL is known to increase blood MEL levels which has been implicated in protective effect against cardiovascular damage, cancer initiation and growth. MEL is also found in wines, green tea, beer, olive oil etc. Moreover, presence of SER and MEL in Coffee beans (green and roasted beans) and decoction has been reported. In this chapter occurrence of indoleamines in edible plants and their implications in human health are presented.

In: Indoleamines
Editor: Ángel Catalá

ISBN: 978-1-63482-097-4

Chapter 1

THE ROLE OF INDOLEAMINES IN REDUCING FREE RADICAL DAMAGE AND OXIDATIVE STRESS: A PHYSICOCHEMICAL PERSPECTIVE

A. Galano*

Departamento de Química. Universidad Autónoma Metropolitana-Iztapalapa,
Iztapalapa, C. P. México D. F. México

ABSTRACT

This chapter presents an overview of the chemical processes involved in the protection against oxidative stress exerted by indoleamines, which are ubiquitous and versatile molecules with important physiological functions. Eight representative indoleamines were chosen to illustrate the various aspects of their protecting effects. They are melatonin, 6-hydroxymelatonin, tryptophan, 5-hydroxytriptophan, serotonin, N-acetylserotonin, tryptamine, and 5-methoxytryptamine. The possible reaction mechanisms for the free radical scavenging activity of these compounds are analyzed, as well the kinetic data gathered so far. Both experimental and theoretical information are taken into account. In addition, the role of free radicals in oxidative damage, as well as the chemical reactivity of different free radicals of biological importance, is discussed. The features desirable for an ideal antioxidant are presented, paying particular attention to the characteristics of indoleamines related to this behaviour. It is concluded that indoleamines are capable of reducing oxidative stress by several pathways including, at least, free radical scavenging and metal chelating processes. This family of compounds fulfills most of the features necessary to be considered ideal protectors against oxidative stress. The literature has been thoroughly reviewed, and the gathered data is expected to contribute to a deeper understanding on the important protective role exerted by indoleamine against oxidative processes in living organisms.

* Corresponding author: agalano@prodigy.net.mx.

I. INTRODUCTION

All matter, including living organisms, is constituted by molecules. This might seem a trivial statement nowadays. However, it has important implications in our understanding of the processes that take place around, and inside, us. Moreover, comprehending chemical reactivity at a molecular level has become an important aspect in the design of pharmacological strategies to decrease, or even prevent, unwanted phenomena with dangerous consequences to our health. As our knowledge in science goes deeper, the necessity of multidisciplinary approaches in research increases. Currently it is common to find investigations where biological and chemical perspectives are successfully combined. Another example is the growing number of joint efforts between experimental and theoretical protocols aiming to solve a particular problem.

Oxidative stress is among the events, concerning all living organisms that has attracted an increasing deal of attention in the last decades. This is fully justified taking into account its dangerous consequences to our health. Therefore, oxidative stress has become the focus of numerous investigations, combined or otherwise, aiming to understand its mechanism of action and to find efficient strategies to decrease its deleterious effects. Moreover, words like "antioxidants", "free radicals", and "oxidative stress" itself are no longer exclusive to the scientific community. On the contrary, it is rather common to hear them in our daily life.

In this chapter, oxidative stress is revised from a physicochemical point of view, devoting particular attention to the chemicals species involved in its development and inhibition. Logically, we are especially interested in the latter, since that is the key to keep a healthy oxidative status. In addition to the enzymatic defense mechanisms within living organisms, molecular defense can be also exerted by a wide variety of chemical species. They may be acquired from endogenous or exogenous sources, and can play single or multiple functions. Indoleamines are particularly interesting species in this regard. Most of them are endogenously produced (although their concentrations can be increased by dietary supplements), and versatile molecules. In fact, they are mainly known for other functions, i.e., non-related to oxidative stress. However, there is increasing evidence on their role in reduced oxidative damage. This chapter is focused just in this aspect.

II. FREE RADICALS AND OXIDATIVE STRESS

Most of the radicals found *in vivo* are reactive oxygen species (ROS) or reactive nitrogen species (RNS). ROS are oxygen-centered free radicals, including the superoxide radical anion ($O_2^{\bullet-}$), hydroxyl ($^{\bullet}OH$), alkoxyl ($RO^{\bullet}$), peroxyl ($ROO^{\bullet}$) and hydroperoxyl ($HOO^{\bullet}$). RNS, on the other hand, are nitrogen-based radicals including peroxynitrite (ONOO−), nitric oxide ($NO^{\bullet}$) and nitrogen dioxide ($NO_2^{\bullet}$). As is the case for almost anything, ROS and RNS can be harmful or beneficial to living systems, depending on their concentration. In healthy organisms there is a fragile balance between the production and the removal of free radicals, which guarantees that they remain in low/moderate concentrations. Under such conditions free radicals have beneficial effects. They are necessary for the maturation process of cellular structures, and for inducing mitogenic responses. Free radicals are also involved in the defense system, participate in cellular signaling systems, and seem to play a role in the regulation of insulin receptor kinase activity and in the apoptosis of defective cells. That is to say, free radicals at low/moderate levels are essential to optimal human health.

On the contrary, at high concentrations free radicals can be harmful to living organisms. Such high concentrations are caused by an imbalance between the production and the consumption of free radicals, which is commonly referred to as oxidative stress. More than 50 years ago Gerschman and coworkers [1] discovered that free radicals are the toxic intermediates associated with oxygen poisoning and ionizing radiation. Surprisingly, despite its important implications, not enough attention was paid to this discovery at the time. Currently, the role of free radicals in the development of a large variety of chronic and degenerative diseases is well documented. Oxidative stress is involved in several neurological disorders such as Parkinson's disease, Alzheimer's disease, multiple sclerosis, memory loss, and depression. It has also been established that oxidative damage to DNA is responsible for cancer development. There is also evidence supporting the role of oxidative stress in several cardiovascular diseases including atherosclerosis, cardiac hypertrophy, ischemia, hypertension, cardiomyopathy, and congestive heart failure, as well as in ocular, renal, and pulmonary diseases; rheumatoid arthritis, fetal growth restriction and pre-eclampsia. Based on such overwhelming data, it is evident that an excess of free radicals can be dangerous to human health.

These species are not only produced by endogenous sources but also by exogenous ones. Endogenous free radicals are generated from immune cell activation, inflammation, mental of physical stress, excessive exercise, ischemia, infection, cancer, and aging. Exogenous free radicals come from environmental pollution, cigarette smoke, alcohol, heavy or transition metals, certain drugs, and radiation. Therefore efficiently detoxifying free radicals to keep them at healthy low/moderate concentrations is a difficult task in the modern world.

II.1. The Chemistry of Free Radicals

From a chemical point of view, free radicals are species containing one or more unpaired electrons. This characteristic makes them highly reactive species that trigger chain reaction mechanisms, which may involve molecules of high biological importance such as DNA, lipids, and proteins.

Because of its high reactivity, and the consequent widespread damage that it can cause, the hydroxyl radical deserves particular attention. It can be formed intracellularly by the Fenton-type reaction, by Haber-Weiss recombination, or from other radicals created from enzyme reactions. It can also be produced by ultraviolet and ionizing radiations. $^{\bullet}$OH is the most electrophilic [2] and reactive of the oxygen-centered radicals, with a half-life of ~10-9 s. [3] It can react though a wide variety of mechanisms and its reactions with a large variety of chemical compounds take place at, or close to, diffusion-controlled rates (rate constants ≥ 109 M-1 s-1). It has been estimated that the $^{\bullet}$OH is responsible for 60%-70% of the tissue damage caused by ionizing radiations. [4] $^{\bullet}$OH is so reactive that it is capable of immediately reacting, after formation, with almost any molecule in its vicinity with little selectivity toward the various possible sites of attack. It has been held responsible for the most important oxidative damage to DNA. [5-7]

Compared to $^{\bullet}$OH, peroxyl radicals are less reactive species, capable of diffusing to remote cellular locations. Their half-lives are in the order of seconds [8] and their electrophilicity is significantly lower than that of $^{\bullet}OH$. [2] However, they can also react with other chemical species through different mechanisms. [9] The simplest of the peroxyl radicals is the hydroperoxyl radical. It is the conjugated acid of the superoxide radical anion, and according to its *p*Ka (4.8) [10] only ~0.3% of any superoxide present in the cytosol of a typical cell is in its protonated form. [11] The reactions of $HOO^{\bullet}/O2^{\bullet}$- with bioorganic compounds are generally quite slow. [10] For example, unsaturated fatty acids react with $HOO^{\bullet}$ at rate constants ~103 M-1 s-1 and with $O2^{\bullet}$- at rate constants about five orders of magnitude slower [9]. However, it has also been demonstrated that $HOO^{\bullet}$ initiates fatty acid peroxidation. [12]

Organic peroxyl radicals are also very important in living organisms. Hydrogen transfer processes from organic compounds or the addition of a radical to a double bond yield carbon-centered radicals, which in the presence of O2 produce peroxyl radicals. [9] In general, they are less reactive than $HOO^{\bullet}$ when R is an alkyl or an alkenyl group. [13,14] However, if R is a more electron-accepting group, for example R= -CC13, the ROO radicals significantly increase their reactivity towards organic molecules. [13] In particular, the rate constants for the electron-transfer reactions involving $ROO^{\bullet}$ strongly depend on the chemical nature of R, increasing to a great extent as the electron-withdrawing character of the substituents increases. [15] Alkoxyl radicals are formed from the reduction of peroxides and are significantly more reactive than ROO radicals, provided that R is the same in both species, [14,16-19] at the same time they are less reactive than $^{\bullet}OH$. [14,16-19]

Regarding RNS, the chemical reactivity of $NO^{\bullet}$ is rather low, and therefore its direct toxicity is actually minor. [20-21] However, it reacts with $O2^{\bullet}$- yielding peroxynitrite, [22] which is a potent oxidant. ONOO− itself is a very damaging species able to react with proteins, lipids, and DNA. [23-25] Therefore the reaction between two rather innocuous free radicals produces a much more reactive one. [26,27] Moreover, this reaction has one of the largest rate constants known among the $NO^{\bullet}$ reactions [28-31], and is believed to occur *in vivo*. [26,32] Nitrogen dioxide is a moderate oxidant, and its reactivity is somewhere between those of $NO^{\bullet}$ and $ONOO^{-}$. [9] NO_2 reacts with organic molecules at rates that range from ~10^4 to 10^6 M^{-1} s^{-1}, depending on the *p*H. [33,34]

III. ANTIOXIDANTS OR FREE RADICAL SCAVENGERS?

The term antioxidant is frequently used in a rather loose way. Therefore it is important to clarify its meaning in the context of this chapter, before starting using it. Here, the term antioxidant refers to the definition made by Halliwel and coworkers, i.e., “any substance that when present at low concentrations compared to that of an oxidizable substrate would significantly delay or prevent oxidation of that substrate” [35-36] In this definition the oxidizable substrate is any biological target that is expected to be protected by the antioxidant, for example DNA, lipids, or proteins.

It is also important to make a distinction between primary (chain breaking, or Type I) and secondary (preventive, or Type II) antioxidants. Even though this classification has been proposed for lipid oxidation[37] it can be extended to any other biological target. Primary antioxidants are chemical species that prevent oxidation by directly reacting with free radicals, turning them into significantly less reactive species, or terminating the radical chain reaction, i.e.,acting as free radical scavengers. Secondary antioxidants, on the other hand, retard oxidation by indirect ways of action including metal chelation, repairing of primary antioxidants by H or electron donation, decomposing hydroperoxide to non-radical species, deactivating singlet oxygen, absorbing ultraviolet radiation, or acting as oxygen scavengers. It should be noted that some antioxidants can act as multiple-function antioxidants, i.e., their protective effects can take place by both, primary and secondary, ways of action. In the next section it will be demonstrated that indoleamines belong to this category.

IV. Indoleamines vs. Oxidative Stress

Indoleamines are characterized by a distinctive bicyclic structure, consisting of a benzene ring fused to a pyrrole ring with a side chain containing an amino group. Since the indoleamine family is rather large, it is not possible to include them all in this chapter. Instead a representative set has been chosen (Scheme 1). They are tryptamine (2-(1H-Indol-3-yl)ethanamine; TA), melatonin (*N*-acetyl-5-methoxytryptamine; MLT), serotonin (5-hydroxytryptamine; 5-HTA), tryptophan (2S-2-amino-3-(1H-indol-3-yl)propanoic acid; TRP), and some of their derivatives: 6-hydroxymelatonin (6-HMLT), *N*-acetylserotonin (NAS), 5-hydroxytryptophan (5-HTP), and 5-methoxytryptamine (5-MTA). In living organisms these indoleamine are interconnected through the metabolic pathways of tryptophan (Scheme 2). There are other metabolites of melatonin that also exert beneficial effects in reducing oxidative stress. This behavior has been well documented at least for cyclic 3-hydroxymelatonin (3OHM), N^1-acetyl-N^2-formyl-5-methoxykynuramine (AFMK), and N^1-acetyl-5-methoxykynuramine (AMK). However, they do not belong to the indoleamine family. That is why they are not included in this chapter.

IV.1. Melatonin (MLT)

MLT is probably the most widely studied indoleamine. The amount of data gathered so far, regarding the protective action of MLT against oxidative stress, is overwhelming. Moreover, based on the ability of melatonin to scavenge a wide variety of ROS and RNS, it has been hypothesized that its primary function in living organisms is to protect them from oxidative stress. [40] The role of MLT regarding inhibition of lipid peroxidation has been well established. [41-45]

The brain is particularly sensitive to free radical damage. This is due to its high utilization of O_2, its relatively poorly developed antioxidant defense, and the fact that it contains large amounts of easily oxidizable fatty acids. [46] Since MLT is generated *in situ* in the brain, and is released into the cerebrospinal fluid, [208] its antioxidant activity may acquire special relevance to protect against neurodegenerative disorders. There are numerous studies documenting the protective effects of MLT against oxidative damage-related disorders affecting the brain. [47] For example, it has been demonstrated that MLT limits ischemia/reperfusion-induced damage in the brain, [48-51] and in other organs. [52] This has been related to the melatonin's ability to scavenge the free radicals produced during such events. [53, 54]

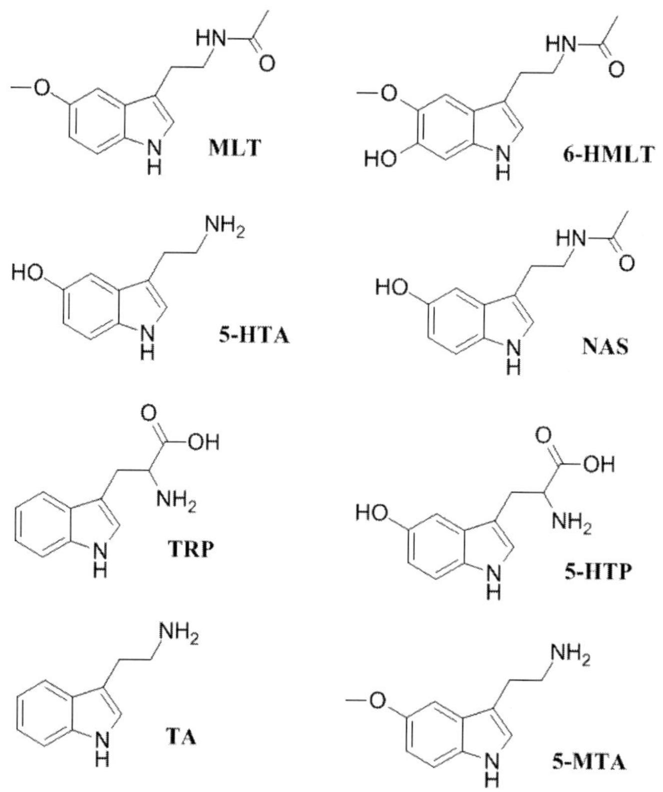

Scheme 1. Chemical structures of some indoleamines.

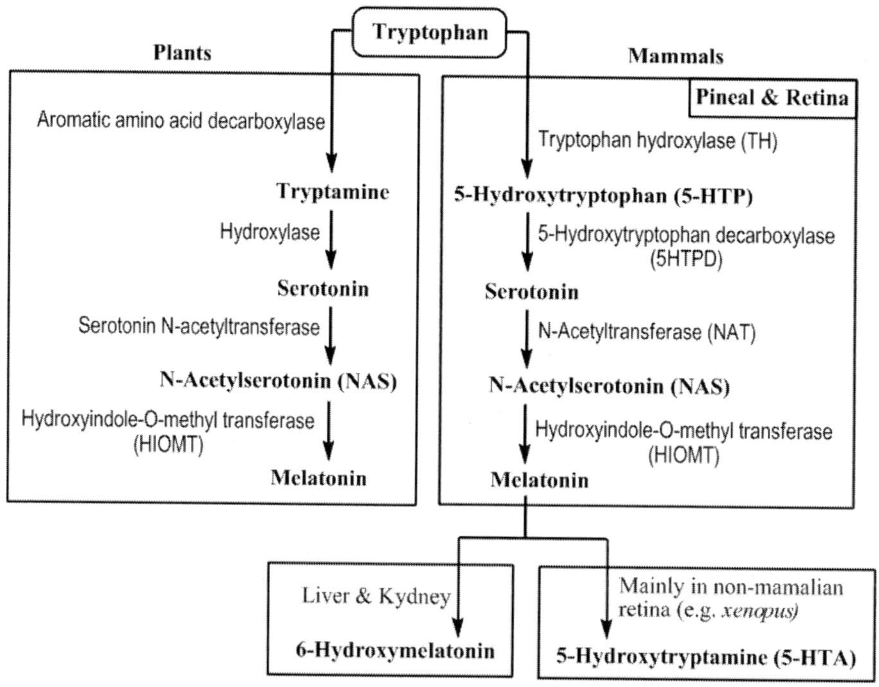

Scheme 2. Metabolic pathways of tryptophan in living organisms (adapted from references 38 and 39).

Experimental models of Alzheimer's disease have been used to evaluate the effect of melatonin, and it has been found that it reduces degenerative changes. [55-63] Moreover, MLT administration to humans has been shown to significantly slow the progression of this disease. [64, 65] There are also numerous studies showing the neuro-protective effect of MLT in Parkinson's disease. [66] This effect has been related to the regulatory role of MLT in the circadian rhythm of dopamine [67] and to the protective action of MLT against autoxidation of dopamine. [68] Moreover, the efficiency of MLT for the latter has been described to be higher than those of vitamins E and C, and that of Ldeprenyl. [69] In addition, it has been reported that the cisplatin-induced $^{\bullet}OH$ production is completely inhibited by MLT. [70]

Whereas the central nervous system is susceptible to be extensively destroyed by free radicals, not only the brain is vulnerable to oxidative stress. On the contrary, all organs suffer from these devastating processes. It has been shown that MLT protects the mucosa of the stomach, [71-73] the liver, [74] and the heart [75,76] from the damaging effects of free radicals generated during ischemia–reperfusion.

In addition, the ability of MLT to scavenge reactive oxygen species (ROS) and reactive nitrogen species (RNS) is shared by its metabolites. This is a very appealing, and rare, characteristic that warranties a continuous action in living organisms. Such an "endless" protection exerted by MLT has been referred to as the free radical scavenging cascade, [77-81] and makes this compound particularly effective, even at low concentrations, in protecting organisms from oxidative damage. Among the melatonin's metabolites, 3-OHM has been proposed to be the most efficient for scavenging peroxyl radicals, [82] and a more potent antioxidant than its precursor, or vitamin C, in terms of its ability to scavenge the hydroxyl radical and to recover oxidized horseradish peroxidase to its ground state. [83]

Regarding its role as secondary antioxidant, there is evidence on the protection exerted by MLT against metal-induced damage. It has been shown that MLT forms complexes with aluminum, cadmium, lead, zinc, iron and copper, [84] which suggests that this indoleamine can play a role in metal detoxification and inhibit the metal-induced radical generation. In fact, it has been demonstrated that MLT protects against copper-mediated lipid peroxidation by binding Cu(II) and Cu(I), which has led to propose that the antioxidant and neuroprotective activities of melatonin can take place by sequestering, and thus removing toxic metals, from the central nervous system. [85] The ability of MLT to counteract oxidative damage to proteins caused by Cu(II)/H_2O_2 has also been proven. [86]

Compared to other antioxidants, it has been documented that the Fe(II) chelating ability of MLT is higher than those of α-tocopherol, butylated hidroxybutilanisol (BHA), and butylated hydroxytoluene (BHT). [87] This behavior was associated with the melatonin's ability to inhibit lipid peroxidation. Moreover, MLT drastically reduces the free radical formation caused by the interaction of Al(III), Zn(II), Cu(II), Mn(II), and Fe(II)) with the amyloid-β peptide. [88] There is also experimental evidence indicating that melatonin inhibits low-density lipoprotein (LDL) oxidation in the serum of postmenopausal women, [89] and protects from copper-mediated free radical damage in rat liver homogenates. [85] The protection exerted by MLT against metal-catalyzed molecular damage has been recently reviewed, [90] and it was proposed that in addition to acting as a potent free radical scavenger, MLT may also bind copper preventing it from participating in free radical generation in vivo.

IV.2. 6-Hydroxymelatonin (6-HMLT)

6-HMLT is the major hepatic metabolite and photo-product of melatonin. This compound shares the protective effects against oxidative stress exerted by its precursor. It has been reported that lipo-peroxidation induced by thiobarbituric acid, which is attributed to the production of ROS, is inhibited by both compounds. [91] Moreover, it was proposed that the antioxidant activity of 6-HMLT is even larger than that of MLT in this context, which may be attributed to the presence of a hydroxyl group in the benzene ring, i.e., a phenol moiety.

It has also been shown that 6-HMLT is capable of protecting rat brain homogenate against iron-induced lipid peroxidation, *in-vitro* [92] and from cyanide induced oxidative stress. [93] The later was rationalize in terms of the 6-HMLT ability to significantly reduce KCN-induced $O_2^{\bullet -}$ generation. It was also found that 6-HMLT protects against quinolinic-acid-induced oxidative neurotoxicity in the rat hippocampus [94] by efficiently scavenging singlet oxygen and $O_2^{\bullet}$. Based on the results from these works it was proposed that 6-HMLT should present neuro-protective effects.

Regarding its activity as secondary antioxidant, it has been reported that 6-HMLT is efficient for protecting against oxidative damage induced by UV irradiation. [95] This metabolite was actually proposed as partially responsible for the potential benefits of incorporating melatonin into sunscreens. In addition there is evidence that this compound can reduce Fe^{2+}-induced lipid peroxidation and necrotic cell damage in the rat hippocampus *in vivo*. [96] Accordingly it seems that 6-HMLT can act as both primary and secondary antioxidant.

On the contrary, it has been suggested that 6-HMLT may induce DNA damage via non-o-quinone type of redox cycle, in the presence of metal ions, leading to carcinogenic effects. [97] Therefore, albeit most of the investigations reported so far, on the oxidative stress related effects of 6-HMLT, indicate that it is beneficial more studies aiming to provide more information on this subject are still necessary.

IV.3. Tryptophan (TRP)

Diets rich in, or enriched with, TRP have been demonstrated to improve the antioxidant status in young, middle-aged, and elderly human individuals. [98-100] However, this beneficial effects are not exclusive for humans. It has been reported that appropriate dietary TRP levels also improves the antioxidant status of some fish. [101] TRP antioxidant effects include inhibiting lipid peroxidation and the lysis of erythrocytes induced by peroxyl radicals. [102] These effects were also evidenced in experiments with rats under endotoxic shock. [103] In this case the protection was explicitly attributed to the free radical scavenging activity of TRP. This activity has also proposed as responsible, at least partially, for the TRP protection against the gastric mucosa from damage by stress and ischaemia reperfusion. [72] In addition it has been proposed that this compound may be a powerful free radical scavenger, naturally present in human milk. [104] However it has been pointed out that its antioxidant effects and potential application in the diets of infants, particularly preterm, must be examined further.

TRP is also capable of inhibiting the 2-deoxy-D-ribose degradation, induced by the Fenton reaction, to a larger extent than mannitol and DMSO. [105] At the same time TRP was proposed as a more efficient inhibitor of the cytochrome P-450-dependent lipid peroxidation than other antioxidants including L-phenylalanine, L-tyrosine and uracil. These results led to propose that TRP is a main antioxidant in human placenta, capable of suppressing lipid peroxidation and increasing the oxidative status. They also suggest that this compound may present both kind of antioxidant activity, primary and secondary.

However, there are some contradictory evidences about the protective vs. promoting effects of TRP regarding oxidative damage. It has been proposed that, when oxidized, TRP radicals may subsequently oxidize urate [106] and the tyrosine side chain. [107] It has also been reported that tryptophan inhibits the *in vivo* activity of the thiol containing enzymes creatine kinase and pyruvate kinase, and induces *in vitro* oxidative stress in rat brain. [108-111] More recently it has been suggested that the observed increase in glutathione levels, after treating young rats with TRP, can be attributed to the tryptophan-induced free radicals generation. [112] Thus, it becomes evident that the antioxidant vs. pro-oxidant effects of TRP deserves further studies.

At this point it seems important to call attention to the fact that TRP is a singular compound regarding oxidative stress. In living organisms it can be found as free TRP, but also as a residue in many proteins. While in the first case the analyses on its role reducing oxidative stress applies, in the second case it becomes a biological target that needs protection against oxidative damage. [113] Therefore, it is not expected to be particularly reactive towards free radicals because in such a case the proteins integrity would be highly compromised. Moreover, it would be desirable that oxidative lesions in TRP residues could be repaired by other antioxidants. Accordingly it should be expected that its antioxidant protection might be mainly attributed to its metabolites, not only those analyzed in this chapter but also those produced along the kynurenine pathway.

There is information available regarding the antioxidant properties of kynurenine-related metabolites of tryptophan. It has been found that they play inhibitory roles on the ROS production in a concentration-dependent way. [114] There is also evidence that these metabolites present a high scavenging ability for reactive oxygen and chlorine species, consequently establishing a self-regulatory mechanism that limits free radicals induced tissue damages. [115] In addition, it has been found that, even at low (micromolar) concentrations, 5-hydroxytryptophan, 3-hydroxykynurenine, xanthurenic acid, and 3-hydroxyanthranilic acid, scavenge peroxyl radicals with high efficiency, while their corresponding non-hydroxylated metabolic precursors do not present such an activity. [116] Moreover, the protective effects of 3-hydroxykynurenine and 3-hydroxyanthranilic acid against peroxyl radical-mediated oxidative damage were reported to be higher than those of equimolar amounts of either ascorbate or Trolox.

IV.4. 5-Hydroxytryptophan (5-HTP)

It has been proposed that the oxidative damage induced by free radicals can be efficiently inhibited by 5-HTP. [117] Its free radical scavenging activity was evidenced by the inhibition of Fe-induced lipid peroxidation processes, [118] and the possible implications of its antioxidant activity for inflammatory diseases has been pointed out. [116] It has also been demonstrated that 5-HTP prevents lipid and protein oxidations, induced by Fe/ascorbate mixtures, in a concentration-dependent manner. [119] Moreover, it has been proposed that 5-

HTP is a more potent hydroxyl radical scavenger, in vitro, than melatonin or vitamin C. [120] In addition it has also been demonstrated that 5-HTP is able of suppressing UV-induced apoptosis in human monocytes [121], and to chelate some divalent metal ions (Ni, Cu, Zn, Cd, Pb). [122]

Accordingly, even though the number of studies on the antioxidant activity of 5-HTP is rather low, compared to MLT or TRP, there seems to be enough evidence supporting its action as both primary and secondary antioxidant. Nevertheless, more studies focused on this compound, regarding its potential roles in oxidative stress processes, are still desirable.

IV.5. Serotonin (5-HTA)

Platelet 5-HTA has been recognized as a locally releasable antioxidant, [123] that acts in a dose-dependent manner. When released from activated thrombocytes at sites of inflammation and endothelial cell damage, 5-HTA acts as a genuine ROS scavenger of reactive oxygen species generated during the respiratory burst of phagocytes. In consequence, it has been proposed that 5-HTA may modulate various aspects of cell-mediated defense reactions. [124] At the same time its oxidation can be prevented by other antioxidants, which is important for preserving its functions as a neurotransmitter.

It has been proposed that 5-HTA protects membranes from lipid peroxidation through its free radical scavenging activity, which is mainly exerted in the aqueous phase, or at the water-membrane interface. [125] This protective activity has been proposed as responsible for the attenuated secondary tissue damage in the central nervous system when serotonin is released from platelets and 5-HTA nerve endings at sites of brain damage or inflammation. [126]

The antioxidant effects of 5-HTA on neuronal tissues have also been examined by studying the oxidative damages of brain synaptosomal components. [127] The oxidation processes induced by Fe^{2+} and ascorbate are efficiently attenuated by 5-HTA, which is also capable of significantly decrease the hydroxyl radical production caused by either Fe^{3+}, EDTA, H_2O_2 and ascorbate or xanthine and xanthine oxidase. 5-HTA was also found to restore synaptosomal Ca^{2+} uptake, decreased by Fe^{2+} and ascorbate. The protective effects of 5-HTA on brain synaptosomes has been attributed to the removing action of reactive oxidants, hydroxyl radicals and probably iron-oxygen complex, without chelating action on iron.

The antioxidant activity of 5-HTA has been also demonstrated by its suppressive effects on phagocytosis-associated, luminol-enhanced chemiluminescence. [128] Its action was rationalized in terms of its reactions with reactive oxygen species, and to be dose-dependent. This behavior is in line with the observation that low blood serotonin correlates with increased rate of infections. In addition serotonin has been reported to have marked *N,N*-dimethyl-p-phenilenediamene (DMPD) radical scavenging and cupric ions reducing ability. [129] In fact the activity of 5-HTA in those two processes was found to be higher than that of MLT, which was attributed to its phenolic group.

Very recently, the in vitro antioxidant and free radical scavenging activities of 5-HTA have been evaluated, with different *in vitro* methodologies, using butylated hydroxyanisole (BHA), butylated hydroxytoluene (BHT), α-tocopherol and trolox, as reference compounds. [130] It was found that 5-HTA is capable of completely inhibiting lipid peroxidation of a linoleic acid emulsion, showing higher efficiency to that purpose than the references. In addition, 5-HTA was effective in scavenging DPPH, $ABTS^{\bullet+}$, and H_2O_2. This indoleamine also showed ferric ions reducing power and ferrous ions chelating activities.

Therefore, based on all the data gathered so far it can be concluded that 5-HTA plays protective roles against oxidative stress by directly scavenging free radicals, and also by sequestering metals, thus inhibiting free radicals production.

IV.6. *N*-acetylserotonin (NAS)

The antioxidant role of NAS has been evidenced in several studies. It has been shown that this molecule protects DNA against damages caused by exposure to chromium III and hydrogen peroxide, [131] and that it inhibits low-density lipoproteins oxidation induced by copper. [132,133] NAS also decreases lipid peroxidation caused by iron, hydrogen peroxide and 2,2′-azobis(2-amidinopropane) (AAPH). [134-138] This compound is also capable of preventing the formation of ultraviolet-radiation-induced cataracts in rats. [46]

The protection exerted by NAS against oxidative damage has been associated with its anti-aging activity. In addition, the antioxidant effects of NAS have been related to the protective effects of MLT, since NAS is both a precursor and a metabolite of MLT. [139] However, despite of their close relationship, the protective effect of NAS against oxidative damage induced by various agents, seems to be independent from the effect of MLT. [135,138] It was even proposed that, taking into account that, *in vivo*, about 15% of MLT is demethylated into NAS, the beneficial effects of melatonin may be mediated by NAS.

It was also reported that administration of NAS during 4 weeks to mice resulted in decreased lipid peroxidation and increased glutathione peroxidase in brain and kidney tissues, and that NAS is more efficient than MLT for protecting against iron and lipopolysaccharide (LPS) induced lipid peroxidation [139]. The same trend in activity has been reported for the autoxidation and iron-induced lipid peroxidation, [140] the copper-mediated oxidation of low density lipoproteins, [132] and the ascorbate-Fe^{2+} induced lipid peroxidation of rod outer segment membranes of bovine retina. [141] It was also reported that in an aqueous cell-free solution, the peroxyl radicals scavenging activity of NAS was higher than that of MLT, and similar to that of Trolox. [142] These findings can be rationalized based on the structures of both indoleamines. While NAS present a phenolic site, in the same position MLT has a methoxy group, the higher reactivity of NAS towards free radicals can be attributed to its phenolic moiety. Detailed, quantitative, studies on the site reactivity of both compounds would be necessary to validate, or reject, this hypothesis.

So far, there is enough evidence supporting the protective effects of NAS against direct free radical, and metal induced, damages. Therefore its activity in reducing oxidative stress seems to be confirmed. While the relative importance of the different mechanisms of action still deserves further studies, it seems that NAS protective effects may be mediated, at least, by direct free radical scavenging and metal deactivating processes. The latter might take place via chelation, but this needs to be confirmed.

IV.7. Tryptamine (TA)

There is rather little information about the direct involvement of TA in oxidative stress related processes. However it is known that this compound is capable of deactivating *tert*-butyl hydroperoxide, [143] peroxynitrite, [144] hydroxyl and peroxyl radicals. [145] The protective effects of TA were compared with those of other indoleamines, and they were found to be lower than those of MLT, NAS and 5-MTA [119]. Based on these results it was proposed that TA itself is a rather poor antioxidant and that its protective effects may be increased by methoxylation, hydroxylation, and *N*-acetylation. Such substitutions are expected to increase not only reactivity towards free radicals, but also the solubility of TA in the incubation medium, which would promote better access to the free radicals.

Since TA functionalization seems to be a practical way of producing better antioxidants, it is logical that producing tryptamine-derivatives with increased activity has become an active area of research in the last years. One example is securinine (TDS), which was found to possess antioxidant and chelating effects, and has been proposed to be of interest as a potential scaffold for the development of new neuro-protective anticonvulsant drugs. [146] Tryptamine triazole derivatives, functionalized with the catechol group, have proven to be multifunctional protectors capable of successfully inhibit β-secretase action and amyloid aggregation, and showing metal chelating and antioxidant effects at micromolar levels. [147] In addition, the recently synthesized tryptamine-gallic acid hybrid was found to prevent indomethacin mediated mitochondrial oxidative stress and dysfunctions in gastric mucosal cells, which play a pathogenic role in inducing gastropathy. [148] It was proposed that these effects are exerted by scavenging mitochondrial superoxide anion and intra-mitochondrial free iron. Accordingly, tryptamine-derivatives seems to be promising agents against oxidative stress, presenting both primary and secondary antioxidant activities.

IV.8. 5-methoxytryptamine (5-MTA)

The information on the possible antioxidant activity of 5-MTA is also rather scarce. It has been reported to be capable of scavenging $^{\bullet}OH$ produced by photolytic rupture of H2O2, [149] reducing lipid peroxidation induced by Fe^{2+} and react with hypochlorous acid [134]. 5-MTA was also proposed to be efficient for inhibiting lipid peroxidation in rat brain, liver and kidney homogenates and hemolysis of rat erythrocytes, as well as for suppressing superoxide and hydroxyl radicals formation. [150] It was proposed to be more efficient for those purposes than other indoles including 5-hydroxytryptophol, 5-methoxytryptophol, 5-methoxyindole-3-acetic acid, 5-hydroxyindole-3-acetic acid, MLT and 5-HTA. Moreover, 5-MTA was proven to be devoided of pro-oxidant effects. There is also evidence that 5-MTA is capable of inhibiting malondialdehyde formation in a concentration dependent manner, carbon tetrachloride induced lipid peroxidation in mice kidneys. [151] Administration of 5-MTA seems to prevents exercise-induced cellular oxidative changes in rats, albeit it is less effective than MLT to that purpose. [152]

Accordingly it can be stated that, even though more studies are still desirable on the potential roles of 5-MTA in oxidative stress processes, it seems to have beneficial effects to the oxidative status of living organisms.

V. REACTION MECHANISMS

Elucidating the main reaction mechanisms involved in the antioxidant activities of chemical compounds may be a challenge. Such action takes place in complex environments, which means that there are large numbers of species present that may be involved in competing chemical reactions. This implies that several chemical processes can take place simultaneously, and the most important ones would depend on the relative reactivity of the reacting species. Moreover, due to the chemical nature of the free radicals, a chain reaction mechanism may also be involved. Therefore, the first oxidation step can be rapidly followed by subsequent chemical processes. Also, different radicals can react in different ways, and the polarity of the environment, as well as the *p*H in the aqueous phase, can also affect the relative importance of the competing mechanisms.

Identifying the main mechanism, or mechanisms, of action can be achieved both experimentally and theoretically. A good experimental strategy may be to perform detailed product analyses. However, this approach also involves a large degree of inference because the processes are usually very fast and composed by several steps. Therefore, the detectable products are frequently those formed after various elementary reactions. Moreover, as mentioned above, different mechanisms might lead to the formation of the same products. Computational strategies also present a series of difficulties, mainly related to the frequent necessity of using simplified models, and to the availability of strategies for properly include solvent effects. Accordingly, combined experimental-theoretical efforts are probably the best way to address this important part of the antioxidant activity related investigations.

V.I. Role of the Indoleamines Structure

Despite of the numerous reports on the free radical scavenging activity of indoleamines, the information on the mechanism, or mechanisms, determining this activity is not yet well defined. Based on their structures, and in what is known about other antioxidants [153-159] it can be assumed that the main reaction mechanisms involved in their protection against free radicals include:

Radical adduct formation (RAF):

$$\text{indoleamine} + {}^{\bullet}R \rightarrow \left[\text{indoleamine} - R\right]^{\bullet}$$

Hydrogen atom transfer (HAT):

$$\text{indoleamine} + {}^{\bullet}R \rightarrow \text{indoleamine}_{(-H)}{}^{\bullet} + HR$$

Single electron transfer (SET):

$$\text{indoleamine} + {}^{\bullet}R \rightarrow \text{indoleamine}^{\bullet+} + R^{-}$$

Additionally, SET can occur after, rapidly followed by, or simultaneously with, proton transfer. These processes are known as sequential proton loss electron transfer (SPLET), sequential electron proton transfer (SEPT) and proton coupled electron transfer (PCET) mechanisms, respectively:

Sequential Proton Loss Electron Transfer (SPLET):

$$\text{Indoleamine} \xrightarrow{-H^+} \text{Indoleamine}_{(-H^+)}^{\ -} + R^{\bullet} \rightarrow \text{Indoleamine}_{(-H)}^{\ \bullet} + HR$$

Sequential Electron Proton Transfer (SEPT):

$$\text{Indoleamine} + {}^{\bullet}R \rightarrow \text{Indoleamine}^{\bullet+} + R^{-} \rightarrow \text{Indoleamine}_{(-H^+)}^{\ \bullet} + HR$$

Proton Coupled Electron Transfer (PCET):

$$\text{Indoleamine} + {}^{\bullet}R \rightarrow \text{Indoleamine}_{(-H)}^{\ \bullet} + HR$$

It seems important to mention that even though SPLET, SEPT and PCET yield the same products as HAT, the influence of the solvent and of the nature of the reacting radical on their feasibility is expected to be different. While SPLET, SET and SEPT are likely to be favored by polar environments that promote solvation of the intermediate ionic species, the PCET mechanism might be also viable in non-polar media since the transfer of the proton and the electron occurs simultaneously in this case and, therefore, no charged intermediaries are formed.

In the particular case of the hypochlorous acid, it has been proposed that the mechanism of reaction of tryptamine and tryptophan derivatives is a complex process involving not only radical but also ionic pathways. [160]

In addition, the above detailed mechanisms correspond only to the free radical scavenging activity, i.e., to the action of indoleamines as primary antioxidants. For their secondary antioxidant activity other mechanisms may be involved. For example metal chelation can take place with indoleamines acting as monodentate or bidentate ligands. In turn, both processes can proceed directly or coupled with indoleamines' depronotation. Singlet oxygen deactivation may also take place through several routes including radiative deactivation and radiation-less deactivation. The later also involve different pathways such as electronic to vibrational energy transfer, charge transfer deactivation, and electronic energy transfer. The interested reader can find more details on this topic in a very thorough review presented by Schweitzer and Schmidt. [161]

SET and HAT have both been identified, from experimental data, as the main mechanisms for the reactions of indole derivatives tryptophan and *N*-methylindole with different free radicals ($N_3^{\bullet}$, $Br_2^{\bullet}$, and $(SCN)_2^{\bullet}$). However, based on structural considerations, it was proposed that because MLT does not have phenolic hydrogens it should react more easily by SET than by HAT. [162] In addition, not all free radicals are good enough electron-acceptors to be involved in the SET mechanisms. Therefore, the nature of the reacting free radical also plays an important role in SET viability. For example, haloperoxyl radicals are highly electron-deficient species and thus capable of easily reacting via SET. [163] It should be kept in mind, however, that while such a mechanism is a likely antioxidant route in aqueous solution, it is highly improbable in the lipid phase, since such environment does not provide enough stabilization for the ionic intermediates.

Based on the structure similarities, SET and HAT mechanisms are also expected to contribute to the overall free radical scavenging activity of other indoleamines, with larger contributions of HAT for the hydroxylated compounds. In addition, also depending on the reacting free radical, RAF at the indole moiety may also contribute to indoleamines free radical scavenging activity. In fact, it has been proposed that this mechanism plays an important role in the reactions of serotonin and tryptamine with the $^{\bullet}OH$ radical [164].

V.II. Role of the Oxidant and *p*H

Based on the above discussion it becomes evident that the chemical nature of the reacting free radical influences the relative importance of competing antioxidant routes. In the case of indoleamines the most abundant data in this regard has been gathered for MLT. Thus it will be used to illustrate this point.

UV-vis spectra data have been collected for the reactions of MLT with $N_3^{\bullet}$ and $Br_2^{\bullet}$ and $ONOO^-$ in aqueous solution. [165,166] From this information it can be inferred that both SET and HAT mechanisms contribute to the scavenging activity of MLT. In addition, at *p*H=7 the main reaction mechanism is expected to be SET, while at *p*H=3 the relative importance of HAT becomes the highest. In line with these findings, for the reaction of MLT with $ONOO^-$ there is evidence for the formation of cyclic 2-hydroxy melatonin, cyclic 3-hydroxymelatonin, and 6-hydroxymelatonin. [167,168] This led to hypothesize that the initial step of the melatonin + $ONOO^-$ reaction involves one electron transfer from melatonin to $ONOO^-$. [167]

On the other hand, for the reaction of MLT with $^{\bullet}OH$, in addition to the products yielded by SET and HAT, there is also evidence of the formation of a product, which was suggested to arise from RAF processes, in particular from $^{\bullet}OH$ additions to the carbon sites in the indole ring. [169] This is a logical finding because, based on the high reactivity of this radical, it is expected to be less selective and form a wider variety of products. It is also in agreement with an early study in which, using different techniques for measurements, other products of melatonin + $^{\bullet}OH$ reaction were identified, and related to OH addition at the C3 site of the indole moiety. [170]

In addition, for the reaction between TRP and 2,2'-azinobis(3-ethylbenzothiazoline-6-sulfonic acid) ($ABTS^{\bullet+}$) it has been found that the reaction rate decreases as the *p*H lowers [171]. Accordingly, it seems that the anionic form of TRP is more reactive towards the $ABTS^{\bullet+}$ radical than the neutral species. This suggests that the main reaction mechanism for this reaction is ruled by electron transfer.

VI. Kinetics

Regarding kinetic data, particularly rate constants for the reactions of indoleamines with different free radicals, the most abundant information gathered so far also corresponds to melatonin, followed by tryptophan and serotonin, in that order (Table 1). These reported rate constants values were obtained using different approaches, including both experimental and theoretical protocols, albeit the former are by far more abundant. In addition, most of the data correspond to reactions taking place in aqueous solution.

Table 1. Rate constants (*k*) of indoleamines as primary antioxidants (free radical scavengers)

Indoleamine	Radical	k (M^{-1} s^{-1})	Approach	Main solvent	Ref.
MLT	$^{\bullet}OH$	2.7×10^{10}	Experimental	Water, *p*H=7	
MLT	$^{\bullet}OH$	6×10^{10}	Experimental	Water	[145]
MLT	$^{\bullet}OH$	4×10^{10}	Experimental	Water	[173]
MLT	$^{\bullet}OH$	1.32×10^{10}	Experimental	Water, *p*H=7	
MLT	$^{\bullet}OH$	1.2×10^{10}	Experimental	Water	[169]
MLT	$^{\bullet}OH$	1.25×10^{10}	Experimental	Water, *p*H=7	[165]
MLT	$^{\bullet}OH$	2.7×10^{10}	Experimental	*ne*	[175]
MLT	$^{\bullet}OH$	1.85×10^{10}	Theoretical	Water	[13]
MLT	$^{\bullet}OH$	2.23×10^{10}	Theoretical	Benzene	[13]
MLT	$Br_2^{\bullet-}$	4.5×10^{8}	Experimental	Water, *p*H=3	[165]
MLT	$Br_2^{\bullet-}$	2.5×10^{8}	Experimental	Water, *p*H=7	[165]
MLT	$H^{\bullet}$	1.3×10^{10}	Experimental	Water, *p*H=7	[165]
MLT	$^{\bullet}OOCCl_3$	6×10^{8}	Experimental	Water, *p*H=7	[165]
MLT	$^{\bullet}OOCCl_3$	2.7×10^{8}	Experimental	Mix	[176]
MLT	$^{\bullet}OOCCl_3$	1.40×10^{9}	Theoretical	Water	[13]
MLT	$^{\bullet}OOCCl_3$	4.40×10^{8}	Theoretical	Benzene	[13]
MLT	$HOO^{\bullet}$	1.99×10^{1}	Theoretical	Water	[13]
MLT	$HOO^{\bullet}$	3.11×10^{2}	Theoretical	Benzene	[13]
MLT	$LOO^{\bullet}$	$\sim 10^{-2}$	Theoretical	Water	[13]
MLT	$LOO^{\bullet}$	~ 2-5	Theoretical	Benzene	[13]
MLT	$LOO^{\bullet}$	$\geq 2.8 \times 10^{4}$	Experimental	Mix	[177]
MLT	$LOO^{\bullet}$	$\geq 9.0 \times 10^{4}$	Experimental	Mix	[178]
MLT	$t\text{-}ButO^{\bullet}$	3.4×10^{7}	Experimental	Acetonitrile	[179]
MLT	$t\text{-}ButO^{\bullet}$	2.8×10^{9}	Experimental	Water	[165]
MLT	$di\text{-}t\text{-}CumO^{\bullet}$	6.7×10^{7}	Experimental	Acetonitrile	[179]
MLT	$O_2^{\bullet-}$	$< 1.0 \times 10^{4}$	Experimental	Water	[174]
MLT	$O_2^{\bullet}$	1.25×10^{3}	Experimental	Water, *p*H=7.4	[180]
MLT	$G^{\bullet}$	3×10^{9}	Experimental	Water, *p*H=7	[165]
MLT	$^{\bullet}N_3$	7.5×10^{9}	Experimental	Water	[165]
MLT	$^{\bullet}N_3$	9.8×10^{9}	Experimental	Water, *p*H=7	[174]
MLT	$^{\bullet}NO$	3×10^{7}	Experimental	Water, *p*H=7	[165]
MLT	$^{\bullet}NO_2$	3.7×10^{6}	Experimental	Water, *p*H=7	[165]
MLT	$SO_4^{\bullet-}$	8.0×10^{9}	Experimental	Water	[181]
TRP	$^{\bullet}OH$	1.2×10^{10}	Experimental	Water, *p*H=8.8	[182]
TRP	$^{\bullet}OH$	1.25×10^{10}	Experimental	Water, *p*H=1	[182]

Indoleamine	Radical	k ($M^{-1} s^{-1}$)	Approach	Main solvent	Ref.
TRP	$Br_2^{\bullet-}$	1.4×10^9	Experimental	Water, pH=3	[162]
TRP	$^{\bullet}H$	1.0×10^9	Experimental	Water, pH=1	[182]
TRP	$^{\bullet}OOCl$	7.6×10^5	Experimental	Water, pH~12	[183]
TRP	$^{\bullet}OOCl$	3.4×10^4	Experimental	Water, pH=7	[184]
TRP	$^{\bullet}NO_2$	$<5 \times 10^5$	Experimental	Water, pH~12	[183]
TRP	$ATRA^{\bullet+}$	1.0×10^8	Experimental	Mix, pH~7	[185]
TRP	$G^{\bullet}$	4.4×10^7	Experimental	Mix, pH=7	[186]
TRP	$G^{\bullet}$	1.0×10^7	Experimental	Mix, pH=7	[187]
TRP	$G^{\bullet}$	1.0×10^9	Experimental	Water, pH=1.3	[188]
TRP	$G^{\bullet}$	2.6×10^8	Experimental	Water, pH=7.5	[188]
TRP	$G^{\bullet}$	2.6×10^8	Experimental	Water, pH=11	[188]
TRP	$G^{\bullet}$	1.2×10^7	Experimental	Water, pH=13.3	[188]
TRP	$dGMP^{\bullet}$	1.2×10^8	Experimental	Water, pH=7	[189]
TRP	$dAMP^{\bullet}$	2.5×10^8	Experimental	Water, pH=7	[189]
TRP	$^{\bullet}N_3$	4.2×10^9	Experimental	Water, pH=3	[162]
TRP	$^{\bullet}N_3$	4.3×10^9	Experimental	Water, pH=7-10	[162]
TRP	$(SCN)_2^{\bullet-}$	3.0×10^8	Experimental	Water, pH=7-10	[162]
5-HTA	$^{\bullet}OH$	1.1×10^{10}	Experimental	*ne*	[175]
5-HTA	$Br_2^{\bullet-}$	3.7×10^9	Experimental	Water, pH=7	[191]
5-HTA	$Br_2^{\bullet-}$	3.5×10^9	Experimental	Water, pH=5	[190]
5-HTA	$Br_2^{\bullet-}$	4.0×10^9	Experimental	Water, pH=9	[190]
5-HTA	$NO_2^{\bullet}$	$\sim 1 \times 10^7$	Experimental	Water, pH=7	[191]
5-HTA	$NO_2^{\bullet}$	1.4×10^8	Experimental	Water, pH=9	[190]
5-HTA	$^{\bullet}OOCCl_3$	1.2×10^8	Experimental	Water, pH=7	[176]
5-HTA	$O_2^{\bullet-}$	2.1×10^4	Experimental	*ne*	[175]
5-HTP	$^{\bullet}OH$	1.2×10^{10}	Experimental	*ne*	[175]
5-HTP	$NO_2^{\bullet}$	9.0×10^5	Experimental	Water, pH=5	[190]
5-HTP	$NO_2^{\bullet}$	5.6×10^7	Experimental	Water, pH=9	[190]
5-HTP	$O_2^{\bullet-}$	1.2×10^4	Experimental	*ne*	[175]
5-HTP	$^{\bullet}DPPH$	$\sim 7 \times 10^{-1}$	Experimental	methanol	[192]
5-MTA	$^{\bullet}OH$	2.3×10^{10}	Experimental	Water, pH=7	[172]
5-MTA	$^{\bullet}OH$	1.7×10^{10}	Experimental	Water, pH=7	[172]
5-MTA	$^{\bullet}OH$	2.3×10^{10}	Experimental	*ne*	[175]
TA	$^{\bullet}OH$	8.6×10^9	Experimental	Water, pH=7	[191]
TA	$^{\bullet}N_3$	3.6×10^9	Experimental	Water, pH=7	[191]
6-HMT	$^{\bullet}OH$	1.1×10^{10}	Experimental	Water, pH=7	[172]
6-HMT	$O_2^{\bullet-}$	2.7×10^4	Experimental	*ne*	[175]
NAS	$^{\bullet}DPPH$	$\sim 2 \times 10^0$	Experimental	methanol	[192]

ne = non specified, $G^{\bullet}$ =guanosyl radical, $ATRA^{\bullet+}$= all-trans retinoic acid radical cation, $LOO^{\bullet}$ = different lipid peroxyl radicals, $dAMP^{\bullet}$ = 2'-deoxyadenosine-5'-monophosphate radical, $dGMP^{\bullet}$ = 2'-deoxyguanosine-5'-monophosphate radical, $^{\bullet}DPPH$ = 1,1-diphenyl-2-picrylhydrazyl radical.

Rate constants are a very useful criterion to establish trends in reactivity and thus to identify the most efficient compounds for scavenging free radicals. The faster the reaction the higher the primary antioxidant activity. Using this kind of comparisons it is possible to analyze the relative activity of the indoleamines presented in this chapter. The gathered kinetic data show that the reactions of all the analyzed indoleamines with the OH radical takes place within the diffusion limited regime. This indicates that these compounds are very efficient for scavenging this, very damaging, free radical. It is interesting to note that the influence of the environment (polarity and *p*H) on the reactivity of indoleamines towards $^{\bullet}OH$ seems to be almost negligible. For example the rate constants for MLT in water and benzene are very similar. This is also the case for the reactions of TRP, in aqueous solution, at *p*H values equal to 1 and 8.8. This allows predicting that indoleamines exert their protective effects against $^{\bullet}OH$ induced damages in all the different regions of living organisms, offering protection under any physiological conditions.

For halogenated peroxyl radicals, on the other hand, the effects of the environment are more important. For example for the reactions of $^{\bullet}OOCCl_3$ with MLT the values of the rate constants indicate that the polarity of the environment increases the scavenging activity, with the largest value corresponding to aqueous solution. A similar behavior appears for the reactions of MLT with the t-ButO radical. In this case changing the solvent from acetonitrile to water increases the rate constant almost 100 times. Regarding the *p*H, the data available for TRP suggest that the reactivity of this compound towards $^{\bullet}OOCl$ increases with the basicity of the environment. In addition there seems to be a direct relationship between the number of chlorine atoms in the reacting radical and the efficiency of indoleamines for scavenging them, being the reactions with $^{\bullet}OOCCl_3$ several orders of magnitude faster than those involving $^{\bullet}OOCl$.

The reactions of indoleamines with $O_2^{\bullet-}$, $LOO^{\bullet}$, and $^{\bullet}DPPH$ are significantly slower than those involving other free radicals such as $^{\bullet}OH$, $^{\bullet}N_3$, $Br_2^{\bullet-}$, $^{\bullet}NO$ and $^{\bullet}NO_2$. This suggests that indoleamines are particular efficient for scavenging this last set of free radicals, while their efficiency for scavenging $O_2^{\bullet-}$, $LOO^{\bullet}$, and $^{\bullet}DPPH$ is only moderate. Unfortunately there is no enough data available to compare the reactivity of all the investigated indoleamines with all the above mentioned free radicals. However some comparisons can still be performed. For example, the reactivity towards $Br_2^{\bullet-}$ seems to be higher for TRP and 5-HTA, when compared with MLT. The rate constants for the reactions with this radical are not significantly influenced by the *p*H. On the contrary, the reactions with $^{\bullet}NO_2$ become faster as the *p*H increases. In the particular case of this radical, 5-HTA seems to be the most efficient scavenger among the analyzed compounds. At neutral *p*H it reacts faster than MLT, and at basic *p*H it reacts faster than 5-HTP and TRP, which has the lowest rate constant. In any case these comparisons should be interpreted with caution since the experimental conditions were not identical in the different studies.

In addition there are rate constants reported for the repair of guanosyl radical ($G^{\bullet}$) by MLT and TRP. In the case of MLT the rate constant corresponds to a reaction that is diffusion controlled, indicating its high efficiency for that purpose. In the case of TRP the rate constants have been measured at different *p*H ranging from 1.3 to 13.3, and all of them are high enough to allow fast repair. Apparently such repairing processes would be close to the diffusion limit at neutral *p*H (near to the physiological one) albeit it is faster as the *p*H lowers. In any case, regardless of the *p*H and the polarity of the solvent, the rate constants for the $G^{\bullet}$

repair by TRP are always higher than 10^7 $M^{-1}s^{-1}$, supporting the efficiency of this indoleamine for repairing oxidated DNA.

VII. Theoretical Investigations

Theoretical, or computational protocols, aiming to elucidate different aspects of the chemical reactions which may be relevant to oxidative stress, are rather recent, but not less important than experimental approaches. They have contributed to provide information on both mechanisms and kinetics by using different strategies. Here some representative examples are presented.

The antioxidant activity of different *ortho-* and *meta-* substituted tryptophan derivatives was theoretically investigated in gas and aqueous phases. [193] Different reaction mechanisms were explored, namely HAT, SPLET and SETPT, while the calculated data included bond dissociation enthalpies, ionization potentials, electron transfer enthalpies, proton affinities and proton dissociation enthalpies. Based on these data it was proposed that substituents in the *ortho* position have the highest potential for developing novel TRP derivatives with enhanced antioxidant activity. In addition, it was also proposed that electron withdrawing groups would be particularly efficient for increasing the activity via SPLET.

The selectivity of the $^{\bullet}OH$ reactions with TA and 5-HTA has been theoretically examined, using the energetics of the radical addition and the water elimination from the adducts. [191] Based on this study, it was possible to proposed the formation of two different types of radical adducts, one involving the H-O bond and the other by C-C linkage. Both dimer products, i.e., yielded by intermolecular H transfer followed by intramolecular reorganization through H shift, and by direct recombination of indoloxyl radicals were described as equally efficient thermodynamic processes, which may interfere with regular metabolic pathways. This theoretical approach was a part of a combined experimental-theoretical investigation.

In another combined study, a theoretical approach was used to calculate several descriptors (first ionization energies, bond dissociation energies, and spin densities) and tested their suitability for evaluating the antioxidant activity of MLT. [194] The possibility of associating these descriptors with different mechanisms of reaction was proposed. The experimental data were found to support the usefulness of the computed scavenging parameters of MLT. This indoleamine is the most studied one also from a theoretical point of view.

The viability of the SET mechanism in its free radical scavenging activity was investigated by first characterizing different conformers of the melatonin radical cation and then calculating the adiabatic ionization energies (IE), which by definition are directly related to the facility for electron donation. [195] The finding that the IE values are small, allowed proposing that the antioxidant properties of MLT are due to the stability of its radical cation. The viability of the HAT mechanism, from the N-atom in the indole moiety, was also investigated and ruled out based on the calculated endothermicity. In the same work, 6-HMLT was also studied, and proposed that the additional electron-donating group in the aromatic moiety should contribute to a better stabilization of the radical cation. For this particular indoleamine the HAT mechanism was found to be viable. These findings are in line with previous experimental results, which had suggested that the antioxidant activity of 6-HMLT is even higher than that of MLT itself.

The competing reaction mechanisms involved in the MLT + $^{\bullet}OH$ reaction was also studied by considering two different reaction channels, one involving the abstraction of an indolic hydrogen (HAT) thus yielding an indolyl neutral radical; and another one involving the $^{\bullet}OH$ addition to the indolic moiety (RAF) to form 2-hydroxymelatonin. [196] The calculated ΔG values were found to be -30 and -31 kcal/mol, respectively. Based on these thermochemical results, it was proposed that MLT can directly scavenge hydroxyl radicals both in vacuum and in aqueous solution. However, the SET mechanism was ruled out because the predicted ΔG values were positive.

The thermochemical feasibility of five different RAF pathways for the reaction of MLT with $^{\bullet}OH$, were investigated in another theoretical work, and it was proposed that all of them can occur easily and with low selectivity. [197] Because of the very reactive nature of $^{\bullet}OH$, attention was called to the possibility that other reaction pathways could also be feasible for the reactions involving this radical.

In a further study, all the reaction paths involved in the reactions of MLT with $^{\bullet}OH$ and a series of peroxyl radicals were included. [13] The mechanisms of reaction included in this study were RAF, HAT, SET, SEPT and PCET. In addition, since MLT is highly soluble in lipids and partially soluble in water, the influence of the environment's polarity was also analyzed. MLT was found to react with $^{\bullet}OH$ in a diffusion-limited way, regardless of the polarity of the environment, which is in line with the experimental kinetic data, and indicates that this molecule is an excellent $^{\bullet}OH$ scavenger. Based on kinetic calculations, MLT was also predicted to be a very good $^{\bullet}OOCCl_3$ scavenger, both in aqueous and in lipid media. However, it was also found that this indoleamine is rather ineffective for scavenging less reactive peroxyl radicals, such as alkenyl peroxyl radicals and $^{\bullet}OOH$. Therefore, it was concluded that the protective effect of MLT against lipid peroxidation does not take place by directly trapping peroxyl radicals, but rather by scavenging more reactive radicals, such as $^{\bullet}OH$, which initiate the degradation process. An estimation of the relative importance of the different mechanisms and reaction paths was also performed in the same work [13]. In aqueous solution, SET was predicted to be the most important mechanism, followed by HAT and RAF (in that order) for the reaction of MLT with $^{\bullet}OH$, while for the MLT + $^{\bullet}OOCCl_3$ reaction RAF was proposed as the main mechanism closely followed by HAT. In non-polar environments, the RAF mechanism was predicted to be the one contributing the most to the overall reactivity of MLT towards $^{\bullet}OOCCl_3$, while HAT was identified as a minor scavenging path. On the other hand, for the reaction with $^{\bullet}OH$, in the same environment, both mechanisms were proposed to be similarly important.

In a very recent study, also using a theoretical approach, it was proposed that melatonin and its metabolites are capable of fully inhibiting the oxidative stress induced by Cu(II)-ascorbate mixtures, via Cu(II) chelation. [198] These compounds were described as capable of preventing the first step of the Haber-Weiss reaction, consequently turning off the $^{\bullet}OH$ production via the Fenton reaction. Therefore, it was proposed that, in addition to the previously reported free radical scavenging cascade, melatonin is also involved in a concurrent "chelating cascade" thereby contributing to reduce oxidative stress.

VIII. Are Indoleamines Ideal Antioxidants?

Many molecules have been reported to exhibit antioxidant activity. However, they are not all equally efficient for that purpose. A series of requirements have been previously proposed [199], which characterize an ideal antioxidant. In this section it is analyzed whether indoleamines fulfill these desired criteria.

VIII.1. Location and Concentrations

Due to their high reactivity, most free radicals have short half-lives in biological systems. This means that they would react with molecules in the vicinity of their site of formation. Accordingly, an efficient antioxidant should be ubiquitous and should also be present in adequate amounts in cells.

MLT is a widespread molecule that can be found in both animal and vegetal kingdoms. In humans it can be found in the pineal gland, serum and cerebrospinal fluids, the aqueous humor of the eyes, plasma, ovarian follicular fluid, saliva, hepato-gastrointestinal tissues, etc. Its concentrations in human serum and cerebrospinal fluid varies widely [200] and in the aqueous humor of the eyes change over the light-dark cycle [201]. Melatonin's concentration in plasma also varies during the day and from childhood to adulthood. In adults, melatonin levels in the blood are usually below 10 pg/ml at night [202,203]. Its concentration has been estimated to be as high as 112 pg/ml in the ovarian follicular fluid [204], 5-15 pg/ml in saliva [205], and ~100 (day) to ~400 (night) pg/mg in the pineal gland [206]. In hepato-gastrointestinal tissues its concentration has been reported to be 80, 176, 163-214, and 214 pg/100mg wet weight in liver, duodenum, colon, and rectum; respectively [207]. While these are measurements from specific reports, the values estimated by others can widely vary [208,209].

5-HTA is also widely distributed in humans. It is present in platelets, the cardiovascular, pulmonary, gastrointestinal, and genitourinary systems, as well as in the central and peripheral nervous systems. [210] Over 95% of serotonin in humans' body is located in the gastrointestinal tract, [211,212] while platelets are considered the major storage site for this indoleamine outside the central nervous system. It has been found that serotonin´s concentrations vary with seasons and sunlight, decreasing in winter and rapidly rising with increased luminosity. [213] The concentrations of 5-HTA in different brain regions have been estimated to be higher in the putamen, amygdala, and caudate nucleus (2.0, 1.8, and 1.2 nmoles/g wet weight, respectively) than in the hippocampus, cingulate cortex and frontal cortex (0.5, 0.5, and 0.1 nmoles/g wet weight). [214] The serotonin levels in healthy human platelets have been estimated in several studies to range from 0.5 to 0.9 $\mu g/10^9$ platelets. [215] In the hypothalamus it is 0.39 μg/g tissue, [216] and ranges from 1.9 to 27.3 nmol/l in ventricular and from 5.7 to 12.0 nmol/l in lumbar cerebrospinal fluids (CSF). [217] However, in a previous study the range of 5-HTA concentrations in cerebrospinal fluid was found to be 0-13 ng/ml. [218] In addition serotonin levels have also been estimated to be 1.61, 2.12 and 2.66 ng/ml in plasma of maternal vein, umbilical vein and umbilical artery, respectively, of pregnant women at the third trimester. [219]

For humans, TRP is an essential amino acid, i.e., it cannot be endogenously synthesized. Fortunately it is present in both vegetal and animal food, and thus its intake through the diet is assured. Within living organisms, including humans, tryptophan is widely distributed and it can be found in both its free form and as a building block in proteins and peptides. Free blood TRP has been reported to be about 1 mg/l in infants of 2 to 6 weeks of age, while total TRP levels range from 8.1 to 10.2 mg/l. [220] The levels of tryptophan were estimated to range from 1593 to 4865 nmol/l in ventricular and from 1257 to 2557 nmol/l in lumbar CSF. [217] In addition for pregnant women at the third trimester, TRP concentrations were found to be 6.48, 12.97, and 12.32 μg/ml. [219]

TA is mainly present in the central nervous system of mammalians, at relatively low concentrations compared to 5-HTA. That is probably why there is much less information on the distribution and levels of TA. Tryptamine's concentrations in human brain have been estimated to be 389, 172, and 150 ng/g in the cerebellum, cerebral cortex, and basal ganglion, respectively. [221] In human lenses an average concentration equal to 0.229 μg/g lens has been reported, which corresponds to the range 0.066-0.488 μg/g lens. [222]

NAS has been isolated form the pineal gland and also from several extra pineal tissues including the brainstem, cerebellum, hippocampus, retina, digestive tract and ovary. In addition, it is produced and released by cultured skin cells and by human peripheral blood lymphocytes. It is present in human serum in nanomolar concentrations, which has been reported to be about 10 to 100-fold higher than those of MLT. [142] The pineal content of NAS has been found to decrease with age, as occurs with MLT. [223]

The data regarding the other indoleamines analyzed in this work is rather scarce. 6-HMT, for example has been used as an indicator of human plasma MLT through the urinary excretion of 6-hydroxymelatonin sulfate. However, to our best knowledge the distribution and concentrations of this indoleamine in human tissues have not been quantified. 5-MTA has been reported to be present, in humans, at concentrations of 12.8, 10.6, 7.9, 7.1, 6.2, ng/g of wet tissue in the pineal gland, brainstem (pons and medulla), striatum, thalamus, hypothalamus, respectively and < 5.0 cerebellum and cerebral cortex. [224] The level of 5-HTP was found to varied from 1.1 to 68.9 nmol/l in ventricular and from 5.3 to 10.8 nmol/l in lumbar CSF. [217]

Thus, according to the above presented data, it can be safely stated that indoleamines are widespread within the human body, at concentrations that should guarantee their availability for scavenging *in situ* produced free radicals.

VIII.2. Versatility

Due to the wide variety of free radicals that can be found in living organisms, a good antioxidant should be versatile, i.e., it should easily react with different free radicals.

As shown in section VI, all the indoleamines analyzed in this chapter are capable of reacting with more than one free radical. Such reactions are fast enough to represent significant sinks for these oxidants, which indicates that indoleamines can act as versatile free radical scavengers. Moreover, these compounds are capable of rapidly repair guanosyl radicals ($G^•$), and at least MLT seems to be capable of inhibiting $^•OH$ production via Fenton reaction, and to quench $^1O_2$, [225,226] thus acting as a multifunctional antioxidant.

VIII.3. Crossing Physiological Barriers

A good antioxidant is expected to have the ability of crossing physiologic barriers and to be quickly transported into the cells. [172, 227-229] Indoleamines, are amphiphilic molecules, i.e., they present both hydrophilic and lipophilic character. In general, they are more soluble in lipids than they are in water, albeit the hydroxylated species are more hydrophilic than the non-hydroxylated ones. Therefore, these molecules can easily penetrate biological membranes. [230-234] Moreover, their intermediate size is optimum for transportation across cellular membranes [199].

VIII.4. Availability

It is also important for an antioxidant to be available when needed. Therefore they should be acquired in the diet, i.e., obtained from exogenous sources; or produced *in situ*, i.e., by endogenous sources.

All the analyzed indoleamines, but TRP, are endogenously produced. Therefore TRP must be acquired through the diet. Main dietary sources of TRP are diverse and include eggs, red meat, fish, poultry, quinoa, soybeans, cheese, milk, yogurt, sesame seeds, chocolate, oats, dried dates, chickpeas, sunflower seeds, pumpkin seeds, spirulina, bananas, cherries, and peanuts.

In addition to their endogenous production, MLT and 5-HTA can also be obtained from the diet. MLT is found in a wide variety of fruits, vegetables, and nuts including cherries, corn, rice, barley grains, and ginger. 5-HTA has also been found in a wide variety of foods, with the main sources being walnut, hickory, chocolate (with diverse cocoa contents), pineapples, banana, kiwi, plums, and tomatoes.

Accordingly, acquired from the diet of metabolically produced, indoleamines are expected to be available, when needed, to offer protection against oxidative stress.

VIII.6. Regeneration

The term regeneration in this context refers to antioxidants that are capable of scavenging several radical equivalents. The reaction between an antioxidant and a free radical yields an oxidized form of the antioxidant. This new species must have, by definition, less scavenging activity than the original compound. However, many antioxidants have physiologically reducing mechanisms that regenerate their original form, which would then be ready to scavenge another free radical. Another way to success in scavenging more than one radical per molecule of antioxidant is that their oxidized forms can still efficiently react with free radicals.

After donating an electron to $^{\bullet}OH$, MLT becomes a free radical itself, the indoyl radical cation [13,235]. However, its reactivity is very low and therefore it is not toxic to cells. [236] In addition, melatonin's metabolites arising from its reaction with ROS, still possess free radical scavenging activities, [237-245] which is referred to as the melatonin scavenging cascade. There is also evidence that MLT can be regenerated after radical quenching by other species such as ascorbate and urate [165]. Oxidized TRP can also be repaired by electron donors such as ascorbate, *N,N,N',N'*-tetramethyl-p-phenylenediamine dihydrochloride (TMPD), and phenolic antioxidants. [246]

Based on the structural similarities among indoleamines, it is expected that their oxidized forms also present rather low reactivity, and that they can be easily repaired via electron transfer from other antioxidants present in their vicinity. It is also possible that other antioxidants are also capable of repairing indoleamines by H transfer processes. However, this hypothesis still remains to be tested.

VIII.7. Toxicity

A very important aspect to consider when evaluating the suitability of a compound as an antioxidant is its toxicity. It should be non-toxic before, and after, the free radical scavenging process takes place.

The information available regarding indoleamines' toxicity is quite varied, and limited in some cases. Probably the most concerning issues, related to indoleamines toxicity, is the excess of 5-HTA in the central nervous system that causes a condition usually referred to as the serotonin syndrome. The associated adverse effects can be can be mild, moderate or even life-threatening. It arises from pharmacological agents increasing 5-HTA synthesis or release, decreasing 5-HTA metabolism, inhibiting 5-HTA reuptake or the direct agonism of the 5-HTA receptors. [247] The symptoms include tremor, hyper-reflexia, fever, tachycardia, agitation, excitement, confusion, and pyramidal rigidity. The etiology is often a consequence of therapeutic drug use, intentional overdose of serotonergic agents, or complex interactions between drugs that directly or indirectly modulate the serotonin system. [247] A wide variety of drugs, and drug combinations, have been associated with the serotonin syndrome. Some examples are: tricyclic antidepressants, opiate analgesics, cough medicines, antibiotics, weight reduction agents, antiemetics, antimigraine agents, drugs of abuse and herbal products. [248] Since 5-HTA is not directly consumed, overdose measurements are taken for the triggering drugs and not for 5-HTA levels.

Because of the tryptophan metabolic pathway (Scheme 2), the highest risk of TRP and 5-HTP overdoses might be related to the serotonin syndrome. On the contrary, there are also several disorders associated with low TRP levels due to poor dietary intake, malabsorption, or high metabolic rates. [249] Different tryptophan dosages have been tested without reaching relevant toxic effects. It has been reported that a moderate dose (3 g per day) produces no more side effects than placebo. [250] Higher doses of TRP such as 9.6g per day in affective disorder patients, and 20g per day in schizophrenic patients, did not produce any adverse effects either. [251,252] Regarding 5-HTP, abundant clinical trials have been carried out to evaluate its efficiency in treating several health disorders in humans. For example, 5-HTP was administered at doses ranging from 50 to 600 mg per day, or even up to 3250 mg per day in some cases, for the treatment of various types of depression, fibomyalgia, obesity, migraines and insomnia. [253] No adverse effects were observed, provided that 5-HTP is pristine (no impurities) and used alone (not combined with other drugs).

On the other hand, it has been reported that the only significant short-term side effect after oral ingestion of ≤ 5 mg of MLT by normal healthy adults is sleepiness. [254] Numerous *in vivo* studies on animals, involving massive doses of melatonin have shown that acute and chronic toxicity of melatonin is extremely low. [255-258] Moreover, oral doses (up to 1g daily) taken by human volunteers resulted in no negative side effects. [259] There is a general agreement that MLT has minimal toxicity over a very wide dose range.

In general it can be stated that, in moderate doses, indoleamines present no health risks. However special attention should be paid to avoid overdoses of serotonin precursors.

CONCLUSION

This chapter presented an overview of several aspects relevant to the protection exerted by indoleaminas against oxidative stress. It is concluded that indoleamines are able of reducing oxidative stress by several pathways including, at least, free radical scavenging and metal chelating processes. This family of compounds fulfills most of the features necessary to be considered ideal protectors against oxidative stress. They are ubiquitous molecules present at enough concentrations in the different regions of living organisms. They are versatile agents capable of acting as both primary and secondary antioxidants; and capable of scavenging a wide variety of free radicals. Their amphiphilic nature, and intermediate size, makes this family of compounds ideal for crossing physiological barriers. They can be regenerated, at least by electron transfer, by other antioxidants. The products of their reactions with free radicals, are expected to be innocuous for biological targets such as lipids, proteins, and DNA. In general they are nontoxic, at moderate levels, provided that they have no impurities and are not combined with other drugs. Special attention should be paid to avoid overdoses of serotonin precursors. In the particular case of melatonin no adverse effects have been reported so far. In summary, in addition to their other important physiological functions, indoleamines are proposed as efficient species for counteracting the deleterious effects caused by oxidative stress.

REFERENCES

[1] Gerschman, R; Gilbert, DL; Nye, SW; Dwyer, P; Fenn, WO. Oxygen poisoning and X-irradiation: A mechanism in common. *Science*, 1954, 119, 623–626.

[2] Pryor, WA. Why is the hydroxyl radical the only radical that commonly adds to DNA? Hypothesis: it has a rare combination of high electrophilicity, high thermochemical reactivity, and a mode of production that can occur near DNA. *Free Radic Biol Med*, 1988, 4, 219–223.

[3] Draganic, IG; Draganic, ZD. The radiation chemistry of water. Academic Press, New York, 1971.

[4] Vijayalaxmi; Reiter, RJ; Tan, DX; Herman, TS; Thomas, CR Jr. Melatonin as a radioprotective agent: A review. *Int J Radiation Oncol Biol Phys*, 2004, 59, 639–653.

[5] Candeias, LP; Steenken, S. Reaction of HO with guanine derivatives in aqueous solution: Formation of two different redox-active OH-adduct radicals and their unimolecular transformation reactions. Properties of G(-H). *Chem Eur J*, 2000, 6, 475–484.

[6] Chatgilialoglu, C; D'Angelantonio, M; Guerra, M; Kaloudis, P; Mulazzani, QG. A reevaluation of the ambident reactivity of the guanine moiety towards hydroxyl radicals. *Angew Chem Int Ed*, 2009, 48, 2214−2217.

[7] Galano, A; Alvarez-Idaboy, JR. Guanosine + OH radical reaction in aqueous solution: A reinterpretation of the UV-Vis data based on thermodynamic and kinetic calculations. *Org Lett*, 2009, 11, 5114−5117.

[8] Pryor, WA. Oxy-radicals and related species: Their formation, life-times, and reactions. *Annu Rev Physiol*, 1986, 48, 657−667.

[9] Huie, RE; Neta, P. Chemistry of reactive oxygen species, in reactive oxygen species in biological systems: An interdisciplinary approach. Ed. Gilbert DL, Colton CA. Kluwer, Academic Publishers, 2002, 33–63.

[10] Bielski, BHJ; Cabelli, DE; Arudi, RL; Ross, AB. Reactivity of HO_2/O_2- radicals in aqueous solution. *J Phys Chem Ref Data*, 1985, 14, 1041–1100.

[11] De Grey, AD. HO2*: The Forgotten Radical. *DNA Cell Biol*, 2002, 21, 251−257.

[12] Aikens, J; Dix, TA. Perhydroxyl radical ($HOO^•$) initiated lipid peroxidation. The role of fatty acid hydroperoxides. *J Biol Chem*, 1991, 266, 15091−15098.

[13] Galano, A. On the direct scavenging activity of melatonin towards hydroxyl and a series of peroxyl radicals. *Phys Chem Chem Phys*, 2011, 13, 7147–7157.

[14] Leon-Carmona, JR; Galano, A. Is caffeine a good scavenger of oxygenated free radicals? *J Phys Chem B*, 2011, 115, 4538–4546.

[15] Neta, P; Huie, RE; Mosseri, S; Shastri, LV; Mittal, JP; Maruthamuthu, P; Steenken, S. Rate constants for reaction of substituted methylperoxyl radicals with ascorbate ions and TMPD. *J Phys Chem*, 1989, 93, 4099–4104.

[16] Galano, A. Relative antioxidant efficiency of a large series of carotenoids: Electron transfer reactions. *J Phys Chem B*, 2007, 111, 12898–12908.

[17] Martinez, A; Vargas, R; Galano, A. What is important to prevent oxidative stress? A theoretical study on electron transfer reactions between carotenoids and free radicals. *J Phys Chem B*, 2009, 113, 12113–12120.

[18] Martinez, A; Vargas, R; Galano, A. Theoretical study on the chemical fate of adducts formed through free radical addition reactions to carotenoids. *Theor Chem Acc*, 2010, 127, 595–603.

[19] Martinez, A; Galano, A. Free radical scavenging activity of ultra short single walled carbon nanotubes with different structures through electron transfer reactions. *J Phys Chem* C, 2010, 114, 8184–8191.

[20] Brunelli, L; Crow, JP; Beckman, JS. The comparative toxicity of nitric oxide and peroxynitrite to Escherichia Coli. *Arch Biochem Biophys*, 1995, 316, 327–334.

[21] Squadrito, GL; Pryor, WA. Oxidative chemistry of nitric oxide: The roles of superoxide, peroxynitrite, and carbon dioxide. *Free Radic Biol Med*, 1998, 25, 392–403.

[22] Radi, R; Peluffo, G; Alvarez, MN; Naviliat, M; Cayota, A. Unraveling peroxynitrite formation in biological systems. *Free Radic Biol Med*, 2001, 30, 463–488.

[23] Douki, T; Cadet, J. Peroxynitrite mediated oxidation of purine bases of nucleosides and isolated DNA. *Free Radic Res*, 1996, 24, 369–380.

[24] Wiseman, H; Halliwell, B. Damage to DNA by reactive oxygen and nitrogen species: Role in inflammatory disease and progression to cancer. *Biochem J*, 1996, 313, 17–29.

[25] Koppal, T; Drake, J; Yatin, S; Jordan, B; Varadarajan, S; Bettenhausen, L; Butterfield, DA. Peroxynitriteinduced alterations in synaptosomal membrane proteins: Insight into oxidative stress in Alzheimer's disease. *J Neurochem*, 1999, 72, 310–317.
[26] Beckman, JS; Beckman, TW; Chen, J; Marshall, PA; Freeman, BA. Apparent hydroxyl radical production by peroxynitrite: Implications for endothelial injury from nitric oxide and superoxide. *Proc Natl Acad Sci USA*, 1990, 87, 1620–1624.
[27] Pryor, WA; Squadrito, GL. The chemistry of peroxynitrite: A product from the reaction of nitric oxide with superoxide. *Am J Physiol (Lung Cell Mol Physiol 12)*, 1995, 268, L699–L722.
[28] Huie, RE; Padmaja, S. The reaction of NO with superoxide. *Free Radic Res* Comm, 1993, 18, 195–199.
[29] Kobayashi, K; Miki, M; Tagawa, S. Pulse-radiolysis study of the reaction of nitric oxide with superoxide. *J Chem Soc Dalton Trans*, 1995, 2885–2889.
[30] Goldstein, S; Czapski, G. The reaction of $NO^{\bullet}$ with $O_2^{\bullet-}$ and $HO_2^{\bullet-}$: A pulse radiolysis study. *Free Radic Biol Med*, 1995, 19, 505–510.
[31] Noda, Y; Mori, A; Liburdy, R; Packer, L. Melatonin and its precursors scavenge nitric oxide. *J Pineal Res*, 1999, 27, 159–163.
[32] Squadrito, GL; Pryor, WA. The formation of peroxynitrite in vivo from nitric oxide and superoxide. *Chem Biol Interact*, 1995, 96, 203–206.
[33] Forni, LG; Packer, JE; Slater, TF; Willson, RL. Reaction of the trichloromethyl and halothane-derived peroxy radicals with unsaturated fatty acids: A pulse radiolysis study. *Chem Biol Interact*, 1983, 45, 171–177.
[34] Prütz, WA; Mönig, H; Butler, J; Land, EJ. Reactions of nitrogen dioxide in aqueous model systems: Oxidation of tyrosine units in peptides and proteins. *Arch Biochem Biophys*, 1985, 243, 125–134.
[35] Halliwell, B; Murcia, MA; Chirico, S; Aruoma, OI. Free radicals and antioxidants in food and in vivo: What they do and how they work. *Crit Rev Food Sci Nutr*, 1995, 35, 7–20.
[36] Halliwell, B; Whiteman, M. Measuring reactive species and oxidative damage in vivo and in cell culture: how should you do it and what do the results mean? *British J Pharmacol*, 2004, 142, 231–255.
[37] Chaiyasit, W; Elias, RJ; McClements, DJ; Decker, EA. Role of physical structures in bulk oils on lipid oxidation. *Crit Rev Food Sci Nutr*, 2007, 47, 299–317.
[38] Yu, HS; Tsin, ATC; Reiter, RJ. Melatonin: History, Biosynthesis, and Assay Metodology, in Melatonin: biosynthesis, physiological effects and clinical applications. Chapter 1, 1–16. CRC Press, Boca Raton., 1992
[39] Murch, SJ; KrishnaRaj, S; Saxena, PK. Tryptophan is a precursor for melatonin and serotonin biosynthesis in in vitro regenerated St. John's wort (Hypericum perforatum L. cv. Anthos) plants. *Plant Cell Rep*, 2000, 19, 698–704.
[40] Tan, DX; Hardeland, R; Manchester, LC; Paredes, SD; Korkmaz, A; Sainz, RM; Mayo, JC; Fuentes-Broto, L; Reiter, RJ. The changing biological roles of melatonin during evolution: from an antioxidant to signals of darkness, sexual selection and fitness. *Biol Rev Camb Philos Soc*, 2010, 85, 607–623.
[41] Zavodnik, IB; Domanski, AV; Lapshina, EA; Bryszewska, M; Reiter, RJ. Melatonin directly scavenges free radicals generated in red blood cells and a cell-free system:

chemiluminescence measurements and theoretical calculations. *Life Sci*, 2006, 79, 391–400.

[42] Gitto, E; Karbownik, M; Reiter, RJ; Tan, DX; Cuzzocrea, S; Chiurazzi, P; Cordaro, S, Corona, G; Trimarchi, G; Barberi, I. Effects of melatonin treatment in septic newborns. *Pediatr Res*, 2001, 50, 756–760.

[43] Longoni, B; Salgo, MG; Pryor, WA; Marchiafava, PL. Effects of melatonin on lipid peroxidation induced by oxygen radicals. *Life Sci*, 1998, 62, 853–859.

[44] Taysi, S; Koc, M; Büyükokuroğlu, ME; Altinkaynak, K; Sahin, YN. Melatonin reduces lipid peroxidation and nitric oxide during irradiation-induced oxidative injury in the rat liver. *J Pineal Res*, 2003, 34, 173–177.

[45] Hong, Y; Palaksha, KJ; Park, K; Park, S; Kim, HD; Reiter, RJ; Chang, KT. Melatonin plus exercise-based neurorehabilitative therapy for spinal cord injury. *J Pineal Res*, 2010, 49, 201–209.

[46] Acuna-Castroviejo, D; Escames, G; Rodriguez, MI; Lopez, LC. Melatonin role in the mitochondrial function. *Front Biosci*, 2007, 12, 947–963.

[47] Samantaray, S; Das, A; Thakore, NP; Matzelle, DD; Reiter, RJ; Ray, SK; Banik, NL. Therapeutic potential of melatonin in traumatic central nervous system injury. *J Pineal Res*, 2009, 47, 134–142.

[48] Manev, H1; Uz, T; Kharlamov, A; Joo, JY. Increased brain damage after stroke or excitotoxic seizures in melatonin-deficient rats. *FASEB J*, 1996, 10, 1546–1551.

[49] Koh, PO. Melatonin attenuates the focal cerebral ischemic injury by inhibiting the dissociation of pBad from 14-3-3. *J Pineal Res*, 2008, 44, 101–106.

[50] Kilic, E; Ozdemir, YG; Bolay, H; Keleştimur, H; Dalkara, T. Pinealectomy aggravates and melatonin administration attenuates brain damage in focal ischemia. *J Cerebr Blood Flow Metab*, 1999, 19, 511–516.

[51] Wakatsuki, A; Okatani, Y; Izumiya, C; Ikenoue, N. Melatonin protects against ischemia and reperfusion induced oxidative lipid and DNA damage in fetal rat brain. *J Pineal Res*, 1999, 26, 147–152.

[52] Cho, EH; Koh, PO. Proteomic identification of proteins differentially expressed by melatonin in hepatic ischemia-reperfusion injury. *J Pineal Res*, 2010, 49, 349–355.

[53] Li, XJ; Zhang, LM; Gu, J; Zhang, AZ; Sun, FY. Melatonin decreases production of hydroxyl radical during cerebral ischemia–reperfusion. *Acta Pharmacol Sinica*, 1997, 18, 394–396.

[54] Cuzzocrea, S; Costantino, G; Gitto, E; Mazzon, E; Fulia, F; Serraino, I; Cordaro, S; Barberi I; De, Sarro, A; Caputi, AP. Protective effects of melatonin in ischemic brain injury. *J Pineal Res*, 2000, 29, 217–227.

[55] Pappolla, MA; Sos, M; Omar, RA; Bick, RJ; Hickson-Bick, DL; Reiter, RJ; Efthimiopoulos, S; Robakis, NK. Melatonin prevents death of neuroblastoma cells exposed to the Alzheimer amyloid peptide. *J Neurosci*, 1997, 17, 1683–1690.

[56] Song, W; Lahiri, DK. Melatonin alters the metabolism of the β-amyloid precursor proteinin the neuroendocrine cell line PL12. *J Mol Neurosci*, 1997, 9, 75–92.

[57] Bozner, P; Grishko, V; LeDoux, SP; Wilson, GL; Chyan, YC; Pappolla, MA. The amyloid β protein induces oxidative damage of mitochondrial DNA. *J Neuropathol Exp Neurol*, 1997, 56, 1356–1362.

[58] Pappolla, M; Bozner, P; Soto, C; Shao, H; Robakis, NK; Zagorski, M; Frangione, B; Ghiso, J. Inhibition of Alzheimer beta-fibrillogenesis by melatonin. *J Biol Chem*, 1998, 273, 7185–7188.

[59] Daniels, WM; van, Rensburg, SJ; van, Zyl, JM; Taljaard, JJ. Melatonin prevents β-amyloidinduced lipid peroxidation. *J Pineal Res*, 1998, 24, 78–82.

[60] Pappolla, MA; Chyan, YJ; Poeggeler, B; Bozner, P; Ghiso, J; LeDoux, SP; Wilson, GL. Alzheimer β protein mediated oxidative damage to mitochondrial DNA: prevention by melatonin. *J Pineal Res*, 1999, 27, 226–229.

[61] Reiter, RJ; Cabrera, J; Sainz, RM; Mayo, JC; Manchester, LC; Tan, DX. Melatonin as a pharmacological agent against neuronal loss in experimental models of Huntington's disease, Alzheimer's disease and parkinsonism. *Ann NY Acad Sci*, 1999, 890, 471–485

[62] Olcese, JM; Cao, C; Mori, T; Mamcarz, MB; Maxwell, A; Runfeldt, MJ; Wang, L; Zhang, C; Lin, X; Zhang, G; Arendash, GW. Protection against cognitive deficits and markers of neurodegeneration by long-term oral administration of melatonin in a transgenic model of Alzheimer disease. *J Pineal Res*, 2009, 47, 82–96.

[63] Dong, W; Huang, F; Fan, W; Cheng, S; Chen, Y; Zhang, W; Shi, H; He, H. Differential effects of melatonin on amyloid-beta peptide 25-35-induced mitochondrial dysfunction in hippocampal neurons at different stages of culture. *J Pineal Res*, 2010, 48, 117–125.

[64] Brusco, LI; Márquez, M; Cardinali, DP. Monozygotic twins with Alzheimer's disease treated with melatonin: case report. *J Pineal Res*, 1998, 25, 260–263.

[65] Wu, YH; Swaab, DF. The human pineal gland and melatonin in aging and Alzheimer's disease. *J Pineal Res*, 2005, 38, 145–152.

[66] Tan, DX; Manchester, LC; Sainz, R; Mayo, JC; Alvares, FL; Reiter, RJ. Read More: http://informahealthcare.com/doi/abs/10.1517/13543776.13.10.1513 Antioxidant strategies in protection against neurodegenerative disorders. *Expert Opin Ther Patents*, 2003, 13, 1513–1543.

[67] Khaldy, H; León, J; Escames, G; Bikjdaouene, L; García, JJ; Acuña-Castroviejo, D. Circadian rhythm of dopamine and their metabolites in mouse striatum: Effects of pinealectomy and melatonin replacement. *Neuroendocrinology*, 2002, 75, 201–208.

[68] Acuña-Castroviejo, D; Coto-Montes, A; Gaia, Monti, M; Ortiz, GG; Reiter, RJ. Melatonin is protective against MPTP-induced striatal and hippocampal lesions. *Life Sci*, 1997, 60, PL23–29.

[69] Khaldy, H; Escames, G; León, J; Vives, F; Luna, JD; Acuña-Castroviejo, D. Comparative effects of melatonin, Ldeprenyl, Trolox and ascorbate in the suppression of hydroxyl radical formation during dopamine autoxidation in vitro. *J Pineal Res*, 2000, 2, 100–107.

[70] Yoshida, M; Fukuda, A; Hara, M; Terada, A; Kitanaka, Y; Owada, S. Melatonin prevents the increase in hydroxyl radical-spin trap adduct formation caused by the addition of cisplatin in vitro. *Life Sci.*, 2003, 72, 1773–1780.

[71] Konturek, PC; Konturek, SJ; Majka, J; Zembala, M; Hahn, EG. Melatonin affords protection against gastric lesions induced by ischemia–reperfusion possibly due to its antioxidant and mucosal microcirculatory effects. *Eur J Pharmacol*, 1997, 322, 73–77.

[72] Konturek, PC; Konturek, SJ; Brzozowski, T; Dembinski, A; Zembala, M; Mytar, B; Hahn EG. Gastroprotective effect of melatonin and its precursor, l-tryptophan, against stress-induced and ischemia-induced lesions is mediated by scavenging of oxygen free radicals. *Scand J Gastroenterol*, 1997, 32, 433–438.

[73] Konturek, PC; Konturek, SJ; Celinski, K; Slomka, M; Cichoz-Lach, H; Bielanski, W; Reiter RJ. Role of melatonin in mucosal gastroprotection against aspirin-induced gastric lesions in humans. *J Pineal Res*, 2010, 48, 318–323.

[74] Sewerynek, E; Reiter, RJ; Melchiorri, D; Ortiz, GG; Lewinski, A. Oxidative damage to the liver induced by ischemia–reperfusion: protection by melatonin. *Hepatogastroenterology*, 1996, 43, 898–905.

[75] Tan, DX; Manchester, LC; Reiter, RJ; Qi, W; Kim, SJ; El-Sokkary, GH. Ischemia/reperfusioninduced arrhythmias in the isolated rat heart: prevention by melatonin. *J Pineal Res*, 1998, 25, 184–191.

[76] Lagneux, C; Joyeux, M; Demenge, P; Ribuot, C; Godin-Ribuot, D. Protective effects of melatonin against ischemia-reperfusion injury in the isolated rat heart. *Life Sci*, 2000, 66, 503–509.

[77] Tan, DX; Manchester, LC; Reiter, RJ; Qi, WB; Karbownik, M; Calvo, JR. Significance of melatonin in antioxidative defense system: reactions and products. *Biol Signals Recept.*, 2000, 9, 137–59.

[78] Rosen, J; Than, NN; Koch, D; Poeggeler, B; Laatsch, H; Hardeland, R. Interactions of melatonin and its metabolites with the ABTS cation radical: extension of the radical scavenger cascade and formation of a novel class of oxidation products, C2-substituted 3-indolinones. *J Pineal Res.*, 2006, 41, 374–81.

[79] Tan, DX; Manchester, LC; Terron, MP; Flores, LJ; Reiter, RJ. One molecule, many derivatives: a never-ending interaction of melatonin with reactive oxygen and nitrogen species? *J Pineal Res.*, 2007, 42, 28–42.

[80] Lopez-Burrillo, S; Tan, DX; Rodriguez-Gallego, V; Manchester, LC; Mayo, JC; Sainz, RM, Reiter, RJ. Melatonin and its derivatives cyclic 3-hydroxymelatonin, N1-acetyl-N2-formyl-5-methoxykynuramine and 6-hydroxymelatonin reduce DNA oxidative damage induced by Fenton eagents. *J. Pineal Res*, 2003, 34, 178–184.

[81] Tan, DX; Hardeland, R; Manchester, LC; Poeggeler, B; Lopez-Burillo, S; Mayo, JC; Sainz RM; Reiter, RJ. Mechanistic and comparative studies of melatonin and classic antioxidants in terms of their interactions with the ABTS cation radical. *J Pineal Res*, 2003, 34, 249–59.

[82] Galano, A; Tan, DX; Reiter, RJ. Cyclic 3-hydroxymelatonin, a key metabolite enhancing the peroxyl radical scavenging activity of melatonin. *RSC Adv*, 2014, 4, 5220–5227.

[83] Tan, DX; Hardeland, R; Manchester, LC; Galano, A; Russel, JR. Cyclic-3-hydroxymelatonin (C3HOM), a potent antioxidant, scavenges free radicals and suppresses oxidative reactions. *Curr Med Chem*, 2014, 21, 1557−1565.

[84] Limson, J; Nyokong, T; Daya, S. The interaction of melatonin and its precursors with aluminium, cadmium, copper, iron, lead, and zinc: An adsorptive voltammetric study. *J Pineal Res*, 1998, 24, 15−21.

[85] Parmar, P; Limson, J; Nyokong, T; Daya, S. Melatonin protects against copper-mediated free radical damage. *J Pineal Res*, 2002, 32, 237–242.

[86] Mayo, JC; Tan, DX; Sainz, RM; Natarajan, M; Lopez-Burillo, S; Reiter, RJ. Protection against oxidative protein damage induced by metal-catalyzed reaction or alkylperoxyl radicals: comparative effects of melatonin and other antioxidants. *Biochim Biophys Acta*, 2003, 1620, 139–150.

[87] Gulcin, I; Buyukokuroglu, ME; Kufrevioglu, OI. Metal chelating and hydrogen peroxide scavenging effects of melatonin. *J Pineal Res*, 2003, 34, 278–281.

[88] Zatta, P; Tognon, G; Carampi, P. Melatonin prevents free radical formation due to the interaction between b-amyloid peptides and metal ions [Al(III), Zn(II), Cu(II), Mn(II), Fe(II)]. *J Pineal Res*, 2003, 35, 98–103.

[89] Wakatsuki, A; Okatani, Y; Ikenoue, N; Izumiya, C; Kaneda, C. Melatonin inhibits oxidative modification of low-density lipoprotein particles in normolipidemic post-menopausal women. *J. Pineal Res.*, 2000, 28, 136–142.

[90] Romero, A; Ramos, E; de, Los, Ríos, C; Egea, J; Del, Pino, J; Reiter, RJ. A review of metal-catalyzed molecular damage: protection by Melatonin. *J Pineal Res*, 2014, 56, 343–370.

[91] Pierrefiche; G; Topall; G; Courboin; G; Henriet; I; Laborit; H. Antioxidant activity of melatonin in mice. *Res Commun Chem Pathol Pharmacol*, 1993, 80: 211–223.

[92] Maharaj, DS; Limson, JL; Daya; S. 6-hydroxymelatonin converts iron (III) to iron (II) and reduces iron-induced lipid peroxidation. *Life Sci.*, 2003, 72, 1367–1375.

[93] Maharaj, DS; Walker, RB; Glass, BD; Daya, S. 6-hydroxymelatonin protects against cyanide induced oxidative stress in rat brain homogenates. *J Chem Neuroanatomy*, 2003, 26, 103–107.

[94] Maharaj, DS; Maharaj, H; Antunes, EM; Maree, DM; Nyokong, T; Glass, BD; Daya, S. 6-hydroxymelatonin protects against quinolinic-acidinduced oxidative neurotoxicity in the rat hippocampus. *J Pharm Pharmacol*, 2005, 57, 877–881.

[95] Maharaj, DS; Anoopkumar-Dukie, S; Glass, BD; Antunes, EM; Lack, B; Walker, RB; Daya S. The identification of the UV degradants of melatonin and their ability to scavenge free radicals. *J Pineal Res.*, 2002, 32, 257–261.

[96] Maharaj, DS; Maharaj, H; Daya, S; Glass, BD. Melatonin and 6-hydroxymelatonin protect against iron-induced neurotoxicity. *J Neurochem*, 2006, 96, 78–81.

[97] Sakano, K; Oikawa, S; Hiraku, Y; Kawanishi, S. Oxidative DNA damage induced by a melatonin metabolite, 6-hydroxymelatonin, via a unique non-o-quinone type of redox cycle. *Biochem Pharmacol*, 2004, 68, 1869–1878.

[98] Garrido, M; Espino, J; González-Gómez, D; Lozano, M; Cubero, J; Toribio-Delgado, AF; Maynar-Mariño, JI; Terrón, MP; Muñoz, JL; Pariente, JA; Barriga, C; Paredes, SD; Rodríguez, AB. A nutraceutical product base don Jerte Valley cherries improves sleep and augments the antioxidant status in humans. *E-SPEN J*, 2009, 4, e321–e323.

[99] Garrido, M; Paredes, SD; Cubero, J; Lozano, M; Toribio-Delgado, AF; Muñoz, JL; Reiter, RJ; Barriga, C; Rodríguez, AB. Jerte Valley cherry-enriched diets improve nocturnal rest and increase 6-sulfatoxymelatonin and total antioxidant capacity in the urine of middle-aged and elderly humans. *J Gerontol A Biol Sci Med Sci*, 2010, 65, 909–914.

[100] Bravo, R; Matito, S; Cubero, J; Paredes, SD; Franco, L; Rivero, M; Rodríguez, AB; Barriga C. Tryptophan-enriched cereal intake improves nocturnal sleep, melatonin, serotonin, and total antioxidant capacity levels and mood in elderly humans. *Age*, 2013, 35, 1277–1285.

[101] Wen, H; Feng, L; Jiang, W; Liu, Y; Jiang, J; Li, S; Tang, L; Zhang, Y; Kuang, S; Zhou, X. Dietary tryptophan modulates intestinal immune response, barrier function, antioxidant status and gene expression of TOR and Nrf2 in young grass carp (Ctenopharyngodon idella). *Fish & Shellfish Immunol*, 2014, 40, 275–287.

[102] Jiang, Y; Ng, TB; Wang, CR; Li, N; Wen, TY; Qiao, WT; Zhang, D; Cheng, ZH; Liu, F. First isolation of tryptophan from edible lotus (Nelumbo nucifera Gaertn) rhizomes and demonstration of its antioxidant effects. *Int J Food Sci Nutr*, 2010, 61, 346–356.

[103] Bitzer-Quintero, OK; Davalos-Marin, AJ; Ortiz, GG; del Angel Meza, AR; Torres-Mendoza, BM; Robles, RG; Huerta, VC; Beas-Zarate, C. Antioxidant activity of tryptophan in rats under experimental endotoxic shock. *Biomed Pharmacother*, 2010, 64, 77–81.

[104] Tsopmo, A; Diehl-Jones, BW; Aluko, RE; Kitts, DD; Elisia, I; Friel, JK. Tryptophan released from mother's milk has antioxidant properties. *Pediatric Res*, 2009, 66, 614–618.

[105] Watanabe, S; Togashi, SI; Takahashi, N; Fukui, T. L-Tryptophan as an antioxidant in human placenta extract. *J Nutr Sci Vitaminol*, 2002, 48, 36–39

[106] Santus, R; Patterson, LK; Filipe, P; Morliere, P; Hug, GL; Fernandes, A; Maziere, JC. Redox reactions of the urate radical/urate couple with the superoxide radical anion, the tryptophan neutral radical and selected flavonoids in neutral aqueous solutions. *Free Radical Res*, 2001, 35, 129–136.

[107] Stuart-Audette, M; Blouquit, Y; Faraggi, M; Sicard-Roselli, C; Houee-Levin, C; Jolles, P. Re-evaluation of intramolecular long-range electron transfer between tyrosine and tryptophan in lysozymes. Evidence for the participation of other residues. *Eur J Biochem*, 2003, 270, 3565–3571.

[108] Feksa, LR; Cornelio, AR; Vargas, CR; Wyse, ATS; Dutra-Filho, CA; Wajner, M; Wannmacher, CMD. Alanine prevents the inhibition of pyruvate kinase activity caused by tryptophan in cerebral cortex of rats. *Metab Brain Dis*, 2003, 18, 129–137.

[109] Cornelio, AR; Rodrigues-Júnior, V; Wyse, ATS; Dutra-Filho, CS; Wajner, M; Wannmacher, CMD. Tryptophan reduces creatine kinase activity in the brain of rats. *Int J Dev Neurosci*, 2004, 22, 95–101.

[110] Feksa, LR; Cornelio, A; Dutra-Filho, CS; Wyse, ATS; Wajner, M; Wannmacher, CMD. The effects of the interactions between amino acids on pyruvate kinase activity from the brain cortex of young rats. *Int J Dev Neurosci*, 2005, 23, 509–514.

[111] Feksa, LR; Latini, A; Rech, VC; Wajner, M; Dutra-Filho, CS; Wyse, ATS; Wannmcher, CMD. Promotion of oxidative stress by tryptophan in cerebral cortex of rats. *Neurochem Int*, 2006, 49, 87–93.

[112] Feksa, LR; Latini, A; Rech, VC; Feksa, PB; Koch, GD; Amaral, MF; Leipnitz, G; Dutra-Filho, CS; Wajner, M; Wannmacher, CM. Tryptophan administration induces oxidative stress in brain cortex of rats. *Metab Brain Dis*, 2008, 23, 221–33.

[113] Perez-Gonzalez, A; Muñoz-Rugeles, L; Alvarez-Idaboy, JR. Tryptophan: Antioxidant or target of oxidative stress? A quantum chemistry elucidation. *RSC Adv*, 2014, asap DOI:10.1039/C4RA11635F.

[114] Genestet, C; Le Gouellec, A; Chaker, H; Polack, B; Guery, B; Toussaint, B; Stasia, MJ. Scavenging of reactive oxygen species by tryptophan metabolites helps Pseudomonas aeruginosa escape neutrophil killing. *Free Radic Biol Med*, 2014, 73, 400–410.

[115] Weiss, G; Diez-Ruiz, A; Murr, C; Theur, I; Fuchs, D. Tryptophan metabolites as scavengers of reactive oxygen and chlorine species. *Pteridines*, 2002, 13, 140–144.

[116] Christen, S; Peterhans, E; Stocker, R. Antioxidant activities of some tryptophan metabolites: Possible implication for inflammatory diseases. *Proc Nati Acad Sci USA*, 1990, 87, 2506–2510.

[117] Simic, MG; al-Sheikhly, M; Jovanovic, SV. Inhibition of free radical processes by antioxidants-tryptophan and 5-hydroxytryptophan. *Bibl Nutr Dieta*, 1989, 43, 288–296.
[118] Cadenas, E; Simic, MG; Sies, H. Antioxidant activity of 5-hydroxytryptophan, 5-hydroxyindole, and dopa against microsomal lipid peroxidation and its dependence on vitamin E. *Free Radic Res*, 1989, 6, 11–17.
[119] Millán-Plano, S; Piedrafita, E; Miana-Mena, FJ; Fuentes-Broto, L; Martínez-Ballarín, E; López-Pingarrón, L; Sáenz, MA; García, JJ. Melatonin and structurally-related compounds protect synaptosomal membranes from free radical damage. *Int J Mol Sci*, 2010, 11, 312–328.
[120] Keithahn, C; Lerchl, A. 5-hydroxytryptophan is a more potent in vitro hydroxyl radical scavenger than melatonin or vitamin C. *J Pineal Res*, 2005, 38, 62–66.
[121] Lysek, N; Kinscherf, R; Claus, R; Lindel, T. L-5-hydroxytryptophan: Antioxidant and anti-apoptotic principle of the intertidal sponge Hymeniacidon heliophila. Zeitschrift fur Naturforschung - *Section C J Biosci*, 2003, 58, 568–572.
[122] Weber, OA; Simeon, VI. Tryptamine, 5-hydroxytryptamine and 5-hydroxytryptophan complexes of proton and some divalent metal ions. *J Inorg Nucl Chem*, 1971, 33, 2097–2101.
[123] Huether, G; Schuff-Werner, P. Platelet serotonin acts as a locally releasable antioxidant. Recent advances in tryptophan research. *Adv Exp Med Biol*, 1996, 398, 299–306.
[124] Schuff-Werner, P; Splettstosser, W; Schmidt, F; Huether, G. Serotonin acts as a radical scavenger and is oxidized to a dimer during the respiratory burst of human mononuclear and polymorphonuclear phagocytes. *Eur J Clin Inv*, 1995, 25, 477–484.
[125] Daniels, WMU; van, Rensburg, SJ; van, Zyl, JM; van, der, Walt, BJ; Taljaard, JJF. Free radical scavenging effects of melatonin and serotonin: possible mechanism. *Neuroreport*, 1996, 7, 1593–1596.
[126] Huether, G; Fettkötter, I; Keilhoff, G; Wolf, G. Serotonin acts as a radical scavenger and is oxidized to a dimer during the respiratory burst of activated microglia. *J Neurochem*, 1997, 69, 2096–2101.
[127] Ham, SS; Kim, DH; Lee, SH; Kim, YS; Lee, CS. Antioxidant effects of serotonin and L-DOPA on oxidative damages of brain synaptosomes. *Korean J Physiol Pharmacol*, 1999, 3, 147–155.
[128] Schuff-Werner, P; Splettstoesser, W. Antioxidative properties of serotonin and the bactericidal function of polymorphonuclear phagocytes. *Adv Exp Med Biol*, 2000, 467, 321–325.
[129] Gülçin, I. Measurement of antioxidant ability of melatonin and serotonin by the DMPD and CUPRAC methods as trolox equivalent. *J Enzyme Inhib Med Chem*, 2008, 23, 871–876.
[130] Sarikaya, SBO; Gülçin, I. Radical scavenging and antioxidant capacity of serotonin. *Curr Bioactive Comp*, 2013, 9, 143–152.
[131] Qi, W; Reiter, RJ; Tan, DX; Manchester, LC; Siu, AW; García, JJ. Increased levels of oxidatively damaged DNA induced by chromium(III) and H_2O_2: Protection by melatonin and related molecules. *J Pineal Res*, 2000, 29, 54–61.
[132] Seegar, H; Mueck, AO; Lippert, TH. Effect of melatonin and metabolites on copper-mediated oxidation of flow density lipoprotein. *Br J Clin Pharmacol*, 1997, 44, 283–284.

[133] Gozzo, A; Lesieur, D; Duriez, P; Fruchart, JC; Teissier, E. Structure-activity relationships in a series of melatonin analogues with the low-density lipoprotein oxidation model. *Free Radic Biol Med*, 1999, 26, 1538–1543.

[134] Chan, TY; Tang, PL. Characterization of the antioxidant effects of melatonin and related indoleamine in vitro. *J Pineal Res*, 1996, 20, 187–191.

[135] Longoni, B; Pryor, WA; Marchiafava, P. Inhibition of lipid peroxidation by N-acetylserotonin and its role in retinal physiology. *Biochem Biophys Res Commun*, 1997, 233, 778–780.

[136] Lezoualc'h, F; Sparapani, M; Behl, C. N-acetyl-serotonin (normelatonin) and melatonin protect neurons against oxidative challenges and suppress the activity of the transcription factor NFkappaB. *J Pineal Res*, 1998, 24, 168–178.

[137] Siu, AW; Reiter, RJ; To, CH. Pineal indoleamines and vitamin E reduce nitric oxide-induced lipid peroxidation in rat retinal homogenates. *J Pineal Res*, 1999, 27, 122–128.

[138] García, JJ; Reiter, RJ; Karbownik, M; Calvo, JR; Ortiz, GG; Tan, DX; Martínez-Ballarín, E; Acuña-Castroviejo, D. N-acetylserotonin suppresses hepatic microsomal membrane rigidity associated with lipid peroxidation. *Eur J Pharmacol*, 2001, 428, 169–175.

[139] Oxenkrug, G; Requintina, P; Bachurin, S. Antioxidant and antiaging activity of N-acetylserotonin and melatonin in the in vivo models. *Ann NY Acad Sci*, 2001, 939, 190–199.

[140] Karbownik, M; Gitto, E; Lewinski, A; Reiter, RJ. Relative efficacies of indole antioxidants in reducing autoxidation and iron-induced lipid peroxidation in hamster testes. *J Cell Biochem*, 2001, 81, 693–699.

[141] Guajardo, MH; Terrasa, AM; Catalá, A. Protective effect of indoleamines on in vitro ascorbate-Fe2 dependent lipid peroxidation of rod outer segment membranes of bovine retina. *J Pineal Res*, 2003, 35, 276–282.

[142] Wolfler, A; Abuja, PM; Schauenstein, K; Liebmann, PM. N-acetylserotonin is a better extra- and intracellular antioxidant than melatonin. *FEBS Lett*, 1999, 449, 206–210.

[143] Stetinova, V; Smetanova, L; Grossmann, V; Anzenbacher, P. In vitro and in vivo assessment of the antioxidant activity of melatonin and related indole derivatives. *Gen Physiol Biophys*, 2002, 21, 153–162.

[144] Nakagawa, H; Takusagawa, M; Arima, H; Furukawa, K; Kinoshita, T; Ozawa, T; Ikota, N. Selective scavenging property of the indole moiety for the nitrating species of peroxynitrite. *Chem Pharm Bull (Tokyo)*, 2004, 52, 146–149.

[145] Poeggeler, B; Reiter, RJ; Hardeland, R; Tan, DX; Barlow-Walden, LR. Melatonin and structurally-related, endogenous indoles act as potent electron donors and radical scavengers in vitro. *Redox Rep*, 1996, 2, 179–184.

[146] Neganova, ME; Blik, VA; Klochkov, SG; Chepurnova, NE; Shevtsova, EF. Investigation of the antioxidant characteristics of a new tryptamine derivative of securinine and its influence on seizure activity in the brain in experimental epilepsy. *Neurochem J*, 2011, 5, 208–214.

[147] Jiaranaikulwanitch, J; Govitrapong, P; Fokin, VV; Vajragupta, O. From BACE1 Inhibitor to Multifunctionality of tryptoline and tryptamine triazole derivatives for Alzheimer's disease. *Molecules*, 2012, 17, 8312–8333.

[148] Pal, C; Bindu, S; Dey, S; Alam, A; Goyal, M; Iqbal, MS; Sarkar, S; Kumar, R; Halder, KK; Debnath, MC; Adhikari, S; Bandyopadhyay, U. Tryptamine-gallic acid hybrid

prevents non-steroidal anti-inflammatory drug-induced gastropathy: Correction of mitochondrial dysfunction and inhibition of apoptosis in gastric mucosal cells. *J Biol Chem*, 2012, 287, 3495–3509.

[149] Tan, DX; Chen, LD; Poeggeler, B; Manchester, LC; Reiter, RJ. Melatonin: A potent, endogenous hydroxyl radical scavenger. *Endocr J.*, 1993, 1, 57–60.

[150] Ng, TB; Liu, F; Zhao, L. Antioxidative and free radical scavenging activities of pineal indoles. *J Neural Transm*, 2000, 107, 1243–1251.

[151] Wang, HX; Liu, F; Ng, TB. Examination of pineal indoles and 6-methoxy-2-benzoxazolinone for antioxidant and antimicrobial effects. *Comp Biochem Physiol C: Toxicol Pharmacol.*, 2001, 130, 379–388.

[152] Hara, M; Iigo, M; Ohtani-Kaneko, R; Nakamura, N; Suzuki, T; Reiter, RJ; Hirata, K. Administration of melatonin and related indoles prevents exercise-induced cellular oxidative changes in rats. *Biol Signals*, 1997, 6, 90–100.

[153] Belcastro, M; Marino, T; Russo, N; Toscano, M. Structural and electronic characterization of antioxidants from marine organisms. *Theor Chem Acc*, 2006, 115, 361–369.

[154] Leopoldini, M; Russo, N; Chiodo, S; Toscano, M. Iron chelation by the powerful antioxidant flavonoid quercetin. *J Agric Food Chem*, 2006, 54, 6343–6351.

[155] Leopoldini, M; Rondinelli, F; Russo, N; Toscano, M. Pyranoanthocyanins: a theoretical investigation on their antioxidant activity. *J Agric Food Chem*, 2010, 58, 8862–8871.

[156] Leopoldini, M; Russo, N; Toscano, M. The molecular basis of working mechanism of natural polyphenolic antioxidants. *Food Chem*, 2011, 125, 288–306.

[157] Perez-Gonzalez, A; Galano, A. OH radical scavenging activity of edaravone: mechanism and kinetics. *J Phys Chem B*, 2011, 115, 1306–1314.

[158] Galano, A. Mechanism and kinetics of the hydroxyl and hydroperoxyl radical scavenging activity of N-acetylcysteine amide. *Theor Chem Acc*, 2011, 130, 51–60.

[159] Galano, A; Francisco-Márquez, M; Alvarez-Idaboy, JR. Mechanism and kinetics studies on the antioxidant activity of sinapinic acid. *Phys Chem Chem Phys*, 2011, 13, 11199–11205.

[160] Carvalho, LC; Estevão, MS; Ferreira, LM; Fernandes, E; Marques, MMB. A new insight on the hypochlorous acid scavenging mechanism of tryptamine and tryptophan derivatives. *Bioorg Med Chem Lett*, 2010, 20, 6475–6478.

[161] Schweitzer, C; Schmidt, R. Physical mechanisms of generation and deactivation of singlet oxygen. *Chem Rev*, 2003, 103, 1685–1757.

[162] Solar, S; Getoff, N; Surdhar, PS; Armstrong, DA; Singh, A. Oxidation of tryptophan and *N*methylindole by $N_3\cdot$, $Br_2\cdot-$, $(SNC)_2\cdot-$ radicals in light and heavy water solutions. A pulse radiolysis study. *J Phys Chem*, 1991, 95, 3639–3643.

[163] Valgimigli, L; Ingold, KU; Lusztyk, J. Antioxidant activities of vitamin E analogues in water and a Kamlet–Taft β-value for water. *J Am Chem Soc*, 1996, 118, 3545–3549.

[164] Hela, PG; Anipindi, NR; Priyadarsini, KI; O'Neill, P. OH radical induced one-electron oxidation of serotonin and tryptamine. *J Phys Chem B*, 1999, 103, 8606–8611.

[165] Mahal, HS; Sharma, HS; Mukherjee, T. Antioxidant properties of melatonin: A pulse radiolysis study. *Free Radic Biol Med*, 1999, 26, 557–565.

[166] Al-Kazwini, AT; O'Neill, P; Adams, GE; Cundall, RB; Jacquet, B; Lang, G; Junino, A. One electron oxidation of methoxylated and hydroxylated indoles by azide. 1 Characterization of primary indolic radicals. *J Phys Chem*, 1990, 94, 6666–6670.

[167] Zhang, H; Squadrito, GL; Pryor, WA. The reaction of melatonin with peroxynitrite: formation of melatonin radical cation and absence of stable nitrated products. *Biochem Biophys Res Commun*, 1998, 251, 83–87.

[168] Zhang, H; Squadrito, GL; Uppu, R; Pryor, WA. Reaction of peroxynitrite with melatonin: a mechanistic study. *Chem Res Toxicol*, 1999, 12, 526–534.

[169] Stasica, P; Ukanski, J; Rosiak, JM. Reactions of melatonin with radicals in deoxygenated aqueous solution. *J Radioanal Nucl Chem*, 1998, 232, 107–113.

[170] Tan, DX; Manchester, LC; Reiter, RJ; Plummer, BF; Hardies, LJ; Weintraub, ST, Vijayalaxmi; Shepherd, AM. A novel melatonin metabolite, cyclic 3-hydroxymelatonin: a biomarker of in vivo hydroxyl radical generation. *Biochem Biophys Res Comm*, 1998, 253, 614–620.

[171] Aliaga, C; Lissi, EA. Reactions of the radical cation derived from 2,2'-azinobis(3-ethylbenzothiazoline-6-sulfonic acid) (ABTS·+) with amino acids. Kinetics and mechanism. *Can J Chem*, 2000, 78, 1052–1059.

[172] Matuszak, ZK; Reszka, J; Chignell, CF. Reaction of melatonin and related indoles with hydroxyl radicals: EPR and spin trapping investigations. *Free Radic Biol Med*, 1997, 23, 367–372.

[173] Chyan, YJ; Poeggeler, B; Omar, RA; Chain, DG; Frangione, B; Ghiso, J; Pappolla, MA. Potent neuroprotective properties against the Alzheimer *β*-amyloid by an endogenous melatonin-related indole structure, indole-3-propionic acid. *J Biol Chem*, 1999, 274, 21937–21942.

[174] Roberts, JE; Hu, DN; Wishart, JF. Pulse Radiolysis Studies of Melatonin and Chloromelatonin. *J Photochem Photobio B: Biology*, 1998, 42, 125–132.

[175] Kawanishi, S; Sakurai, H. Differential anti-lipid peroxidative activity of melatonin. *Naturwissenschaften*, 2002, 89, 31–33.

[176] Marshall, KA; Reiter, RJ; Poeggeler, B; Aruoma, OI; Halliwell, B. Evaluation of the antioxidant activity of melatonin in vitro. *Free Radic Biol Med*, 1996, 21, 307–315.

[177] Mekhloufi, J; Bonnefont-Rousselot, D; Yous, S; Lesieur, D; Couturier, M; Therond, P; Legrand, A; Jore, D; Gardes-Albert, M. Antioxidant activity of melatonin and a pinoline derivative on linoleate model system. *J Pineal Res*, 2005, 39, 27–33.

[178] Mekhloufi, J; Vitrac, H; Yous, S; Duriez, P; Jore, D; Gardes-Albert, M; Bonnefont-Rousselot, D. Quantification of the water/lipid affinity of melatonin and a pinoline derivative in lipid models. *J Pineal Res*, 2007, 42, 330–337.

[179] Scaiano, JC. Exploratory laser flash photolysis study of free radical reactions and magnetic field effects in melatonin chemistry. *J Pineal Res*, 1995, 19, 189–195.

[180] Zang, LY; Cosma, G; Gardner, H; Vallyathan, V. Scavenging of reactive oxygen species by melatonin. *Biochim Biophys Acta*, 1998, 1425, 469–477.

[181] Zhu, H; Zhang, Z; Zhao, H; Wang, W; Yao, S. Transient species and its properties of melatonin. *Science in China Series B: Chemistry*, 2006, 49, 308–314.

[182] Armstrong, RC; Swallow, AJ. Pulse- and gamma-radiolysis of aqueous solutions of tryptophan. *Rad Res*, 1969, 40, 563–579.

[183] Alfassi, ZB. Selective oxidation of tyrosine-oxidation by NO_2 and ClO_2 at basic pH. *Radiat Phys Chem*, 1987, 29, 405–406.

[184] Stewart, DJ; Napolitano, MJ; Bakhmutova-Albert, EV; Margerum, DW. Kinetics and mechanisms of chlorine dioxide oxidation of tryptophan. *Inorg Chem*, 2008, 47, 1639–1647.

[185] Li, K; Wang, M; Wang, T; Sun, D; Zhu, R; Sun, X; Wu, X; Wang, SL. Interaction of retinoic acid radical cation with lysozyme and antioxidants: Laser flash photolysis study in microemulsion. *Photochem Photobiol*, 2013, 89, 1064–1070.
[186] Milligan, JR; Aguilera, JA; Ly, A; Tran, NQ; Hoang, O; Ward, JF. Repair of oxidative DNA damage by amino acids. *Nucl Ac Res*, 2003, 31, 6258–6263.
[187] Milligan, JR; Tran, NQ; Ly, A; Ward, JF. Peptide repair of oxidative DNA damage. *Biochemistry*, 2004, 43, 5102–5108.
[188] Morozova, OB; Kiryutin, AS; Yurkovskaya, AV. Electron transfer between guanosine radicals and amino acids in aqueous solution. II. Reduction of guanosine radicals by tryptophan. *J Phys Chem B*, 2008, 112, 2747–2754.
[189] Pan, J; Lin, W; Wang, W; Han, Z; Lu, C; Yao, S; Lin, N; Zhu, D. A kinetic study on the interaction of deprotonated purine radical cations with amino acids and model peptides. *Biophys Chem*, 2001, 89, 193–199.
[190] Gaikwad, P; Naik, GH; Priyadarsini, KI; Mohan, H; Rao, BSM. Radiation induced oxidation of hydroxyl indoles by $NO_2^{\bullet}$ and $Br_2^{\bullet-}$ radicals: effect of pH. *J Phys Org Chem*, 2011, 24 657–662.
[191] Gaikwad, P; Priyadarsini, KI; Naumov, S; Rao, BSM. Oxidation of tryptamine and 5-hydroxytryptamine: A pulse radiolysis and quantum chemical study *J Phys Chem* A, 2009, 113, 8249–8257.
[192] Rodriguez-Naranjo, MI; Moya, ML; Cantos-Villar, E; Garcia-Parrilla, MC. Comparative evaluation of the antioxidant activity of melatonin and related indoles. *J Food Compos Anal*, 2012, 28, 16–22.
[193] Farmanzadeh, D; Najafi, M. On the antioxidant activity of the tryptophan derivatives. *Bull Chem Soc Jpn*, 2013, 86, 1041–1050.
[194] Velkov, ZA; Velkov, YZh; Galunska, BT; Paskalev, DN; Tadjer, AV. Melatonin: Quantum-chemical and biochemical investigation of antioxidant activity. *Eur J Med Chem*, 2009, 44, 2834–2839.
[195] Migliavacca, E; Ancerewicz, J; Carrupt, PA; Testa, B. Theoretical parameters to characterize antioxidants. Part 2. The cases of melatonin and carvedilol. *Helv Chim Acta*, 1998, 81, 1337–1348.
[196] Turjanski, AG; Rosenstein, RE; Estrin, DA. Reactions of melatonin and related indoles with free radicals: a computational study. *J Med Chem*, 1998, 41, 3684–3689.
[197] Stasica, P; Paneth, P; Rosiak, JM. Hydroxyl radical reaction with melatonin molecule: A computational study. *J Pineal Res*, 2000, 29, 125–127.
[198] Galano, A; Medina, ME; Tan, DX; Reiter, RJ. Melatonin and its metabolites as copper chelating agents and their role in inhibiting oxidative stress: A physicochemical analysis. In press *J. Pineal Res.*, 2014.
[199] Rose, RC; Bode, AM. Biology of free radical scavengers: an evaluation of ascorbate. *FASEB J*, 1993, 7, 1135–1142.
[200] Rousseau, A; Petrén, S; Plannthin, J; Eklundh, T; Nordin, C. Serum and cerebrospinal fluid concentrations of melatonin: a pilot study in healthy male volunteers. *J Neural Transm*, 1999, 106, 883–888.
[201] Chiquet, C; Claustrat, B; Thuret, G; Brun, J; Cooper, HM; Denis, P. Melatonin concentrations in aqueous humor of glaucoma patients. *Am J Ophthalmol*, 2006, 142, 325–327.

[202] Waldhauser, F; Dietzel, M. Daily and annual rhythms in human melatonin secretion. Role in puberty control. *Ann NY Acad Sci*, 1985, 453, 205–214.

[203] Srinivasan, V. Melatonin, oxidative stress and ageing. *Curr Sci*, 1999, 76, 46–54.

[204] Tamura, H; Takasaki, A; Miwa, I; Taniguchi, K; Maekawa, R; Asada, H; Taketani, T; Matsuoka, A; Yamagata, Y; Shimamura, K; Morioka, H; Ishikawa, H; Reiter, RJ; Sugino, N. Oxidative stress impairs oocyte quality and melatonin protects oocytes from free radical damage and improves fertilization rate. *J Pineal Res*, 2008, 44, 280–287.

[205] Higuchi, S; Motohashi, Y; Liu, Y; Ahara, M; Kaneko, Y. Effects of VDT tasks with a bright display at night on melatonin, core temperatura, heart rate, and sleepinesws. *J Appl Physiol*, 2003, 94, 1773–1776.

[206] Skene, DJ; Vivien-Roels, B; Sparks, DL; Hunsaker, JC; Pévet, P; Ravid, D; Swaab, DF. Daily variation in the concentration of melatonin and 5-methoxytryptophol in the human pineal gland: effect of age and Alzheimer's disease. *Brain Res*, 1990, 528, 170–174.

[207] Messner, M; Huether, G; Lorf, T; Ramadori, G; Schwörer, H. Presence of melatonin in the human hepatobiliary-gastrointestinal tract. *Life Sci*, 2001, 69, 543–551.

[208] Reiter, RJ. Normal patterns of melatonin levels in the pineal gland and body fluids of humans and experimental animals. *J Neural Transm* Suppl, 1986, 21, 35–54.

[209] Tan, DX; Manchester, LC; Sanchez-Barcelo, E; Mediavilla, MD; Reiter, RJ. Significance of high levels of endogenous melatonin in mammalian cerebrospinal fluid and in the central nervous system. *Curr Neuropharmacol*, 2010, 8, 162–167.

[210] Berger, M; Gray, JA; Roth, BL. The expanded biology of serotonin. *Annu Rev Med*, 2009, 60, 355–66.

[211] Gershon, MD; Tack, J. The serotonin signaling system: from basic understanding to drug development for functional GI disorders. *Gastroenterology*, 2007, 132, 397–414.

[212] Al-Tikriti, MS; Khamas, W; Chebolu, S; Darmani, NA. Distribution of serotonin-immunoreactive enterochromaffin cells in the gastrointestinal tract of the least shrew (Cryptotis parva). *Int J Morphol*, 2012, 30, 916–923.

[213] Lambert, GW; Reid, C; Kaye, DM; Jennings, GL; Esler, MD. Effect of sunlight and season on serotonin turnover in the brain. *Lancet*, 2002, 360, 1840-1842.

[214] Wester, P; Hardy, JA; Marcusson, J; Nyberg, P; Winblad, B. Serotonin concentrations in normal aging human brains: Relation to serotonin receptors. *Neurobiol Aging*, 1984, 5, 199–203.

[215] Crosti, PF; Lucchelli, PE. An easy method to determine the serotonin content of human platelets. *J Clin Pathol*, 1962, 15, 191–193 (and references therein).

[216] Shannak, K; Rajput, A; Rozdilsky, B; Kish, S; Gilbert, J; Hornykiewicz, O. Noradrenaline, dopamine and serotonin levels and metabolism in the human hypothalamus: observations in Parkinson's disease and normal subjects. *Brain Res*, 1994, 639, 33–41.

[217] Koskiniemi, M; Laakso, J; Kuurne, T; Laipio, M; Härkönen, M. Indole levels in human lumbar and ventricular cerebrospinal fluid and the effect of L-tryptophan administration. *Acta Neurol Scandin*, 1985, 71, 127–132.

[218] Kemerer, VF; Lichtenfeld, KM; Koch, TR. A column chromatographic method for the determination of 5-hydroxytryptamine (Serotonin) and 5-hydroxyindole acetic acid in cerebrospinal fluid. *Clini Chim Acta*, 1979, 92, 81–85.

[219] Sagara, Y; Okatani, Y; Yamanaka, S; Kiriyama, T. Determination of plasma 5-hydroxytryptophan, 5-hydorxytryptamine, 5-hydorxyindoleacetic acid, tryptophan and melatonin by high-performance liquid chromatography with electrochemical detection. *J Chromatogr B: Biomed Sci Appl*, 1988, 431, 170–176.
[220] Zanardo, V; D'Aquino, M; Stocchero, L; Biasiolo, M; Allegri, G. Serum total and free tryptophan levels in term infants fed cow's milk formula or human milk. *Eur J Pediatr.*, 1989,148, 781–783.
[221] Martin, WR; Sloan, JW; Christian, ST; Clements, TH. Brain levels of tryptamine. *Psychopharmacologia*, 1972, 24, 331–346.
[222] Candia, OA; Alvarez, LJ; Lanzetta, PA; Cook, P. Tryptamine in the vertebrate lens. *Biochim Biophys Acta - Mol Cell Res*, 1983, 762, 232–240.
[223] Pang, SF; Tang, F; Tang, PL. Negative correlation of age and the levels of pineal melatonin, pineal N-acetyl serotonin, and serum melatonin in male rats. *J Exp Zool*, 1984, 229, 41–47.
[224] Prozialeck, WC; Boehme, DH; Vogel, WH. The fluorometric determination of 5-methoxytryptamine in mammalian tissues and fluids. *J Neurochem*, 1978, 30, 1471–1477.
[225] Cagnoli, CM; Atabay, C; Kharlamova, E; Manev, H. Melatonin protects neurons from singlet oxygen-induced apoptosis. *J Pineal Res*, 1995, 18, 222–226.
[226] Matuszak, Z1; Bilska, MA; Reszka, KJ; Chignell, CF; Bilski, P. Interaction of Singlet Molecular Oxygen with Melatonin and Related Indoles. *Photochem Photobiol*, 2003, 78, 449–455.
[227] Karbownik, M; Reiter, RJ. Antioxidative effects of melatonin in protection against cellular damage caused by ionizing radiation. *Proc Soc Exp Biol Med*, 2000, 225, 9–22.
[228] Melchiorri, D; Reiter, RJ; Attia, AM; Hara, M; Burgos, A; Nistico, G. Potent protective effect of melatonin on in vivo paraquat-induced oxidative damage in rats. *Life Sci*, 1994, 56, 83–89.
[229] Tan, DX; Reiter, RJ; Chen, LD. Both physiological and pharmacological levels of melatonin reduce DNA adduct formation induced by the carcinogen safrole. *Carcinogenesis*, 1994, 15, 215–218.
[230] Shida, CS; Castrucci, AML; Lamy-Freund, MT. High melatonin solubility in aqueous medium. *J Pineal Res*, 1994, 16, 198–201.
[231] Menendez-Pelaez, A; Reiter, RJ. Distribution of melatonin in mammalian tissues: The relative importance of nuclear versus cytosolic localization. *J Pineal Res*, 1993, 15, 59–69.
[232] Ceraulo, L; Ferrugia, M; Tesoriere, L; Segreto, S; Livrea, MA; Turco, Liveri, V. Interactions of melatonin with membrane models: Portioning of melatonin in AOT and lecithin reversed micelles. *J Pineal Res*, 1999, 26, 108–112.
[233] Menendez-Pelaez, A; Poeggeler, B; Reiter, RJ; Barlow-Walden, L; Pablos, MI; Tan, DX. Nuclear localization of melatonin in different mammalian tissues: Immunocytochemical and radioimmunoassay evidence. *J Cell Biochem*, 1993, 53, 373–382.
[234] Leon, J; Acuña-Castroviejo, D; Sainz, RM; Mayo, JC; Tan, DX; Reiter, RJ. Melatonin and mitochondrial function. *Life Sci*, 2004, 75, 765–790.
[235] Reiter, RJ; Pablos, MI; Agapito, TT; Guerrero, JM. Melatonin in the context of the free radical theory of aging. *Ann NY Acad Sci*, 1996, 786, 362–378.

[236] Lewis, AJ; Kerenyi, NA; Feuer, G. Neuropharmacology of pineal secretions. *Rev Drug Metab Drug Interact*, 1990, 8, 247–312.

[237] Tan, DX; Manchester, LC; Burkhardt, S; Sainz, RM; Mayo, JC; Kohen, R; Shohami, E; Huo YS; Hardeland, R; Reiter, RJ. N1-acetyl-N2-formyl-5-methoxykynuramine, a biogenic amine and melatonin metabolite, functions as a potent antioxidant. *FASEB J*, 2001, 15, 2294–2296.

[238] León, J; Escames, G; Rodríguez, MI; López, LC; Tapias, V; Entrena, A; Camacho, E; Carrión, MD; Gallo, MA; Espinosa, A; Tan, DX; Reiter, RJ; Acuña-Castroviejo, D. Inhibition of neuronal nitric oxide synthase activity by N1-acetyl-5-methoxykynuramine, a brain metabolite of melatonin. *J Neurochem*, 2006, 98, 2023–2033.

[239] Manda, K; Ueno, M; Anzai, K. AFMK, a melatonin metabolite, attenuates X-ray-induced oxidative damage to DNA, proteins and lipids in mice. *J Pineal Res*, 2007, 42, 386–393.

[240] Hardeland, R; Backhaus, C; Fadavi, A. Reactions of the NO redox forms NO+, ˙NO and HNO (protonated NO–) with the melatonin metabolite N1-acetyl-5-methoxykynuramine. *J Pineal Res*, 2007, 43, 382–388.

[241] Ressmeyer, AR; Mayo, JC; Zelosko, V; Sáinz, RM; Tan, DX; Poeggeler, B; Antolín, I; Zsizsik, BK; Reiter, RJ; Hardeland, R. Antioxidant properties of the melatonin metabolite N1-acetyl-5-methoxykynuramine (AMK): scavenging of free radicals and prevention of protein destruction. *Redox Rep*, 2003, 8, 205–213.

[242] Manda, K; Ueno, M; Anzai, K. Space radiation-induced inhibition of neurogenesis in the hippocampal dentate gyrus and memory impairment in mice: ameliorative potential of the melatonin metabolite, AFMK. *J Pineal Res*, 2008, 45, 430–438.

[243] Hardeland, R; Tan, DX; Reiter, RJ. Kynuramines, metabolites of melatonin and other indoles: the resurrection of an almost forgotten class of biogenic amines. *J Pineal Res*, 2009, 47, 109–126.

[244] Schaefer, M; Hardeland, R. The melatonin metabolite N1-acetyl-5-methoxykynuramine is a potent singlet oxygen scavenger. *J Pineal Res*, 2009, 46, 49–52.

[245] Tapias, V; Escames, G; López, LC; López, A; Camacho, E; Carrión, MD; Entrena, A; Gallo, MA; Espinosa, A; Acuña-Castroviejo, D. Melatonin and its brain metabolite N1-acetyl-5-methoxykynuramine prevent mitochondrial nitric oxide synthase induction in Parkinsonian mice. *J Neurosci* Res, 2009, 87, 3002–3010.

[246] Slobodan, V. Jovanovic, Michael G. Simic. Repair of tryptophan radicals by antioxidants. *J Free Radic Biol Med*, 1985, 1, 125–129.

[247] Dvir, Y; Smallwood, P. Serotonin syndrome: a complex but easily avoidable condition. *General Hospital Psychiatry*, 2008, 30, 284–287 (and references therein).

[248] Boyer, EW; Shannon, M. The serotonin syndrome. *N Engl J Med*, 2005, 352, 1112–1120.

[249] Sainio, EL; Pulkki, K; Young, SN. L-Tryptophan: Biochemical, nutritional and pharmacological aspects. *Amino Acids*, 1996, 10, 21–47.

[250] Thomson, J; Rankin, H; Ashcroft, GW; Yates, CM; McQueen, JK; Cummings, SW. (1982). The treatment of depression in general practice: a comparison of L-tryptophan, amitriptyline; and, a, combination, of, L-tryptophan, and, amitriptyline, with, placebo. *Psychol Med*, 12, 741–751.

[251] Murphy, DL; Baker, M; Goodwin, FK; Miller, H; Kotin, J; Bunney, WE. LTryptophan in affective disorders: indoleamine changes and differential clinical effects. *Psychopharmacologia*, 1974, 34, 11–20.
[252] Gillin, JC; Kaplan, JA; Wyatt, RJ. Clinical effects of tryptophan in chronic schizophrenic patients. *Biol Psychiatry*, 1976, 11, 635–639.
[253] Das, YT; Bagchi, M; Bagchi, D; Preuss, HG. Safety of 5-hydroxy-l-tryptophan. *Toxicol Lett*, 2004, 150, 111–122 (and references therein).
[254] Arendt, J. Safety of melatonin in long-term use. *J Biol Rhythms*, 1997, 12, 673–681.
[255] Vijayalaxmi; Meltz, ML; Reiter, RJ; Herman, TS. Melatonin and protection from genetic damage in blood and bone marrow: whole-body irradiation studies in mice. *J Pineal Res*, 1999, 27, 221–225.
[256] Jahnke, G; Marr, M; Myers, C; Wilson, R; Travlos, G; Price, C. Maternal and developmental toxicity evaluation of melatonin administered orally to pregnant Sprague-Dawley rats. *Toxicol Sci*, 1999, 50, 271–279.
[257] Vijayalaxmi; Meltz, ML; Reiter, RJ; Herman, TS; Kumar, KS. Melatonin and protection from whole-body irradiation: survival studies in mice. *Mutat Res*, 1999, 425, 21–27.
[258] Kaya, H; Delibas, N; Serteser, M; Ulukaya, E; Ozkaya, O. The effect of melatonin on lipid peroxidation during radiotherapy in female rats. Strahlenther. *Onkol*, 1999, 175, 285–288.
[259] Nordlund, JJ; Lerner, AB. The effects of oral melatonin on skin color and on the release of pituitary hormones. *J Clin Endocrinol Metab*, 1977, 45, 768–774.

In: Indoleamines
Editor: Ángel Catalá

ISBN: 978-1-63482-097-4

Chapter 2

MELATONIN: ENDOGENOUS SOURCES AND ROLE IN THE REGULATION OF FISH REPRODUCTION

Saumen Kumar Maitra**, *Sourav Mukherjee and Kazi Nurul Hasan
Department of Zoology, Visva-Bharati University, Santiniketan, India

ABSTRACT

Melatonin (*N*-acetyl-5-methoxytryptamine), since first isolation, purification and chemical characterization from the extracts of bovine pineal gland, has been detected in several extra-pineal organs, more specifically in retina, Harderian gland, gut, and ovary in vertebrates as well as in a wide variety of organisms, ranging from invertebrates to various plants and even in microorganisms. This tiny, but versatile, indoleamine hormone is synthesized from amino acid tryptophan following four enzymatic reactions. The regulatory mechanisms of melatonin synthesis in the pineal and retina are synchronized with the environmental light-dark cycle, with a nocturnal peak and nadir at midday, irrespective of the habit and habitat of studied animals. Conversely, so far no significant influences of environmental light-dark conditions have been evident in the regulation of melatonin synthesis in any extra-pineal or extra-retinal tissues. Hence, the daily and seasonally changing melatonin rhythm patterns in serum are depicted as the result of varying melatonin synthesis in the pineal and/or retina, which are involved in transmission of information on 'time of day' and 'time of year' and thereby serve as a biological clock and/or a biological calendar. Past few decades have witnessed an enormous progress in understanding the sources and mechanism of melatonin actions in the regulation of diverse physiological functions especially seasonal reproduction in a wide range of vertebrates including fish. However, relative contribution of different melatonin producing tissues to circulating profiles of melatonin and their temporal relationship with reproductive events in an annual cycle in any vertebrates remained speculative. Accordingly, findings on a carp *Catla catla* appeared important, as this is the single fish species that has been used for collection of data on the daily and seasonal profiles of melatonin in the pineal, retina, gut and serum as well as the density of arylalkylamine *N*-acetyltransferase (AANAT) protein, the key regulator of melatoninergic pathway, in the pineal, retina and gut under natural and diverse

* Corresponding author: Professor Saumen Kumar Maitra, Department of Zoology, Visva-Bharati University, Santiniketan, 731235, India. Fax, 91-3463-261079, Email: dgp_skmaitra@yahoo.co.in.

experimental conditions as well. Notably, daily concentrations of melatonin and the band intensity of a ~23kDa AANAT protein in the pineal and serum titers of melatonin showed parallel changes in a daily cycle with a peak in late dark-phase (during the preparatory phase) or at midnight (in the remaining parts of annual reproductive cycle). On the other hand, levels of melatonin and AANAT protein in the gut, irrespective of sampling months, exhibited a peak at midday, and thereby depicted a negative correlation between the melatoninergic system in gut and pineal in the same species. Likewise, the influences of altered photoperiods on the activity of both the testis and the ovary as well as the levels of melatonin and the density of AANAT protein in the pineal varied in relation to the reproductive seasons, though daily rhythmic profiles of gut melatonin remained identical irrespective of the reproductive status of the fish and the light-dark regimens to which the carp were held. The obvious outcome of such study is that both the pineal and gut appeared as major sites of melatonin synthesis, but their environmental as well as physiological synchronizer(s) might not be identical. Nonetheless, convincing data gathered from a number of carefully controlled studies employing exogenous administration of melatonin to several fish species, especially carp, implicated melatonin as a potent candidate in the regulation of seasonal reproduction in fish. Remarkably, an identical regimen of exogenous melatonin treatment led to stimulation or inhibition or no influences at all on the functions of gonads depending on the reproductive status of concerned fish. Cross-talk between the ovarian steroids and melatonin has also been demonstrated by an *in vitro* study, in which melatonin accelerated the action of 17α, 20β-dihydroxy-4-pregnen-3-one or maturation inducing hormone (MIH) on meiotic cell cycle resumption in carp oocytes by formation of maturation promoting factor (MPF) - a complex of two proteins, cyclin B and cyclin dependant kinase (Cdk1). While several lines of evidence suggest melatonin actions on the hypothalamo-hypophyseal-gonadal axis, localization and dynamics of a 37-kDa melatonin receptor protein in carp oocytes argued in favor of extra-hypothalamic action of melatonin on fish reproduction. The carp study also indicated that response of ovary to photoperiods might be related to varied profiles of serum melatonin, but not to any rhythm parameters of MT1 or MT2 receptors. Very recently, melatonin has been detected and measured in the carp ovary. Though it remained unclear whether measured melatonin is synthesized within the ovary itself or is derived from the circulation, ovarian melatonin concentrations displayed significant seasonal variations with a peak at spawning and nadir during post-spawning, and thereby underlined its proximity to the course of ovarian development in carp. Moreover, search for a correlation between melatonin concentrations, oxidative status and oocyte dynamics in the carp ovary led to emergence of an idea that ovarian melatonin may be associated with an enhanced activity/level of antioxidative agents for reduction in oxidative stress to augment ovarian functions during spawning. This article summarizing available information on the endogenous sources of melatonin in fish deals with external factor(s) that may act as synchronizer of rhythmic synthesis of melatonin in different organs and finally discusses the modalities by which this indoleamine hormone impacts on physiological regulations of fish reproduction mainly focusing the recent findings on carp.

Keywords: Fish, melatonin, ovary, pineal, photoperiod, reproduction

1. INTRODUCTION

Isolation, purification, and characterization of an indole compound (*N*-acetyl-5-methoxy-tryptamine) from the extracts of cattle pineal gland (Lerner et al., 1958) has been a major breakthrough in indoleamines research. In recognition to an initial finding that treatment with crude acetone extract of bovine pineal glands to tadpole of *Rana pipens* caused lightening (blanching) of skins resulting in clear visibility of their larger viscera (McCord and Allen, 1917), the purified pineal indole substance was designated as 'melatonin'. Subsequent studies reveal that melatonin is indeed present in several non-pineal organs/tissues like, retina (Iuvone and Besharse, 1983) and gut (Bubenik, 1980) in most of the vertebrates; Harderian gland (Menendez-Pelaez et al., 1987), gonads (Itoh et al., 1999) and placenta (Iwasaki et al., 2005) in mammals; as well as in a wide variety of organisms ranging from invertebrates to various plants and even in microorganisms (Chowdhury and Maitra, 2012). Nonetheless, it remains doubtful whether melatonin detected in these tissues is produced within the tissue itself or somewhere else and incorporated from the circulation thereafter.

Existing information on the biosynthesis of melatonin and its regulation is based mostly on the studies on the pineal, where synthesis of melatonin is a four-step phenomenon. First, the precursor L-tryptophan is taken up by the pinealocyte from the circulation (blood) and is converted to 5-hydroxy-tryptophan (5-HTP) in the mitochondria by tryptophan-5-mono-oxygenase/hydroxylase before undergoing decarboxylation to form serotonin (5-hydroxytryptamine, 5-HT) by L-aromatic amino acid decarboxylase in the cytosol. The most physiologically important metabolic pathway of serotonin leads to the synthesis of melatonin (Axelrod et al., 1969). Serotonin is first acetylated (*N*-acetylation) by arylalkylamine-*N*-acetyltransferase (AANAT) into *N*-acetyl serotonin (NAS), which plays a key role in melatonin synthesis. Finally, *N*-acetyl serotonin is O-methylated by hydroxyindole-O-methyltransferase (HIOMT) to form melatonin (Axelrod and Weissbach, 1960). AANAT is considered as the rate limiting enzyme regulating melatonin biosynthesis (Klein et al., 1997). One of the unique properties of melatonin in the pineal is that it is synthesized rhythmically with a nocturnal peak in synchronization with the environmental light-dark cycles (Collin et al., 1989). Considering this conserved feature, melatonin appears to be the signal of darkness and the time keeping molecule of the organisms (Arendt and Deacon, 1997). Despite this apparent homogeneity, it remains largely unknown whether biosynthesis of melatonin in other non-pineal tissues follows the same pattern and the mechanisms by which melatonin synthesis have been greatly modified from fish to mammals (Falcón et al.,1992).

The chronobiotic property of melatonin underlines its significant role in the regulation of diverse rhythmic body functions, most important of which is seasonal reproduction in a wide range of vertebrates, including fish (Okimoto and Stetson, 1999; Amano et al.,2000; Falcón et al., 2010; Maitra et al., 2013). However, the question whether temporal pattern and, more importantly, regulatory role of melatonin on seasonal reproduction, vary with its source has not yet been duly addressed in any group of vertebrates. In recent years, a tropical carp (*Catla catla*), which because of its surface dwelling habit maintains close proximity to environmental lighting conditions- an attribute especially important for understanding the role of photic information in the regulation of different physiological functions, has been used to gather basic information on the diurnal profiles of melatonin in circulation as well as in the pineal, retina, gut and ovary during different phases of an annual reproductive cycle (Table

1). Additionally, the mechanism of actions of melatonin in the regulation of oocyte maturation has been critically investigated in this fish species. This chapter is thus an attempt to bring together the available information on the major sources of endogenous melatonin and shape the current working hypotheses on its physiological significance in the regulation of reproduction supported by recent findings under natural as well as experimental conditions following exogenous administration of melatonin in carp and other fish species.

2. Endogenous Sources of Melatonin

2.1. The Pineal: A Major Source of Endogenous Melatonin

The pineal organ or epiphyses cerebri, as the major source of melatonin, is located at the roof of the diencephalon or third ventricle of the brain. This organ appears to have undergone a clear transformation from being primarily a sensory organ in lower group of vertebrates to a major endocrine gland in birds and mammals (Vollrath, 1981). More specifically, the pineal in higher vertebrates, like mammals and birds, acts as a typical endocrine gland, while in lower groups of vertebrates, especially in fishes, is known to perform both sensory and secretory functions (Korf and Oksche, 1986; Falcón, et al., 2010; Seth and Maitra, 2010a,b). The pineal in carp *Catla catla* is represented by a complex of three distinct structures: an antero-laterally elongated, dorso-ventrally flattened vesicular body - the 'end vesicle' located close to the cranium, connected to the brain by a very long 'pineal stalk', which at cerebral end is encircled by highly folded plexiform saccular - 'dorsal sac' (Bhattacharya et al., 2003). However, immunocytochemical study demonstrated the presence of melatonin secretory cells only in the end vesicle of the pineal complex (Maitra et al., 2005).

Table 1. Major characteristics of melatonin detected and measured in different organs in carp (*Catla catla*)

Attributes	***Source of Measured Melatonin***			
	Pineal organ	***Retina***	***Gastro-Intestinal Tract (Gut)***	***Ovary***
Site of Synthesis	Pinealocytes	Photoreceptor cells	Enterochromafin cells	Unknown
Types of AANAT	AANAT2 (~23kDa)	AANAT1 (~26 / ~90 / ~98kDa)	AANAT2 (~23kDa)	Undetected
Possible Synchronizer	Environmental light-dark cycle	Environmental light-dark cycle	Non-photic environmental cue(s)	Ovarian growth and development
Daily Profiles	Rhythmic	Rhythmic	Rhythmic	No rhythms
Seasonal Profiles	Rhythmic	Rhythmic	No rhythms	Rhythmic
Relation to Serum Melatonin	Positive	Positive	Uncertain	Unknown
Nature of Action	Endocrine	Neuroendocrine/ autocrine/paracrine	Endocrine/ paracrine/luminal	Antioxidative agent

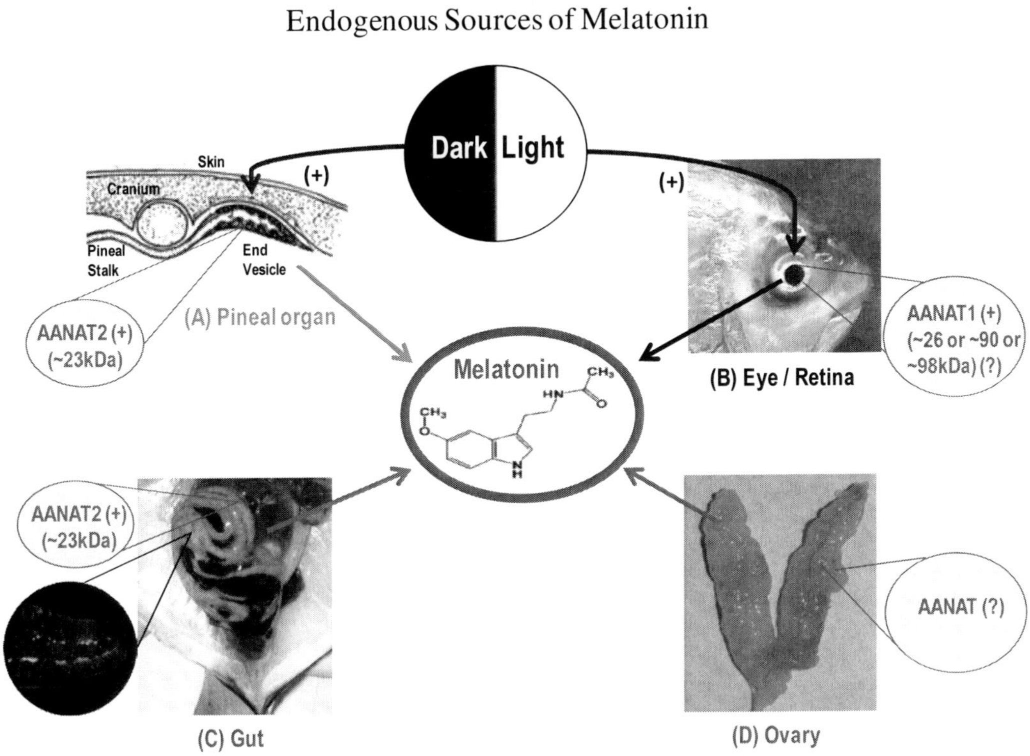

Figure 1. Schematic presentation of the summary of information gathered on the endogenous sources of melatonin like, (A) the pineal, (B) the retina, (c) the gut and (D) the ovary in carp (*Catla catla*). Melatonin is produced in the pineal and retina in response to light-dark conditions of the environment and is released in blood. The gut tissues also synthesize melatonin, but independent of the influences of environmental light-dark cycle. In inset, melatonin immune-positive reactions are shown at the junction between the basal part of epithelial lining and the outer margin of lamina propria of mucosal layer in a section (× 1000) of carp gut collected at midday. The reactions do not differ with the anatomical segments of gut. Melatonin concentrations have also been measured in the ovary. Notably, arylalkylamine *N*-acetyltransferase (AANAT) has been detected in the pineal, retina and gut, but not in the ovary in carp. Two AANATs have been identified in fish: AANAT1 (~26 or ~90 or ~98kDa) is specifically expressed in the retina; while AANAT2 (~23kDa) is specifically expressed in the pineal organ and gut.

Arylalkylamine-*N*-acetyltransferase (AANAT), the rate limiting enzyme in melatonin biosynthetic pathway (Klein, 2007), is a photosensitive protein that determines melatonin rhythms in the pineal organ in a number of fish species (like, goldfish, carp, pike, trout, and white sucker). Characteristically, the fish pineal is a photoreceptive organ that directly receives the light information from the environment and conveys this message by varying the amount of melatonin synthesis within the pinealocytes, and acts to correlate the activity of the animals with the conditions of the external environment, irrespective of the habit and habitat of the species. However, the temporal organization of AANAT synthesis in the pineal of different fish is not identical. In osmerids (e.g., pike *Esox lucius*, zebrafish *Danio rerio* and seabream *Sparus aurata*), the endogenous clock in the pineal contributes to the rhythmic expression of AANAT mRNA. Conversely, clock-driven changes in AANAT mRNA in the pineal of salmonids (e.g., rainbow trout *Oncorhynchus mykiss*) do not play a role in melatonin synthesis (Bégay et al., 1998; Iigo et al., 2007a,b). The circadian organization of melatoninergic system in carp (*Catla catla*) pineal is also dependent on the external light-dark conditions (Figure 1), rather than an endogenous circadian oscillator in the pineal (Seth and Maitra, 2010a,b). Thus it is evident that the mechanisms involved in the regulation of melatonin synthesis in the pineal through the circadian axis differ with the teleostean species, and this variation is possibly a reflection of the photic environment in which they have evolved (Migaud et al., 2007).

2.2. Extra-Pineal Sources of Melatonin

With the use of highly sensitive antibodies against melatonin and application of different molecular biology techniques, melatonin has been detected in several extra-pineal tissues, most important of which are retina, gut and ovary (Figure 1), where often high concentrations of this indoleamine have been measured than in the pineal and circulation (Figures 2,3).

Daily and Seasonal Profiles of Melatonin in the Pineal and Serum in Carp

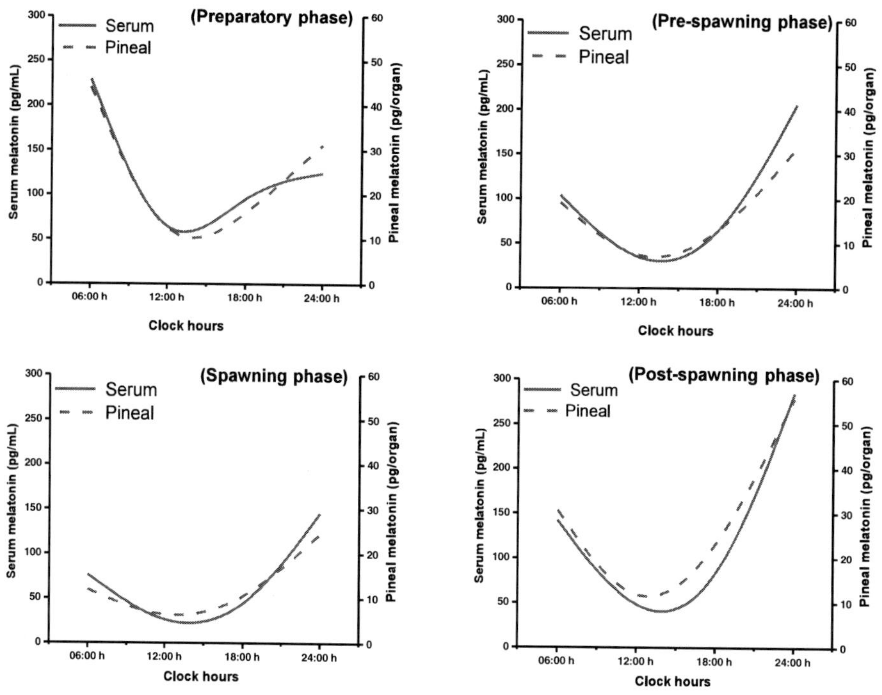

Figure 2. Graphical representation of the diurnal profiles of melatonin concentrations in the pineal and serum at four different time points (06:00, 12:00, 18:00, and 24:00 h) in a 24 h cycle during four different reproductive phases (preparatory, pre-spawning, spawning and post-spawning) in an annual cycle in carp *Catla catla*.

Daily and Seasonal Profiles of Extra-Pineal (retina, gut and ovary) Melatonin in Carp

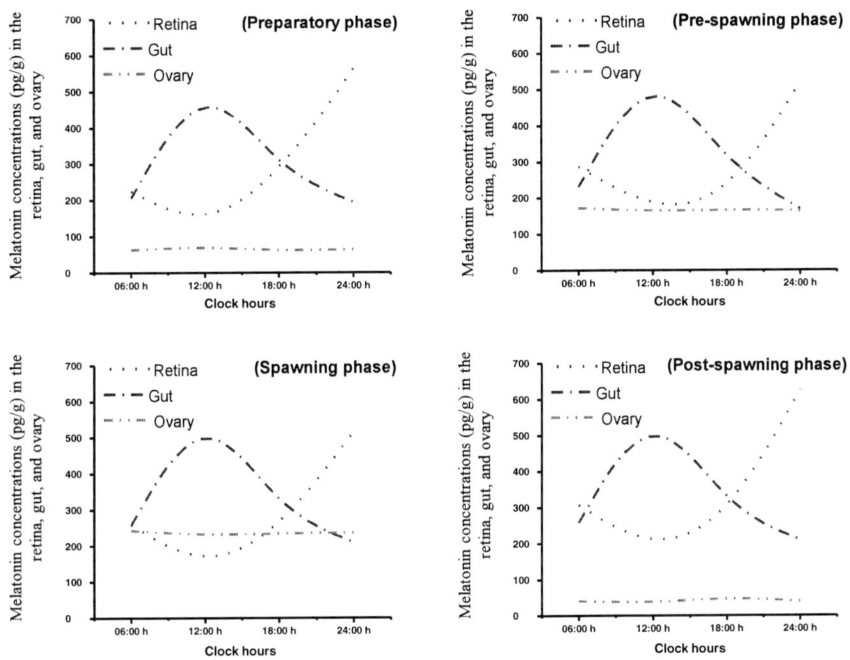

Figure 3. Graphical representation of the diurnal profiles of melatonin concentrations measured in different non-pineal tissues like retina, gut (anterior region) and ovary at four different time points (06:00, 12:00, 18:00, and 24:00 h) in a 24 h cycle during four different reproductive phases (preparatory, pre-spawning, spawning and post-spawning) in an annual cycle in carp *Catla catla.*

2.2.1. The Retina: An Extra-Pineal Source of Melatonin

The retina is one of the major sites of melatonin formation in many vertebrates, where it exerts multiple regulatory functions (Wiechmann and Summers, 2008; Tosini et al., 2012). Presence of melatonin and existence of its daily rhythm were reported first in the retina of 2-day-old rat pups (Bubenik et al., 1978), in which melatonin levels increased with age up to day 20. Since then several studies demonstrated rhythmic production and regulation of retinal AANAT and melatonin in different mammals, including human (Miller et al., 1980; Tosini and Menaker, 1996, 1998). Melatonin in the retina, as in pineal, is produced during darkness. However, *in vitro* studies on golden hamster retina demonstrate that melatonin rhythm may be affected by altered light-dark cycles, continuous darkness, and circadian short-period mutation 'tau' (Tosini and Menaker, 1996). The rhythm, but not the synthesis, of melatonin is related to the integrity of rod photoreceptors in the retina. AANAT is detected in the inner nuclear layer and ganglion cell layer of the rat retina, but only rhythmic AANAT transcription is evident in the outer nuclear layer (Liu et al., 2004). The retinal pigment epithelium seems to be another source of melatonin synthesis in the eye. Several melatonin biosynthetic enzymes are known to express in cultured human retinal pigment cell line (Zmijewski et al., 2009). On the basis of the study showing that exogenous melatonin administration increased the content of melatonin in the rat retina (Yu et al., 1981), any control derived from the pineal gland over extra-pineal melatonin synthesis may not be ruled out. However, the findings that melatonin rhythms in the rat retina are not affected by pinealectomy argue in favor of an independent retinal oscillator controlling melatonin rhythms in the studied tissue (Reiter et al., 1983; Acuña-Castroviejo et al., 2014).

Melatonin producing cells have also been identified in teleostean retina. Immune-reactive melatonin cells have been localized in the outer nuclear layer (ONL), outer plexiform layer (OPL) and inner nuclear layer (INL) of the retina (Falcón and Collin, 1991). Two cell types are involved in melatonin synthesis; photoreceptor cells and Müller cells. Immuno-fluorescence study indicated that perikarya of the ONL, the synaptic processes of photoreceptor cells of the OPL are rich in melatonin content (Falcón and Collin, 1991). However, no apparent homogeneity has been found in the rhythmic pattern of melatonin in fish retina. A nocturnal retinal melatonin pattern, as seen in zebrafish and goldfish (Cahill et al., 1991; Cahill, 1996; Iigo et al., 1997a) as well as in the carp (Mukherjee, S., unpublished data), may not be the general rule. In the concentrations of melatonin retina in different fish, either no rhythm is detected, or the peak is noted at different times of the light-dark cycle, including day (Gern et al., 1978; Iigo et al., 1997b; Besseau et al., 2006). The phase of the retinal rhythm in the sea bass (García-Allegue et al., 2001) changed throughout seasons. Such variations in the retinal melatonin rhythm patterns may be related to the expression of enzymes controlling melatonin synthesis in different retinal cell types like photoreceptors, inter-neurons and ganglion cells (Besseau et al., 2006; Vuilleumier et al., 2007). Notably, non-visual photoreceptors have been identified in the inner nuclear and ganglion cell layers of the fish retina (Cahill and Besharse, 1995; Foster and Bellingham, 2004). Thus, contrary to existing belief, the photoperiodic control of melatonin production may not be exclusive feature of photoreceptor cells. Alternatively, the retinal/pineal differences may be explained from the unique feature of teleost fish, that express one or two AANAT1 (1a, 1b) genes in the retina (Bégay et al., 1998; Falcón et al., 2003; Coon and Klein, 2006). Despite these differences, melatonin synthesis in the retina, as in the pineal, appears to be under circadian clock control in some, but not all species (Iuvone et al., 2005; Falcón et al., 2010). Overall, AANAT is responsible for the rhythmic secretion of melatonin in the retina, where the expression of AANAT1 (Falcón et al., 2001) exhibits a circadian rhythm similar to that in the AANAT2 in pineal (Binkley et al., 1980). The activity of AANAT shows a peak at the mid dark-phase followed by a decrease thereafter. The study showing a persistent daily rhythm of AANAT, at least for one 24h-cycle under constant dark conditions, suggests that circadian oscillator drive the rhythmic activity of AANAT in the retina, as in the pineal organ. It is argued that melatonin originated from the retina might act as modulator of neurotransmission and neuronal excitability of this tissue (Dubocovich et al., 1985). Retinal melatonin may have autocrine, paracrine or neuro-endocrine functions (Binkley, 1986; Gern et al., 1986; Wiechmann, 1986).

2.2.2. The Gut: An Extra-Pineal Source of Melatonin

In addition to pineal and retina, the gut or gastro-intestinal tract (GIT) is a major site of melatonin synthesis in vertebrates (Bubenik, 2002). Available literature on localization, distribution and functional aspect of gastrointestinal melatonin is based largely on mammalian and avian studies. Presence of melatonin in the gastrointestinal tract was first reported in rat (Raikhlin and Kvetnoy, 1974). Further studies demonstrated that GIT serves as a source of melatonin in different groups of vertebrates ranging from mammals (Bubenik and

Brown, 1997), birds (Vakkuri et al., 1985a), to reptiles, amphibians and fish (Bubenik and Pang, 1997; Mukherjee et al., 2014a). Pinealectomy has no effects on the melatonin profile of the GIT in pigeon (Vakkuri et al., 1985b) and in rat (Bubenik and Brown, 1997) indicating that the gut melatonin titers are independent of the activity of the pineal in concerned animals. In mammals, the concentrations of melatonin vary in between different parts of the gut, but such variations are not identical among the studied species. Melatonin in rat is detected low in the jejunum and ileum, high in the caecum and highest in the colon and rectum (Bubenik et al., 1977), while in mouse the levels of same hormone are highest in the stomach, lower in the ileum and lowest in the jejunum and colon (Bubenik et al., 1992).

Considering *Aanat*-2 mRNA expression as an indicator of melatonin synthesis in the gut of goldfish (*Carassius*), diurnal rhythms in melatoninergic system with a peak at subjective night was found only in the hind gut and such rhythmicity persisted under continuous light (LL; LD 24:00) and continuous darkness (DD; LD 00:24) as well (Velarde et al., 2010). On the other hand, a recent study on a surface dwelling tropical carp (*Catla catla*) demonstrated a daily peak at midday in both the concentrations of melatonin and in the abundance of a ~23kDa AANAT protein irrespective of the anatomical segments of gut (Mukherjee et al., 2014a). Extensive experimental studies on carp gut melatoninergic system under long (LD; 16:08) or short (LD; 08:16) photoperiod as well as under continuous light (LD; 24:00) and darkness (LD; 00:24), revealed that none of the employed photo-schedules exert any significant influences on the daily profiles of melatonin as well as the AANAT protein in gut as both the gut variables exhibit a midday peak, irrespective of the photoperiodic history (Mukherjee et al., 2014b). Thus it appears reasonable to suggest that the environmental light-dark conditions may not be the external synchronizer for the daily rhythmic profiles of gut melatoninergic system in carp and thereby warrants search for any non-photic variable which may provide necessary external cue for attaining a daily peak at midday. Moreover, the differences between the functional characteristics of the pineal derived and the GIT derived melatonin may not be the same, as the nature of release and the functions of the melatonin from this extra-pineal organ are not yet fully known. Most likely, pineal melatonin acts as a hormone, whereas GIT melatonin may perform the functions in endocrine and/or autocrine and/or paracrine manner (Bubenik, 2008).

2.2.3. The Ovary: An Extra-Pineal Source (?) of Melatonin

Presence of melatonin in the ovary is known for quite a long time in mammals (Brzezinski et al., 1987; Ronnberg et al., 1990). Very recently, melatonin has been detected and measured also in the carp ovary (Hasan et al., 2014). Melatonin measured in the human ovarian follicular fluid (FF) exceeds those in simultaneously obtained blood samples (Reiter et al., 2013), though it is not yet clear whether melatonin detected in the FF is synthesized within the ovary itself or, is derived from the circulation. Since the rat and cat ovaries have the ability to take up and retain circulating [^{3}H]-melatonin (Tamura et al., 2008), melatonin concentrations measured in the follicular fluid may be exclusively a result of its uptake, against a gradient, from the blood (Tamura et al., 2009). However, it is possible that there are other ovarian cells which could synthesize melatonin and release it into the follicular fluid. On the other hand, occurrence of melatonin and its precursors, serotonin and *N*-acetyl-serotonin, in extracts of human ovary (Itoh et al., 1999) underlines the possibility that ovary itself may be the site of melatonin synthesis. Additional support to the hypothesis that ovary is a melatonin synthesizing organ is provided by the identification of acetylserotonin O-methyltransferase (ASMT), the final enzyme in the melatonin synthesis pathway, in the cumulus cells of the bovine follicle (Salhab et al., 2013). The follicular fluid melatonin concentrations in humans have been found two times higher in large follicles just prior to ovulation compared to those in smaller antral, immature follicles (Nakamura et al., 2003, Reiter et al., 2013). Although a debate on the source of melatonin detected in the ovary still continues, the study on carp demonstrated temporal pattern of ovarian melatonin for the first time in any animal species (Hasan et al., 2014). The results of such study have two major implications: one is the information on the daily profiles and another is on the seasonal fluctuations (Figure 3). Notably, ovarian melatonin levels, regardless of fluctuations in environmental light and temperature experienced by carp in natural habitat, do not show any significant daily variations and thereby rules out possible role of environmental photo-thermal conditions in determining its temporal pattern. However, significant seasonal variations in the amplitude of ovarian melatonin titers parallel with the development of ovary may underline its physiological role in the regulation of several ovarian functions (oogenesis, vitellogenesis, steroidogenesis) and associated hormones in an annual cycle.

3. Physiology of Melatonin in the Regulation of Fish Reproduction

Most fish species are seasonal breeders, in which the reproductive cycle comprises of sequential events in the gonads. Primarily, on the basis of germ cell profiles of gonads, the annual breeding cycle in fish is divided into four different phases, like (a) preparatory phase, (b) pre-spawning phase, (c) spawning phase, and (d) post-spawning phase. Each reproductive phase has a specific duration and is characterized by distinct germ cell profiles in the ovary (Dey et al., 2004) and the testis (Bhattacharyya and Maitra, 2006) as well. The ovaries exhibit mostly oogonia and few stage-I primary oocytes during the preparatory phase; stage-II oocytes as the most advanced germ cells in the pre-spawning phase; yolk-laden stage-III and stage-IV oocytes in the spawning phase; and a large number of atretic or post-ovulatory follicles during the post-spawning phase. On the other hand, in males, recrudescence of gametogenesis occurs with stage-I spermatids as the most advanced germ cell during preparatory phase, while advancement of germ cell development beyond spermatid-I stage occurs during pre-spawning phase. The peak in gametogenesis is attained during spawning phase, followed by a sudden fall in post-spawning phase.

Reproductive events in a seasonal breeder are often synchronized with periodic changes in one or more environmental components in such a way that spawning occurs in the most favourable part of the year (Foster and Kreitzman, 2005). Temporal organization of seasonal reproduction, which is mostly a species specific phenomenon, may be the function of a self-sustained circannual rhythm (Bromage and Duston, 1986) or, influenced by one or more components of the environment (Bromage et al., 2001). Most teleosts, which breed at mid- and high latitudes (Porter et al., 2000; Bromage et al., 2001; Bayarri et al., 2009; Falcón et al., 2010; Zohar et al., 2010), and even in few species at low latitudes (Dey et al., 2004, 2005; Bhattacharya et al., 2005; Bhattacharyya and Maitra, 2006; Maitra et al., 2013), use the information on changes in the duration of the solar day or photoperiod, individually or in combination with water temperature or other component(s) of the environment, as the 'driving force' to determine temporal pattern of seasonal reproduction. Processing of light information at physiological level, thus, appears to be the most important event that occurs at neuroendocrine system to coordinate and/or synchronize different body functions associated with the seasonality of reproduction. Here is the physiological significance of melatonin, which following synthesis in a photo-sensitive organ, like pineal and retina, is released in the circulation to reach the target organs and act on them through specific G-protein coupled receptors. Obviously, demonstration of melatonin receptors in several tissues and organs, especially gonads in fish (Chattoraj et al., 2009a), has been a major breakthrough in understanding the physiology of melatonin in the regulation of fish reproduction, as most of the earlier studies were confined to the serum profiles of melatonin in relation to growth and development of gonads in an annual reproductive cycle, or to response at target sites to melatonin administration, leaving the endogenous source of melatonin in circulation mostly undefined. Thus, we are compelled to discuss the role of melatonin as a potent indoleamine candidate in the regulation of seasonal reproduction in fish using the data on profiles of melatonin in various organs as well as in circulation during different reproductive phases of an annual cycle, but more importantly, the data gathered from *in vivo* and *in vitro* studies on the mechanism of gonadal response and its regulatory axis to exogenous melatonin in respective species.

3.1. Endogenous Melatonin Profiles in Relation to Fish Reproduction

3.1.1. Pineal and Serum Melatonin Profiles in relation to Seasonality of Reproduction

Melatonin concentrations in the pineal and serum exhibit parallel rhythmic changes with a nocturnal peak in a 24 h light-dark cycle and this rhythmic pattern is conserved across all vertebrates. However, three variants of nocturnal melatonin peak, namely type-A, -B and -C profiles have been identified in different vertebrates, including fish (Falcón et al., 2010; Migaud et al., 2010). The A-type profiles are characterized by a discrete peak in late dark phase e.g., Atlantic cod *Gadus morhua* (Porter et al., 2000), haddock *Melanogrammus aeglefinus* (Davie et al., 2007), whereas B-type profiles are characterized by a discrete peak in the mid dark phase e.g., Nile tilapia *Oreochromis niloticus niloticus* (Martinez-Chavez et al., 2008), and the C-type is characterized by a rapid rise in melatonin immediately after the onset of darkness e.g., Atlantic salmon *Salmo salar*, rainbow trout *Onchorhynchus mykiss,* Atlantic halibut *Hippoglossus hippoglossus* and most teleosts (Falcón et al., 2010; Migaud et al., 2010). Though the etiology of different types of melatonin profile is far from being clearly understood, it might be linked to the ability or not to anticipate photic signals under the control of circadian clocks (Martinez-Chavez et al., 2008). However, the study on carp *Catla catla* revealed for the first time that the nocturnal peak pattern of both serum (Maitra et al., 2005) and pineal (Seth and Maitra, 2010a) melatonin in the same species may vary from A-type (during the preparatory phase) to B-type (in the remaining parts of reproductive cycle) in different parts of an annual cycle (Figure 2). Likewise, a shift of nocturnal peak in melatonin from midnight to the end of the night was found in Atlantic cod (Porter et al., 2000) and in Senegal sole (Vera et al., 2007). A shift of the melatonin offset in mammals is generally attributed to the influences of light (Liu and Borjigin, 2005). Though due to lack of any direct evidence the mechanism of a seasonal shift of nocturnal peak of melatonin in teleosts remains unknown, carp study indicated that attainment of a daily peak in melatonin levels may not be unrelated to the reproductive status of the fish. Notably, minimum diurnal value of melatonin during the midday does not vary with the reproductive seasons, but the daily peak may vary from midnight (during the pre-spawning, spawning, and post-spawning phases) to late-night (in preparatory phase) in an annual cycle (Figure 2) (Maitra et al., 2005; Chattoraj et al., 2009b). Since natural day-length (about 12L:12D) during the preparatory phase (February-March) and the post-spawning phase (September-October) appears mostly identical, the study on carp suggests a possible relationship between the functions of pineal organ and reproductive organs (Randall et al., 1995), rather than a function of photoperiod (Porter et al., 1999). Data on the peak values of serum melatonin in different reproductive seasons showing maximum value during the post-spawning phase and minimum during the spawning phase may also indicate a functional interplay between the levels of melatonin and the development of gonads in an annual cycle in carp and other fish species (Bornestaf et al., 2001; Maitra et al., 2013).

3.1.2. Gut Melatonin in Relation to Seasonality of Reproduction

Concentrations of gut melatonin have been measured in several vertebrates (Bubenik, 2002), but information on their daily profiles during different reproductive phases in an annual cycle is limited only to the study on carp (Mukherjee et al., 2014a). In this fish, concentrations of gut melatonin undergo significant variations in a 24 h light-dark cycle, but the daily peak is shown at midday, unlike a nocturnal peak as shown in the pineal and retina (Seth and Maitra, 2010a; Moniruzzaman and Maitra, 2012; Mukherjee, S., unpublished data). Strikingly, the pattern of this daily rhythm of gut melatonin persists throughout the annual reproductive cycle without any variations in its any rhythm features (Figure 3). The seasonal profiles of gut melatonin titers also do not show any significant correlation with the fluctuations of photoperiod and water temperature, which appears to act as most important synchronizer of annual reproductive events in carp (Dey et al., 2004; Bhattacharya and Maitra, 2006). The supposition also earned support from study the experimental conditions, where no significant influences of altered photoperiods have been noted on the daily or seasonal profiles of melatonin and AANAT in carp gut (Mukherjee et al., 2014b). Thus, it appears logical to surmise that the control mechanisms of melatoninergic system in the gut and the pineal are not only different, but their physiological implications, especially to the regulation of reproductive events, may not be identical as well. However, the discrepancy between rhythm features of serum, pineal, retina and gut melatonin under identical conditions may underline the importance of different endogenous (Velarde et al., 2010) or external signals, like quality, quantity and even timing of food supply (Huether et al., 1992; Herrero et al., 2007; Konturek et al., 2011) in determining the daily rhythmic patterns of gut melatonin. Obviously, further experimental studies are warranted before possible influences of gut melatonin in the regulation of seasonal reproductive events in any fish are finally ruled out.

3.1.3. Ovarian Melatonin in Relation to Seasonality of Reproduction

The endocrine status of ovary in carp varies with reproductive seasons. More specifically, under natural conditions, the serum levels of two major ovarian steroids, i.e., estradiol (E_2) and 17α, 20β-dihydroxy-4-pregnen-3-one (17α, 20β DHP) or maturation inducing hormone (MIH), show seasonal peak during spawning and sudden fall thereafter (Chattoraj et al., 2009b). A recent study on carp (Hasan et al., 2014) showing that intra-ovarian melatonin titres during spawning is ~5 times higher than measured during post-spawning phase (Figure 3) promotes the idea of a possible relationship between the steroidogenic status of the ovary and seasonal profiles of ovarian melatonin titers in carp.These findings are consistent with the report on the human follicular fluid melatonin levels, which are two times higher in large follicles just prior to ovulation compared to those in smaller antral, immature follicles (Nakamura et al., 2003). Likewise, ram seminal plasma melatonin exhibit marked seasonal variations with low levels during non-breeding phase and high levels during breeding (Casao et al., 2010). Collectively, available data underlines physiological importance of ovarian melatonin as the functional correlate of seasonal changes in ovarian hormones as well as vitellogenin incorporation in the oocytes, and thereby opens up the possibility that ovarian melatonin performs local actions to regulate activities of ovary itself in an annual cycle.

3.2. Influence of Exogenous Melatonin on Fish Reproduction

3.2.1. Influence of Exogenous Melatonin on Growth and Development of Testis

Melatonin, due to its transducing property of photoperiodic information (Falcón et al., 1994), flourished the immergence to investigate the effects of exogenously administered melatonin on the activities of gonads. Initial studies implicating melatonin as a hormonal candidate in the regulation of reproduction were based on the results derived from experiments in which the tryptophan derivative melatonin was administered exogenously (Vaughan et al., 1987) that displayed a pattern to mimic the effects of endogenous melatonin on animal physiology (Smith, 1981). Therefore, study of the testicular response under experimental conditions emerged as effective following modulation of endogenous melatonin levels. At this point, manipulation of light-dark conditions to extreme range was taken into account due to its pre-established property that the amplitude of pineal melatonin secretion is dependent on the period of experienced darkness (Reiter, 1991). Such studies revealed that a schedule of continuous darkness (DD) enhances melatonin synthesis in the pineal (Okimoto and Stetson, 1999), while continuous illumination (LL), a regimen known as "functional pinealectomy" (Carter et al., 1982), diminishes melatonin secretion. The potent regulatory functions of melatonin on annual testicular events in carp has been thoroughly studied employing diverse experimental regimens such as (a) exogenous administration of melatonin at variable doses (25-, 50-, or 100μg/100g body weight) and durations (1-, 15-, or 30 days), and (b) exposure of fish to LL or to DD for 30 days during four different reproductive phases in an annual cycle (Bhattacharya et al., 2007). The findings collectively revealed for the first time in any tropical teleost that LL and DD modulate the endogenous system of melatonin (Bhattacharya et al., 2007). In fact, single administration of melatonin does not exert any significant changes in weight and spermatogenetic status of the testis compared to those in the respective control fish in any reproductive phase, but serum testosterone levels displays significant decline during the pre-spawning and spawning phases of the annual cycle. On the contrary, treatment of melatonin for consecutive 15- and 30 days causes dose dependent precocious maturation during the preparatory phase and anti-gonadal effects during the pre-spawning and spawning phases. Exogenous melatonin administration resulted in a sharp decline in the number of testicular lobules containing spermatocytes and stage II spermatids during the pre-spawning, and stage II spermatids during the spawning phase followed by an inhibition of 17β-HSD and $\Delta^5$3β-HSD activity. However, melatonin administered at higher doses for longer duration has a different response. No significant changes are noted in the values of testicular weight, spermatogenetic index, activity of 17β-HSD and $\Delta^5$3β-HSD in the testis, and serum level of testosterone in any group of carp following administration of melatonin at the dose of 25 μg-, or 50 μg-, or 100 μg /100g body wt. for 15 or 30 days. Earlier studies in different teleosts have also shown that melatonin may have pro-, or anti-gonadal influences depending upon the dose of hormone (Borg and Ekström, 1981). But in *Catla catla* the inhibitory (during pre-spawning and spawning phases), or stimulatory (during preparatory phase), or no influences (during post-spawning phase) of melatonin on the gonads are noted not only in relation to the dose and duration of hormone treatment, but to the reproductive status of the recipient fish as well (Bhattacharya et al., 2007).

3.2.2. Influences of Exogenous Melatonin on Growth and Development of Ovary

Exogenous administration of melatonin in salmonid fish depicted that melatonin controls the reproductive seasonality by stimulating the final stages of maturation and by synchronizing the oocyte maturity with optimal timing of spawning (Bromage et al., 1994). However, the study on the effects of melatonin administration during different phases of reproductive cycle in female carp (*Catla catla*) demonstrated that influences of exogenous melatonin vary in relation to the dose of hormone, duration of treatment as well as the reproductive status of the fish (Maitra et al., 2005; Maitra et al., 2013). Information on the the effects of the exogenous melatonin in adult carp is provided following administration of different doses (25, or 50, or 100 μg /100 g body weight) at the on-set of darkness for 1 day, 15-, and 30 days (Maitra et al., 2005). Exogenous melatonin, irrespective of its dose and duration of treatment, fails to elicit any ovarian response during the post-spawning phase, while no remarkable impact of single administration of melatonin are observed in any part of the annual cycle, though in laboratory rodents single injection of melatonin is known to exert significant influences on the functions of hypophyseal-gonadal axis (Maitra and Ray, 2000). Each dose of melatonin injected daily for 15 or 30 days during the preparatory phase causes precocious growth of oocytes (formation of stage-II oocytes) compared to the ovaries in control fish showing stage-I oocytes as the most advanced follicles. Conversely, melatonin at each dose causes a significant decrease in the number of stage-II and stage-III oocytes during the pre-spawning phase and spawning phase respectively, with subsequent increase in the number of degenerated follicles relative to their number in the ovaries of control fish. It may thus be concluded that exogenous melatonin administration may result in variable influences, like pro-gonadal or anti-gonadal or no subtle effects on gonads depending on the reproductive status of fish in an annual cycle (Maitra et al., 2005). The anti-gonadal effects of melatonin are also known for other tropical fish like *Heteopneustes fossilis* (Sundararaj and Keshavanath, 1976; Joy and Agha, 1991), *Mystus tengara* (Saxena and Anand, 1977), and *Clarias batrachus* (Singh and Lal, 1994), though a pro-gonadal action of exogenous melatonin is also known in fish (Singh and Lal, 1994; Bromage et al., 1995). The effects of exogenous melatonin may also depend on photoperiod as well as reproductive phase of the fish. Melatonin administration in *Heteopneustes fossilis* for 20 days at 6-8 h after the onset of light phase inhibits vitellogenesis, induces follicular atresia and lowers the number of pituitary gonadotrops during the pre-spawning phase and induces ovarian regression and reduces the number of pituitary gonadotrophs in the spawning phase (Sundararaj and

Keshavanath, 1976). In *Mystus tengara* melatonin injection 2 h after the onset of light phase during the phase of ovarian recrudescence arrests vitellogenesis and increases the frequency of atretic follicle (Saxena and Anand, 1977). The study on *Clarias batrachus* reveals that melatonin administration prior to darkness at the dose of 100 and 200 μg/fish decreases testosterone level, while at higher dose (400 μg/fish) increases the testosterone level (Singh and Lal, 1994). In another study on the same species demonstrates that melatonin administration at lower doses of 25- and 50 μg/fish significantly lowers the 17α-hydroxyprogesterone levels and at higher doses (100-, 200- and 400 mg/fish) further reduces the hormone to non-detectable levels (Singh and Lal, 1994). Implantation of melatonin in *Fundulus similes* induces spawning in rainbow trout in an asynchronous pattern (Bromage et al.,1995), while the fecundity in zebrafish females is affected positively by exogenous melatonin administration (Carnevali et al., 2010). This is likely that melatonin action interacts with the brain-pituitary-gonadal axis (Amano et al., 2000), or with one or more of a variety of peripheral and/or central sites, including the brain (Khan and Thomas, 1996), the pituitary gland (Gaildrat and Falcón, 2000) and also on the gonads (Amano et al., 2000; Maitra et al., 2013).

3.2.3. Influences of Exogenous Melatonin on Oocyte Maturation

Fish oocyte maturation is a naturally occurring phenomenon, in which two successive meiotic divisions take place producing two polar bodies, when the enlarged nucleus of the primary oocyte, which is arrested at the diplotene stage, shifts towards a more peripheral position, followed by the breakdown of its membrane and with the extrusion of a polar body that ultimately mark completion of the first meiotic division (Nagahama, 1994). At the transition of prophase/metaphase, oocyte nuclear envelope denatures or germinal vesicle breakdown (GVBD) takes place indicating completion of the penultimate step in the progression of oocyte maturation (Tokumoto et al., 2004). Second meiotic division starts immediately thereafter, but progression continues only up to the metaphase stage during which a distinct animal pole differentiates from the vegetal pole followed by yolk maturation, hydration, resulting in less opaqueness. The ovulated eggs are spawned in water and are fertilized by the male immediately after the mature oocyte is ovulated out of its follicular envelope. In vertebrates, pre-ovulatory gonadotropin surge stimulates final oocyte maturation or resumption of meiosis. The promoting effects of gonadotropin on oocyte of most teleosts involve two major mechanisms in two temporally different developmental stages (Patiño et al., 2001): (a) a steroid-independent priming stage that increases the maturational competence of the oocytes, and (b) a maturational stage that depends on the gonadotropin stimulated synthesis of maturation-inducing steroid (MIS) or maturation-inducing hormone (MIH), involving steroidogenic pathways in the ovarian follicular cells (Patiño and Thomas, 1990). In most fish, final step of oocytes maturation is initiated by the potent MIH that induces the activation of the maturation- promoting factor (MPF) in the oocyte cytoplasm through action on the oocyte membrane. MPF comprises of two subunits- (i) cyclin B, a regulatory subunit and (ii) cyclin-dependent kinase (Cdk1, or Cdc2, or p34 kinase), a catalytic subunit of which the activation of cdc2 is related with its mobility shift from 35 to 34 kDa (Nagahama and Yamashita, 2008).

Though several hormones are known to act as inducer of the MPF synthesis (Tokumoto et al., 2004), influences of any indoleamine on the steroid-induced maturation of oocytes and/or formation of MPF in fish oocytes has been reported first in carp (Chattoraj et al.,

2005). Such *in vitro* study reveals that the rate of GVBD in the oocytes increases by the formation of MPF when denuded oocytes are incubated with melatonin especially 4 h prior to addition of MIH in the medium. Densitometric analysis of the immunoblot data collected from the melatonin pre-treated MIH incubated oocytes shows that cyclin B level continues to increase even after 4 h of incubation to reach the peak after 12 h. Moreover, determination of H1 kinase activity as an indicator of MPF activity in oocytes reveals that melatonin pre-incubation increases MIH stimulation of histone H1 phosphorylation as compared to MIH alone. Thus the study showing that prior incubation with melatonin accelerates the action of MIH on oocyte maturation presents first evidence of extra-hypothalamic action of melatonin in the maturation of fish oocytes (Chattoraj et al., 2005). The hypothesis earns further support from another *in vitro* study on *Danio rerio* in which melatonin is known to exert a direct action on follicles (as shown by the increase of the oocytes undergoing to GVBD) and modulate mpr α and β gene (whose proteins are involved in oocyte maturation) expression (Carnevali et al., 2011). Apart from interactions between melatonin and MIH during fish oocyte maturation, actions of other agents like serotonin, another indoleamine and precursor of melatonin, are also gaining attention. Incubation with serotonin alone or, serotonin and melatonin, or melatonin and MIH respectively (Chattoraj et al., 2008) reduced stimulatory actions of melatonin alone, or melatonin and MIH on the rate of GVBD.

4. Mechanism of Melatonin Actions in the Regulation of Fish Reproduction

The mechanism of action of melatonin in the regulation of reproduction is one of the major areas that received wide attention in studies on different animals, including fish (Kennaway and Rowe, 1995; Wojtowicz and Jakiel, 2002; Maitra and Chattoraj, 2007; Reiter et al., 2009; Tamura et al., 2009; Maitra et al., 2013). There is possibility that melatonin interacts with the brain-pituitary-gonadal axis (Amano et al., 2000), or with one or more of a variety of peripheral and/or central sites, including the brain (Khan and Thomas, 1996), the pituitary (Gaildrat and Falcón, 2000), also on the gonads (Figure 4) (Amano et al., 2000; Maitra et al., 2013).

4.1. Actions on the Hypothalamo-Hypophyseal-Gonadal Axis

The hypothesis that emerged from such studies in common, states that actions of melatonin on the reproductive system are mediated by an interaction with the hypothalamic control of pituitary function (Figure 4). The hypothalamo-hypophyseal system in periodic breeders in general plays a major role in the photoperiodic regulation of reproduction, where melatonin provides temporal information related to day length and season (Falcón et al., 2007). However, pathways of melatonin action through hypothalamus in different vertebrate groups vary with the organization of their hypothalamic neurosecretory system. In fish, the preoptic area (POA) of hypothalamus is known to integrate photoperiodic information from different sources because it receives nervous inputs from the retina and pineal organ, and hormonal (melatonin) input from the pineal organ (Ekström and Vanecek, 1992; Ekström and Meissl, 1997; Vernadakis et al., 1998; Mazurais et al., 1999) and thereby occupies a pivotal position in photoneuroendocrine regulation of reproduction. Neurons from the POA transmit monoaminergic (i.e., dopamine, 5-hydroxytryptamine) and/or peptidergic information to the pituitary gland, which corresponds to peptides that reach the neurohypophysis (e.g., isotocin, arginine vasotocin) or releasing factors that control the activity of the pituitary cells for synthesis of a number of hormones associated with the seasonal growth and development of gonad. Thus the effects of melatonin on reproductive organs result from an interaction with the hypothalamic control of pituitary function. Melatonin administration in the masu salmon (*Oncorhynchus masou*) reduced pituitary luteinizing hormone (LH) content, but stimulated pituitary follicle stimulating hormone (FSH) content (Amano et al., 2004). Intraventricular administration of melatonin in the POA of Atlantic croaker (*Micropogonias undulatus*), with fully developed gonads, also resulted in reduced LH release by the pituitary gland (Khan and Thomas, 1996). Moreover, the detection of melatonin receptors (both MT1 and MT2 subtypes) and display of binding of ^{125}I Mel to membrane preparations in pike and trout pituitary glands provide evidence that melatonin might modulate neuroendocrine functions by targeting the pituitary gland itself (Gaildrat and Falcón, 1999; Falcón et al., 2003). Thus it seems likely that variable influences of melatonin under different experimental conditions and even during different parts of the reproductive cycle depend on the type of melatonin receptors expressed in the pituitary glands in the concerned animals (Falcón et al., 2007). Obviously, melatonin is *per se* neither anti-gonadotropic nor pro-gonadotropic, rather the changing duration of the nocturnal melatonin message is a passive signal that provides the hypothalamic-pituitary-gonadal axis information as to the time of year (calendar information) (Hazlerigg, 2001). The reproductive axis uses the season dependent melatonin rhythm to adjust time of reproduction accordingly. However, considering hypothalamus as the highest centre of endocrine regulation of pituitary functions, more convincing explanation for the mechanism of action of melatonin on this part of brain deserves special attention.

Gonadotropin-releasing hormone (GnRH) is a decapeptide that is synthesized and released by hypothalamic neurosecretory cells in a pulsatile pattern into the hypothalamo-hypophyseal portal circulation and plays a pivotal role in the regulation of reproduction. This peptide binds and activates its cognate receptor (GnRH receptor) on the pituitary gonadotrope cells, and in turn stimulates the synthesis and secretion of gonadotropins (GtHs) by acting through G-protein coupled transmembrane receptor having extracellular amino terminal and intra-cellular carboxy terminal domains (Weil et al., 1992). Eleven GnRH structures have been elucidated from different vertebrates and proto-chordates, among these four different isoforms are isolated from fishes (Bhattacharya, 1999). GnRH in fish structurally varies from other vertebrates, but the amino acid length and in NH_2 terminus and COOH terminus are conserved. Several studies in sea bass provided evidence in favour of expression of three different GnRH forms, i.e., GnRH-1, GnRH-2 and GnRH-3 (González-Martínez et al., 2002a,b; Zmora et al., 2002; Servili et al., 2010) that could act through five different GnRH receptors (González-Martínez et al., 2004; Lethimonier et al., 2004; Moncaut et al., 2005; Zohar et al., 2010). Existence of day-night variations as well as melatonin inhibitory effects on the expression of GnRH-1, GnRH-3 and several GnRH receptors in the brain are reported in the same fish (Servili et al., 2013). Obviously, these interactions between melatoninergic and GnRH systems could represent a substrate of photoperiod effects on reproductive and other rhythmic physiological events in European sea bass and other fish species, in which melatonin is directly implicated to photoperiodic regulation of seasonal reproduction.

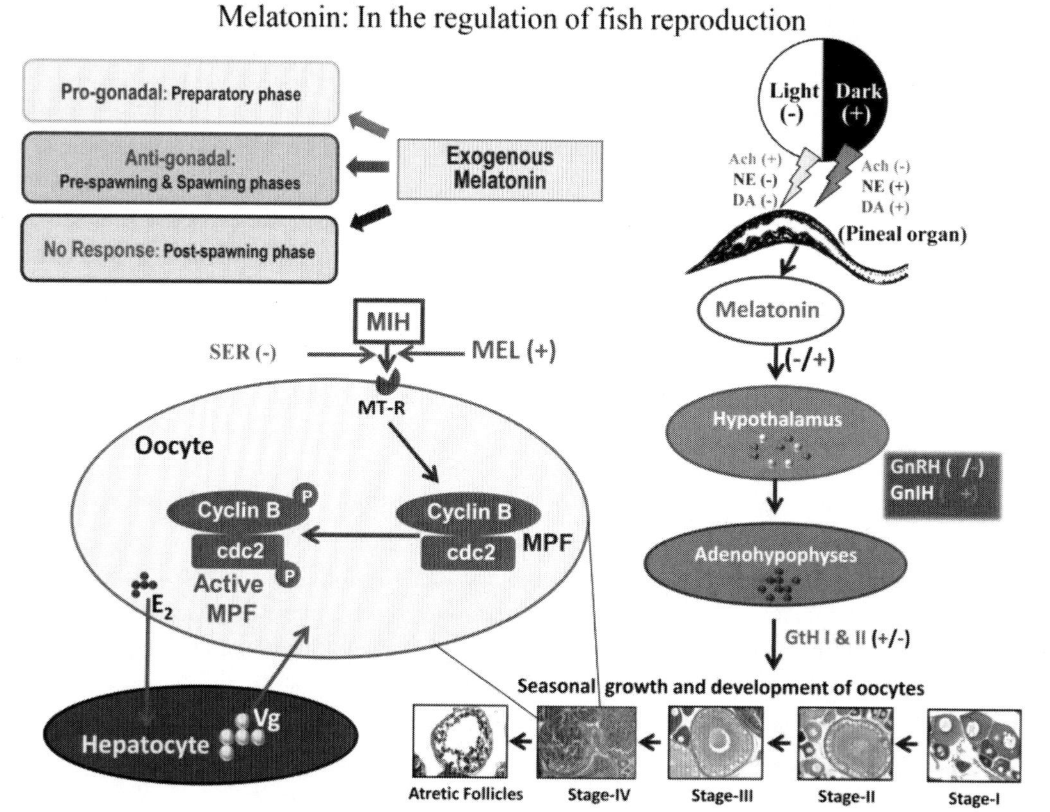

Figure 4. Diagrammatic presentation of the summary of information gathered from different studies on fish to explain possible role of melatonin in the regulation of seasonal events in the ovary and oocyte maturation. Melatonin, once synthesized in the pineal in response to light-dark conditions of the environment and released in blood, may act through hypothalamo-hypophyseal gonadal axis (to regulate oocyte development from stage-I to stage-IV) and/or directly on the gonad [to interact with other hormones like maturation inducing hormone (MIH) and serotonin (SER) via respective hormonal receptors to induce final stage of maturation by the active formation of maturation promoting factor (MPF)- a complex of cyclin B and cyclin-dependent kinase, Cdk1]. Remarkably, exogenous administration of melatonin leads to acceleration of the mechanisms responsible for transformation of stage-I to stage-II oocytes during the preparatory phase (pro-gonadal), but retardation of oocyte development during the pre-spawning and spawning phases (anti-gonadal) or no subtle changes during the post-spawning phase. Ach = acetyl choline; DA = dopamine; E_2 = 17β estradiol; GnIH = gonadotropin inhibitory hormone; GnRH = gonadotropin releasing hormone; GtH = gonadotropic hormone; MEL = melatonin; MT-R = melatonin receptors; NE = norepinephrine; Vg = Vitellogenin.

The recent discovery of a novel hypothalamic gonadotropin-inhibitory hormone (GnIH) in different vertebrates, including fish, has added a new dimension to the current understanding on the mechanism of hypothalamic control of reproduction (Tsutsui, 2009). GnIH inhibits gonadotropin release and synthesis, thus appears to inhibit gonadal development and maintenance. The GnIH receptor, comprising of seven transmembrane domains that specifically bind to GnIH, is expressed in the pituitary and several brain regions including hypothalamus. Anatomically, GnIH fibres remain in close proximity to GnRH neurons in the preoptic area (Ubuka et al., 2008), while GnIH and GnRH are connected in the median eminence (Bentley et al., 2003). GnIH appears to influence the GnRH system at the neuron and fiber terminal levels. As the GnIH receptors (GnIH-R) are expressed in the pituitary gonadotropes, it seems likely that GnIH acts directly on the gonadotrope cells and inhibits the synthesis and release of gonadotropin. GnIH may also act on GnRH neurons in the preoptic area to inhibit GnRH release. Recent studies provide evidence that melatonin induces GnIH expression in GnIH neurons (Tsutsui et al., 2012). Due to the expression of melatonin receptor (Mel1c) in the GnIH neurons, melatonin can act directly on GnIH neurons to induce GnIH expression and release (Tsutsui, 2009). An *in vitro* study on the effects of melatonin demonstrated a dose-dependent increase in GnIH release from the hypothalamus (Chowdhury et al., 2010). The expression of GnIH in brain is increased under short day, when the nocturnal duration of melatonin release is also increased, indicating that melatonin may directly induce synthesis of GnIH (Ubuka et al., 2005). Experimental study suggests that GnIH via altered melatonin signal may transducer photoperiodic information to the avian reproductive axis (Tsutsui et al., 2010). However, it remains to be known whether such mechanism also operates in seasonally reproducing fish, including carp (Maitra et al., 2013).

4.2. Direct Actions of Melatonin on the Gonads

Demonstration of melatonin receptors in isolated human oocytes (Woo et al., 2001) and in rat ovaries (Soares et al., 2003) supported the notion of a direct effect of melatonin in ovarian functions in mammals. However, no information on melatonin receptors on the gonads of any fish species was available till use of an anti-MT1 goat polyclonal antibody in Western blot analysis detected a band of 37-kDa corresponding to Mel_{1a} melatonin receptor ($Mel_{1a}R$) proteins in the carp ovarian homogenate and provided first data on their localization in different cellular fractions of ovary and their diurnal profiles in different parts of an annual reproductive cycle focusing temporal relation with the titers of serum melatonin (Chattoraj et al., 2009a). Mel_{1a} melatonin receptor proteins are detected in both the membrane and cytosolic fractions of the carp ovarian homogenate, though the relative band intensity (a ratio of the band intensity of $Mel_{1a}R$ to β-actin) of ovarian $Mel_{1a}R$ is always found greater in the membrane fraction than in its cytosolic counterpart. Functional role of membrane bound $Mel_{1a}R$ in the mediation of intra-cellular effects of melatonin is well known (Reppert et al., 1996; Gaildrat et al., 2002; Dubocovich et al., 2003), but the significance of cytosolic $Mel_{1a}R$ remains obscure. In a diurnal cycle, immunoreactivity of $Mel_{1a}R$ protein in the carp ovary is highest at midnight and lowest at midday. The pattern of day-night rhythmicity in ovarian $Mel_{1a}R$ does not vary in relation to the reproductive status of the carp or associated environmental variables like photoperiod and water temperature in an annual cycle. However, the nocturnal expression of ovarian $Mel_{1a}R$ varies in relation to the reproductive stages in an annual cycle, as it reaches peak during the spawning phase and becomes lowest during the post-spawning phase. The profiles of serum melatonin and $Mel_{1a}R$ in the ovary of carp exhibits an identical pattern of diurnal variations in each reproductive phase, but not in the preparatory phase when diurnal peak values of serum melatonin (in late dark phase) and of ovarian $Mel_{1a}R$ (at midnight) are completely dissociated from each other (Moniruzzaman and Maitra, 2012). Such findings may point to an unexplained mechanism of action of exogenous melatonin on carp gonads. For example, daily administration of melatonin at 25 μg/100 g body wt for 30 days in male (Bhattacharya et al., 2007) and female (Maitra et al., 2005) carp leads to a pro-gonadal response during the preparatory phase, but anti-gonadal or no response in the remaining parts of the reproductive cycle. Thus, it may be argued that reproductive phase-dependent effects of exogenous melatonin on gonads in carp might be related to the respective season-related diurnal profiles of endogenous melatonin, rather than to its receptor on the ovary. Moreover, demonstration of melatonin receptors in the carp ovary would obviously contribute to our knowledge on the mechanism of action of melatonin on the gonads and its interactions with other hormones in the regulation of reproduction in fish. Nonetheless, supporting data from more carefully controlled extensive studies are needed to justify this hypothesis. The *in vitro* modeling may be used to solve one of the central questions of melatonin physiology: how the rhythmic patterns of melatonin and its receptors on the ovary are associated with the photoperiodic response. The location of the melatonin target for photoperiodic regulations is not determined quite precisely, but according to the available data, melatonin might act through more than one structure even within one species (Vanecek, 1998). The experimental strategy testing most directly which features of the melatonin signal and/or of the ovarian melatonin receptor trigger photoperiodic responses in carp would be daily melatonin infusion. In essence, available data on the localization and dynamics of melatonin receptors on the carp oocytes offer interesting perspectives to explain the control mechanisms of its rhythmicity and response to external cues like, photoperiods and other hormones. In view of the data discussed here, it appears that the results obtained in carp might permit insights into the physiology of melatonin in the control of reproduction in fish in general and in carp in particular.

4.3. Receptor-Independent Actions of Melatonin in the Regulation of Fish Reproduction

A novel finding, that melatonin, due to its amphiphilic nature, can easily cross the cellular barrier and may act as a highly ubiquitous direct free radical scavenger and indirect antioxidant (Tan et al., 1993), has opened up a new possibility that this indoleamine hormone, in addition to its receptor-mediated actions, may perform a receptor-independent action in the regulation of a number of physiological functions, including reproduction. Since melatonin is capable of donating electrons to reduce the reactivity of molecules with an unimpaired electron in their valance orbital, i.e., free radicals, it may apparently directly interact with potentially damaging agents without the necessity of first binding to a receptor.

Follicle development in the mammalian ovary is a complex physiological process during which active synthesis of steroids is accompanied by production of large amount of free-radicals, especially reactive nitrogen species (RNS) and reactive oxygen species (ROS) (Agarwal et al., 2005). Excessive amount of free radicals, especially ROS, cause oxidative stress resulting in damage of oocytes (Tamura et al., 2013), accelerate oocyte aging and deteriorate oocyte quality (Goud et al., 2008). The study on a carp (*Cyprinus carpio*) also provides indication that steroidogenesis in the ovarian follicular cells is related to generation of a large quantity of free radicals (Behl, 2006). One of the alleged roles of melatonin detected in the mammalian pre-ovulatory follicles is the protection of oocyte against free radicals by playing a potent role in the physiological cascade of oocyte maturation (Tamura et al., 2014). *In vitro* study on isolated oocytes revealed that elevation of melatonin level in the ovary depends upon follicular growth during the complex process of ovulation (Nakamura et al., 2003; Tamura et al., 2014). Available data also suggest that melatonin may act on various cells of ovarian follicles as an indirect antioxidant to activate major antioxidative enzymes like superoxide dismutase (SOD), catalase (CAT), and glutathione peroxidase (GPx) which metabolize free radicals to reduce oxidative stress *in vivo* (Reiter et al., 2000). SODs represent a class of closely related enzymes that catalyze the breakdown of the superoxide anion into oxygen and hydrogen peroxide (H_2O_2) (Zelko et al., 2002). GPx and catalase then independently convert H_2O_2 to water (Gaeta et al., 2002). SOD together with GPx and CAT form the main enzyme defense mechanism against harmful effects of free-radicals. GST (glutathione transferase), like GPx, is an important enzyme in the glutathione redox cycle catalyzing the conjugation of reduced glutathione via a sulfhydryl group to xenobiotic substrates for detoxification purpose. Reduced glutathione (GSH), a non-enzymatic antioxidative agent, participates in removal of toxic peroxides and aldehydes (Samiec et al., 1998). Unfortunately, the hypothesis was not tested in any fish species, in which poor egg quality is one of the major constrains for induced breeding.

Melatonin, by playing the role of a free radical scavenger (Hardeland et al., 1993), may minimize the free radical damage in the ovary and ultimately lead to improve the quality of oocytes (Reiter et al., 2013). Since comparable data from the studies on any fish species were not known, a recent study on carp *Catla catla* (Hasan et al., 2014) aimed at presentation of basic information on the profiles of melatonin, enzymatic as well as non-enzymatic antioxidative agents, and malondialdehyde (MDA), a specific marker of intra-cellular oxidative stress, in relation to dynamics of oocyte development in the ovary during four distinct reproductive phases in an annual cycle. Moreover, possible physiological relationship has been sought between the profiles of considered ovarian variables in carp under natural photo-thermal conditions. The study showing significant seasonal variations in MDA levels with a peak in the post-spawning phase and gradual decrease thereafter towards seasonal minimum value during the spawning phase denotes ovarian oxidative stress as very high during the post-spawning season and low during the spawning phase. A significant negative correlation between the concentrations of melatonin and levels of MDA in the ovary points to a possible antioxidative role of ovarian melatonin in carp. The study also reveals that the activity of SOD, CAT, as well as GPx in the ovary varies significantly in between different reproductive phases in an annual cycle. However, the temporal pattern of their activity in a seasonal cycle does not appear identical. While the activity of both SOD and CAT follows an identical pattern of seasonal variations with a peak during the spawning, seasonal maximum values of GST and GPx are noted during the pre-spawning and the post-spawning phases respectively. The levels of reduced glutathione (GSH), unlike the activity of other studied antioxidative enzymes, in the ovary do not show any significant variations in between different sampling points in an annual reproductive cycle. However, melatonin levels in the ovary exhibit a positive correlation with the values of GSI as well as the number of advanced stages (stage-II or, stage-III or, stage-IV) of oocyte in an annual cycle. Correlation coefficient analysis indicating a positive correlation between the seasonal values of melatonin and the activity of SOD, CAT and GST in the ovary underlines their functional interplay as well as importance of ovarian melatonin in reducing oxidative stress (Figure 5). Taken together, obtained data are consistent with the findings on human granulosa cells where a direct stimulatory effect of melatonin on the activity of antioxidative enzymes like SOD and CAT is suggested to improve the quality of oocytes (Rodriguez et al., 2004; Tamura et al., 2013). It seems possible that physiological actions of melatonin in the reduction of oxidative stress in the carp ovary may be related to the activity of antioxidative enzymes like SOD, CAT and GPx (Rikans and Hornbrook, 1997), but not to the levels of non-enzymatic antioxidant (GSH). The possibility also remains open that ovarian melatonin may act at more than one points in the antioxidant defence system, rather than exclusively on the activity of a particular antioxidative enzyme/agent. Thus, it appears reasonable to argue that concentrations of ovarian melatonin in carp may be one of the physiological variables which can reduce oxidative stress in the course of seasonal growth and development of ovary and maturation of oocytes as well (Hasan et al., 2014). However, demonstration of individual role of different antioxidative enzymes/agents in the melatonin-induced ovarian growth and maturation in carp as well as in other fish species remains as an exciting topic of future research.

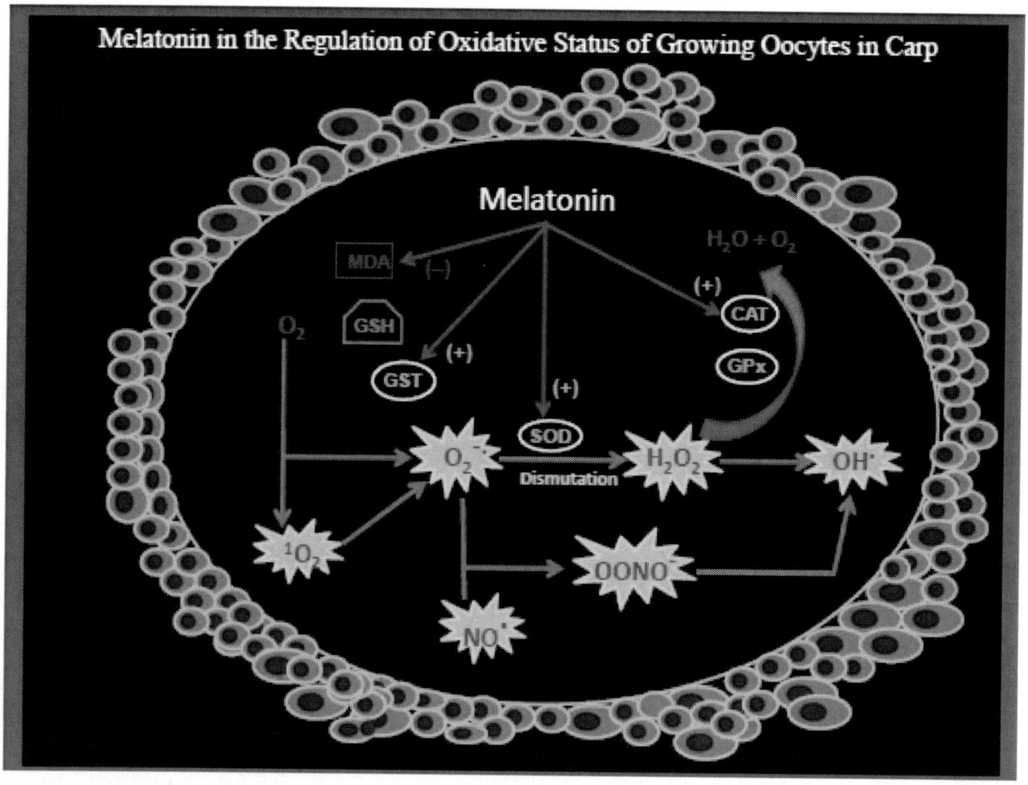

Figure 5. Schematic presentation of possible interactions between melatonin, antioxidative enzymes and stress inducing free-radicals [$^1O_2$, O_2^-, NO·, ONOO$^-$, OH· (marked as explosion)] within the carp oocyte. Recent findings indicate that melatonin present in the ovary induces (+) the activity of SOD, CAT and GST, but reduce the level of MDA, faithful marker of stress. However, intra-ovarian melatonin concentrations do not show any significant relation with GPx activity or GSH level. CAT = catalase; GPx = glutathione peroxidase; GSH = reduced glutathione; GST = glutathione-S-transferase; H_2O_2 = hydrogen peroxide; MDA = malondialdehyde; NO. = Nitrogen oxide; $^1O_2$ = singlet oxygen; O_2. = superoxide anion; OH. = hydroxyl radical; ONOO$^-$ = peroxinitrite anion; SOD = superoxide dismutase.

CONCLUSION

The existence of multiple sites in the body, where melatonin is synthesized from tryptophan through serotonin, may reflect an adaptive mechanism throughout the living system. Synthesis of this indoleamine in phylogenetically very ancient organisms, like cyanobacteria, algae and protozoa to highly evolved mammals (Poeggeler and Hardeland, 1994; Macias et al., 1999), suggests its physiological importance for the functions of cell. Convincing data are available to show that melatonin has a double role in the body: synchronization of rhythmic physiological functions with the changing environment and protection of cells from oxidative stress relating mostly to infections and thereto inflammatory damage. Remarkably, the study on fish pineal and retina by showing parallel rhythmic changes in melatonin concentrations with the same in circulation provides the basis of a belief that both the organs/tissues are the major sources of circulating melatonin. While pineal appears to be the primary source of serum melatonin, the retina, due to common intracellular mechanisms of photoreception with the pineal, may play complementary or supplementary role to the functions of pineal. Gastrointestinal tract is indeed a major site of melatonin synthesis and release, but its rhythm features being independent of the influences of environmental light-dark conditions bespeak of a hitherto unknown non-photic circadian function. On the other hand, a direct evidence of melatonin synthesis in the ovary of any fish is still lacking and thus melatonin detected and measured in the fish ovary leaves doubt on its cellular site of synthesis. The possibility does exist that the ovary, in which concentrations of melatonin undergo seasonal, but not diurnal, variations, may synthesize its own melatonin or incorporate this indoleamine hormone from the circulation. Obviously, the topic should remain speculative until further investigations. The physiological functions in fish, as in other animals, require a homeostatic equilibrium among all organs and tissues, in which the chronobiotic system uses melatonin as the main endogenous synchronizer of environmental variables in the regulation of different rhythmic body functions. In consideration to both chronobiotic and free radical scavenging functions, melatonin appears to act as an essential candidate involved in the regulation of reproduction in fish. This explains why extra-pineal/retinal melatonin acts in a different manner than that of the pineal and retina, and they have different regulatory mechanisms. In all probability, pineal/retina-produced melatonin seems to play a critical role in synchronization of body functions with reliable environmental variable, like duration of solar day or photoperiods, to perform reproductive functions at a time when possibility of survival of offspring is optimum, i.e., determination of temporal pattern of seasonal reproduction. Conversely, ovarian melatonin appears to be associated with additional mechanisms regulating the redox status of the ovary associated with the attainment of seasonal peak in gonadal growth and development leading to oocyte maturation during spawning. The conjecture, that antioxidant and anti-inflammatory properties of melatonin require higher concentrations than those produced by the pineal gland and released in circulation to play a chronobiotic role in the regulation of seasonal reproduction, may also explain the higher concentrations of melatonin in the gut and ovary compared to its serum levels. Finally, it appears logical to support the connotation that the synthesis of melatonin by the same cells expressing its membrane and nuclear receptors contributes to the intra-, auto-, and paracrine actions of this indoleamine, and thereby refining its homeostatic potential (Acuña-Castroviejo et al., 2014).

ACKNOWLEDGMENTS

We thankfully acknowledge financial assistance from the Department of Biotechnology, (BT/PR2842/AAQ/O3/131/2001 and BT/PR11423/AAQ/03/421/2008), Council of Scientific and Industrial Research (37(1220)/05/EMR-II and 37(1390)/09/EMR-II), and Department of Science and Technology (SR/SO/AS/72/2006), Government of India. Active research input from Drs. S. Bhattacharya, R. Dey, A. Chattoraj, M. Seth and M. Moniruzzaman in the current program is deeply appreciated.

REFERENCES

Acuña-Castroviejo, D., Escames, G., Venegas, C., Díaz-Casado, M. E., Lima-Cabello, E., López, L. C., Rosales-Corral, S., Tan, D. X. & Reiter, R. J. (2014). Extrapineal melatonin: sources, regulation, and potential functions. *Cell Mol Life Sci*, *71*, 2997-3025.

Agarwal, A., Gupta, S. & Sharma, R. K. (2005). Role of oxidative stress in female reproduction. *Reprod Biol Endocrinol*, *3*, 28.

Amano, M., Iigo, M., Ikuta, K., Kitamura, S., Yamada, H. & Yamamori, K. (2000). Roles of melatonin in gonadal maturation of underlying precocious male masu salmon. *Gen Comp Endocrinol*, *120*, 190-197.

Amano, M., Iigo, M., Ikuta, K., Kitamura, S., Okuzawa, K., Yamada, H. & Yamamori, K. (2004). Disturbance of plasma melatonin profile by high dose melatonin administration inhibits testicular maturation of precocious male masu salmon. *Zoolog Sci*, *21*, 79-85.

Arendt, J. & Deacon, S. (1997). Treatment of circadian rhythm disorders with melatonin. *Chronobiol Int*, *14*, 185-204.

Axelrod, J. & Weissbach, H. (1960). Enzymatic O-methylation of N-acetylserotonin to melatonin.*Sci*, *131*,1312.

Axelrod, J., Shein, H. M. & Wurtman, R. J. (1969). Stimulation of C14-melatonin synthesis from C14-tryptophan by noradrenaline in rat pineal in organ culture. *Proc Natl Acad Sci USA*, *62*, 544-549.

Bayarri, M. J., Zanuy, S., Yilmaz, O. & Carrillo, M. (2009). Effects of continuous light on the reproductive system of European sea bass gauged by alterations of circadian variations during their first reproductive cycle. *Chronobiol Int*, *26*, 184-199.

Bégay, V., Falcón, J., Cahill, G. M., Klein, D. C. & Coon, S. L. (1998). Transcripts encoding two melatonin synthesis enzymes in the teleost pineal organ: circadian regulation in pike and zebrafish, but not in trout. *Endocrinol*, *139*, 905-912.

Behl, R. (2006). Stimulation of catalase activity in carp (*Cyprinus carpio*) ovarian follicles by gonadotropin *in vitro*. *Fish Physiol Biochem*, *32*, 203-207.

Bentley, G. E., Perfito, N., Ukena, K., Tsutsui, K. & Wingfield, J. C. (2003). Gonadotropin-inhibitory peptide In song sparrows (*Melospiza melodia*) in different reproductive conditions, and in house sparrows (*Passer domesticus*) relative to chicken gonadotropin-releasing hormone. *J Neuroendocrinol*, *15*, 794-802.

Besseau, L., Benyassi, A., Moller, M., Coon, S. L., Weller, J. L., Boeuf, G., Klein, D. C. & Falcón, J. (2006). Melatonin pathway: breaking the 'high-at-night' rule in trout retina. *Exp Eye Res*, *82*, 620-627.

Bhattacharya, S. (1999). Recent advances in the hormonal regulations of gonadal maturation and spawning in fish. *Current Sci*, *76*, 342-349.

Bhattacharyya, S. & Maitra, S. K. (2006). Environmental correlate of the testicular events in a major carp *Catla catla* in an annual reproductive cycle. *Biol Rhythm Res*, *37*, 87-110.

Bhattacharya, S., Dey, R., Basu, A., Maitra, S. K. & Banerji, T. K. (2003). The structure of the pineal complex in a common Indian teleost, *Catla catla*: evidence for pinealinduced inhibition of testicular function within an annual reproductive cycle. *Endocr Res*, *29*, 141-156.

Bhattacharya, S., Dey, R. & Maitra, S. K. (2005). Photoperiodic regulation of annual testicular events in the Indian major carp *Catla catla*. *Acta Zoologica*, *86*, 71-79.

Bhattacharya, S., Chattoraj, A. & Maitra, S. K. (2007). Melatonin in the regulation of annual testicular events in carp *Catla catla*: evidence from the studies on the effects of exogenous melatonin, continuous light, and continuous darkness. *Chronobiol Int*, *24*, 629-650.

Binkley, S. (1986). Melatonin and N-acetyltransferase rhythms in pineal and retina. In: O' Brien PJ, Klein DC. (Eds.) Pineal and retinal relationships. Academic press, New York. pp. 185-196.

Binkley, S., Reilly, K. B. & Hryshchyshyn, M. (1980). N-acetyltransferase in chick retina: I. Circadian rhythms controlled by environmental lighting are similar to those in the pineal gland. *J Comp Physiol*, *139*, 103-108.

Borg, B. & Ekström, P. (1981). Gonadal effects on melatonin in the three-spined stickleback, *Gasterosteus aculeatus* L. during different seasons and photoperiods. *Reprod Nutr Development*, *21*, 919-927.

Bornestaf, C., Mayer, I. & Borg, B. (2001).Melatonin and maturation pace in female three-spined stickleback *Gasterosteus aculeatus*. *Gen Comp Endocrinol*, *112*, 341-348.

Bromage, N. & Duston, J. (1986). The control of spawning in the rainbow trout using photoperiod techniques. *Rep Inst Fres Res Drottn*, *63*, 26-35.

Bromage, N. R., Randall, C., Davies, B. & Portar, M. (1994). Photoperiod, melatonin and fish reproduction. In: Mackinlay DD. (ed) Proc. Intern. Fish Physiol. Symp. Vancouver, BC Canada Fish Physiol. Asso. Canada. 137-142.

Bromage, N. R., Randall, CF., Porter, M. J. R. & Davis, B. (1995). How do photoperiod, the pineal, melatonin and circannual rhythms interact to co-ordinate seasonal reproduction in salmonid fish? In: Goetz F, Thomas O. (eds.) *Proc Fifth Intl. Symp. Reprod. Physiol. Fish Symp. (Austin)*, *95*, 164-166.

Bromage, N. R., Porter, M. J. R. & Randall, C. F. (2001). The environmental regulation of maturation in farmed finfish with special reference to the role of photoperiod and melatonin. *Aquaculture*, *197*, 63-98.

Brzezinski, A., Seibel, M. M., Lynch, H. J., Deng, M. H. & Wurtman, R. J. (1987). Melatonin in human preovulatory follicular fluid. *J Clin Endocrinol Metab*, *64*, 865-867.

Bubenik, G. A. (1980). Localization of melatonin in the digestive tract of the rat. Effect of maturation, diurnal variation, melatonin treatment and pinealectomy. *Hormone Res*, *12*, 313-323.

Bubenik, G. A. (2002). Gastrointestinal melatonin localization, function, and clinical Relevance. *Dig Dis Sci*, *47*, 2336-2348.

Bubenik, G. A. (2008). Thirty four years since the discovery of gastrointestinal melatonin. *J Physiol and Pharmacol*, *2*, 33-51.

Bubenik, G. A. & Brown, G. M. (1997). Pinealectomy reduces melatonin levels in the serum but not in the gastrointestinal tract of the rat. *Biol Signals*, *6*, 40-44.

Bubenik, G. A. & Pang, S. F. (1997). Melatonin level in the gastrointestinal tissue of fish, amphibians, and a reptile. *Gen Comp Endocrinol*, *106*, 415-419.

Bubenik, G. A., Brown, G. M. & Grota, L. J. (1977). Immunohistological localization of melatonin in the rat digestive tract. *Experientia*, *33*, 662-663.

Bubenik, G. A., Purtill, R. A., Brown, G. M. & Grota, L. J. (1978). Melatonin in the retina and the Harderian gland. Ontogeny, diurnal variations and melatonin treatment. *Exp Eye Res*, *27*, 323-333.

Bubenik, G. A., Ball, R. O. & Pang, S. F. (1992). The effect of food deprivation on brain and gastrointestinal tissue levels of tryptophan, serotonin, 5-hydroxyindoleacetic acid, and melatonin. *J Pineal Res*, *12*, 7-16.

Cahill, G. M. (1996). Circadian regulation of melatonin production in cultured zebrafish pineal and retina. *Brain Res*, *708*, 177-181.

Cahill, G. M. & Besharse, J. C. (1995). Circadian rhythmicity in verterate retinas: Regulation by a photoreceptor oscillator. *Prog Retina Eye Res*, *14*, 267-291.

Cahill, G. M., Grace, M. S. & Besharse, J. C. (1991). Rhythmic regulation of retinal melatonin: Metabolic pathways, neurochemical mechanisms, and the ocular circadian clock. *Cell Mol Neurobiol*, *11*, 529-560.

Carnevali, O., Gioacchini, G., Piccinetti, CC., Maradonna, F., Lombardo, F., Giorgini, E. & Tosi, G. (2010). Melatonin control of oogenesis and metabolic resources in Zebrafish. *J Appl Ichthyol*, *26*, 826-830.

Carnevali, O., Gioacchini, G., Maradonna, F., Olivotto, I. & Migliarini, B. (2011). Melatonin induces follicle maturation in *Danio rerio*. *PLoS One* 6 e19978. http://dx.doi.org/ 10.1371/ journal.pone.001997.

Carter, D. S., Hall, V. D., Tamakin, L. & Goldman, B. D. (1982). Pineal is required for testicular maintenance in the Turkish hamster (*Mesocricetus brandti*). *Endocrinol*, *111*, 863-871.

Casao, A., Vega, S., Palacin, I., Perez-Pe, R., Laviña, A., Quintin, F.J., Sevilla, E., Abecia, J.A., Cebrián-Pĕrez, J.A., Forcada, F. & Muiño-Blanco, T. (2010). Effects of melatonin implants during non-breeding season on sperm motility and reproductive parameters in *Rasa aragonesa* rams. *Reprod Domest Anim*, *45*, 425-432.

Chattoraj, A., Bhattacharya, S., Basu, D., Bhattacharya, S., Bhattacharya, S. & Maitra, S. K. (2005). Melatonin accelerates maturation inducing hormone (MIH): induced oocyte maturation in Carps. *Gen Comp Endocrinol*, *140*, 145-155.

Chattoraj, A., Seth, M. & Maitra, S. K. (2008). Influence of serotonin on the action of melatonin in MIH-induced meiotic resumption in the oocytes of carp *Catla catla*. *Comp Biochem Physiol A*, *150*, 301-306.

Chattoraj, A., Seth, M. & Maitra, S. K. (2009a). Localization and dynamics of Mel_{1a} melatonin receptor in the ovary of carp *Catla catla* in relation to serum melatonin levels. *Comp Biochem Physiol A*, *152*, 327-333.

Chattoraj, A., Seth, M., Basu, A., Srivastava, T. G., Porta, S. & Maitra, S. K. (2009b). Temporal relationship between the circulating profiles of melatonin and ovarian steroids under natural photo-thermal conditions in an annual reproductive cycle in carp *Catla catla*. *Biol Rhythm Res*, *40*, 347-359.

Chowdhury, I. & Maitra, S. K. (2012). Melatonin in the Promotion of Health. In: Watson RR. (Ed.) Chapter 1, Melatonin time line: From discovery to therapy. Taylor and Francis, Boca Raton, Florida, USA. pp. 1-60.

Chowdhury, V. S., Yamamoto, K., Ubuka, T., Bentley, G. E., Hattori, A. & Tsutsui, K. (2010). Melatonin stimulates the release of gonadotropin inhibitory hormone by the avian hypothalamus. *Endocrinology*, *151*, 271-280.

Collin, J. P., Voisil, P., Falcón, J., Faure, J. P., Brisson, P. & Defaye, J. R. (1989). Pineal transducers in the course of evolution: molecular organization, rhythmic metabolic activity and role. *Arch Histol Cytol*, *52*, 441-449.

Coon, S. L. & Klein, D. C. (2006). Evolution of arylalkylamine N-acetyltransferase: emergence and divergence. *Mol Cell Endocrinol*, *252*, 2-10.

Davie, C. M., Mazzora, N., Bromage, J., Treasurer, H. & Migaud, H. (2007). Inhibition of sexual maturation in tank reared haddock (*Melanogrammus aeglefinus*) through the use of constant light photoperiods. *Aquaculture*, *270*, 379-389.

Dey, R., Bhattacharya, S. & Maitra, S. K. (2004).Temporal pattern of ovarian activity in a major carp *Catla catla* and its possible environmental correlate in an annual cycle. *Biol Rhythm Res*, *35*, 329-353.

Dey, R., Bhattacharya, S. & Maitra, S. K. (2005). Importance of photoperiods in the regulation of ovarian activities in Indian major carp *Catla catla* in an annual cycle. *J Biol Rhythm*, *20*, 145-158.

Dubocovich, M. L. (1985). Characterization of a retinal melatonin receptor. *J Pharmacol Exp Ther*, *234*, 395-401.

Dubocovich, M. L., Rivera-Bermudez, M. A., Gerdin, M. J. & Masana, M. I. (2003). Molecular pharmacology, regulation and function of mammalian melatonin receptors. *Front Biosci*, *8*, 1093-1108.

Ekström, P. & Vanecek, J. (1992). Localization of 2-[^{125}I]iodomelatonin binding sites in the brain of the Atlantic salmon, *Salmo salar* L. *Neuroendocrinol*,*55*, 529-537.

Ekström, P. & Meissl, H. (1997). The pineal organ of teleost fishes. *Rev Fish Biol Fisheries*, *7*, 199-284.

Falcón, J. & Collin, J. P. (1991). Pineal-retinal relationships: rhythmic biosynthesis and immunocytochemical localization of melatonin in retina of the pike (*Esox lucius*). *Cell Tissue Res*, *265*, 601-609.

Falcón, J., Thiabault, C., Bégay, V., Zachmann, A. & Collin, J. P. (1992). Regulation of the rhythmic melatonin secretion by fish pineal photoreceptor cells. In: Ali MA (Ed.) Rhythms in Fish. Plenum Press, New York. pp. 167-198.

Falcón, J., Begay, V., Goujon, J. M., Voisin, P., Guerlotte, J. & Collin, J. P. (1994). Immunocytochemical localization of hydroxyindole-O-methyltransferase in pineal photoreceptor cells of several fish species. *J Comp Neurol*, *341*, 559-566.

Falcón, J., Galarneau, K. M., Weller, J. L., Ron, B., Chen, G., Coon, S. L. & Klein, D. C. (2001). Regulation of arylalkylamine *N*-acetyltransferase-2 (AANAT2, EC 2.3.1.87) in the fish pineal organ: Evidence for a role of proteasomal proteolysis. *Endocrinol*, *142*, 1804-1813.

Falcón, J., Gothilf, Y., Coon, S. L., Boeuf, G. & Klein, D. C. (2003). Genetic, temporal and developmental differences between melatonin rhythm generating systems in the teleost fish pineal organ and retina. *J Neuroendocrinol*, *15*, 378-382.

Falcón, J., Besseau, L., Sauzet, S. & Boeuf, G. (2007). Melatonin effects on the hypothalamopituitary axis in fish. *Trends Endocrinol Metab*, *18*, 81-88.

Falcón, J., Migaud, H., Muñoz-Cueto, J. A. & Carrillo, M. (2010). Current knowledge on the melatonin system in teleost fish. *Gen Comp Endocrinol*, *165*, 469-482.

Foster, R. G. & Bellingham, J. (2004). Inner retinal photoreceptors (IRPs) in mammals and teleost fish. *Photochem Photobiol Sci*, *3*, 617-627.

Foster, R. G. & Kreitzman, L. (2005). Rhythms of Life: The biological clocks that control the daily lives of every living thing. Profile Books Ltd., London.

Gaeta, L. M., Tozzi, G., Pastore, A., Federici, G., Bertini, E. & Piemonte, F. (2002). Determination of superoxide dismutase and glutathione peroxidise activities in blood of healthy pediatric subjects. *Clin Chim Acta*, *322*, 117-120.

Gaildrat, P. & Falcón, J. (1999). Expression of melatonin receptors and 2-[^{125}I]iodomelatonin binding sites in the pituitary of a teleost fish. *Adv Exp Biol Med*, *460*, 61-72.

Gaildrat, P. & Falcón, J. (2000). Melatonin receptors in the pituitary of a teleost fish: mRNA expression, 2-[^{125}I]iodomelatonin binding and cyclic AMP response. *Neuroendocrinol*, *72*, 57-66.

Gaildrat, P., Becq, F. & Falcón, J. (2002). First cloning and functional characterization of a melatonin receptor in fish brain: a novel one? *J Pineal Res*,*32*, 74-84.

García-Allegue, R., Madrid, J. A. & Sánchez-Vázquez, F. J. (2001). Melatonin rhythms in European sea bass plasma and eye: influence of seasonal photoperiod and water temperature. *J Pineal Res*, *31*, 68-75.

Gern, W. A., Owens, D. W.&Ralph, C. H. L. (1978). The synthesis of melatonin by the trout retina.*J Exp Zool*, *206*, 263-270.

Gern, W. A., Duvall, D. & Nervina, J. M. (1986). Melatonin: a discussion of its evolution and actions in vertebrates. *Am Zool*, *26*, 985-996.

González-Martínez, D., Zmora, N., Mañanos, E., Saligaut, D., Zanuy, S., Zohar, Y., Elizur, A., Kah, O. & Muñoz-Cueto, J. A. (2002a). Immunohistochemical localization of three different prepro-GnRHs in the brain and pituitary of the European sea bass (*Dicentrarchus labrax*) using antibodies to the corresponding GnRH-associated peptides. *J Comp Neurol*, *446*, 95-113.

González-Martínez, D., Zmora, N., Zanuy, S., Sarasquete, C., Elizur, A., Kah, O. & Muñoz-Cueto, J. A. (2002b). Developmental expression of three different prepro-GnRH (gonadotrophin-releasing hormone) messengers in the brain of the European sea bass (*Dicentrarchus labrax*). *J Chem Neuroanat*, *23*, 255-267.

González-Martínez, D., Madigou, T., Mañanos, E., Cerdá-Reverter, J. M., Zanuy, S., Kah, O. & Muñoz-Cueto, J. A. (2004). Cloning and expression of gonadotropin-releasing hormone receptor in the brain and pituitary of the European sea bass: An *in situ* hybridization study. *Biol Reprod*, *70*, 1380-1391.

Goud, A. P., Goud, P. T., Diamond, M. P., Gonik, B. & Abu-Soud, H. M. (2008). Reactive oxygen species and oocyte aging: role of superoxide, hydrogen peroxide, and hypochlorous acid. *Free Radical Biol Med*, *44*,1295-1304.

Hardeland, R., Reiter, R. J., Poeggeler, B. & Tan, D. X. (1993). The significance of the metabolism of the neurohormone melatonin: Antioxidative protection and formation of bioactive substances. *Neurosci Biobehav Rev*, *17*, 347-357.

Hasan, K. N., Moniruzzaman, M. & Maitra, S. K. (2014). Melatonin concentrations in relation to oxidative status and oocyte dynamics in the ovary during different reproductive phases of an annual cycle in carp *Catla catla. Theriogenology*, *82*, 1173-1185.

Hazlerigg, D. G. (2001). What is the role of melatonin within the anterior pituitary? *J Endocrinol*, *170*, 493-501.

Herrero, M. J., Martínez, F. J., Míguez, J. M. & Madrid, J. A. (2007). Response of plasma and gastrointestinal melatonin, plasma cortisol and activity rhythms of European sea bass (*Dicentrarchus labrax*) to dietary supplementation with tryptophan and melatonin. *J Comp Physiol*, *177*, 319-326.

Huether, G., Poegeller, B., Reimer, R. & George, A. (1992). Effect of tryptophan administration on circulating melatonin levels in chicks and rats: evidence for stimulation of melatonin synthesis and release in the gastrointestinal tract. *Life Sci*, *51*, 945-953.

Iigo, M., Furukawa, K. H. A., Ohtani-Kaneko, R., Hara, M., Suzuki, T., Tabata, M. & Aida, K. (1997a). Ocular melatonin rhythms in the goldfish, *Carassius auratus*. *J Biol Rhythms*, *12*, 182-192.

Iigo, M., Sánchez-Vázquez, F. J., Madrid, J. A., Zamora, S. & Tabata, M. (1997b). Unusual responses to light and darkness of ocular melatonin in European sea bass. *Neuroreport*, *8*, 1631-1635.

Iigo, M., Abe, T., Kambayashi, S., Oikawa, K., Masuda, T., Mizusawa, K., Kitamura, S., Azuma, T., Takagi, Y., Aida, K. & Yanagisawa, T. (2007a). Lack of circadian regulation of *in vitro* melatonin release from the pineal organ of salmonid teleosts. *Gen Comp Endocrinol*, *154*, 91-97.

Iigo, M., Azuma, T. & Iwata, M. (2007b). Lack of circadian regulation of melatonin rhythms in the sockeye salmon (*Oncorhynchus nerka*) *in vivo* and *in vitro*. *Zool Sci*, *24*, 67-70.

Itoh, M. T., Ishizuka, B., Kuribayashi, Y., Amemiya, A. & Sumi, Y. (1999). Melatonin, its precursors, and synthesizing enzyme activities in the human ovary. *Mol Hum Reprod*, *5*, 402-408.

Iuvone, P. M. & Besharse, J. C. (1983). Regulation of indoleamine N-acetyltransferase activity in the retina: effects of light and dark, protein synthesis inhibitors and cyclic nucleotide analogs. *Brain Res*, *273*, 111-119.

Iuvone, P. M., Tosini, G., Pozdeyev, N., Haque, R., Klein, D. C. & Chaurasia, S. S. (2005). Circadian clocks, clock networks, arylalkylamine N-acetyltransferase, and melatonin in the retina. *Prog Retin Eye Res*, *24*, 433-456.

Iwasaki, S., Nakazawa, K., Sakai, J., Kometani, K., Iwashita, M., Yoshimura, Y. & Maruyama, T. (2005). Melatonin as a local regulator of human placental function. *J Pineal Res*, *39*, 261-265.

Joy, K. P. & Agha, A. K. (1991). Seasonal effect of administration of melatonin and 5-methoxytrytophol on ovarian activity in the catfish *Heteropneustes fossilis* (Bloch). *J Pineal Res*, *10*, 65-70.

Kennaway, D. J. & Rowe, S. A. (1995). Melatonin binding sites and their role in seasonal reproduction. *J Reprod Fertil Suppl*, *49*, 423-435.

Khan, I. A. & Thomas, P. (1996). Melatonin influences gonadotropin II secretion in the Atlantic croaker, (*Micropogonia sundulatus*). *Gen Comp Endocrinol*, *104*, 231-242.

Klein, D. C. (2007). Arylalkylamine *N*-acetyltransferase: "the timezyme". *J Biol Chem*, *282*, 4233-4237.

Klein, D. C., Coon, S. L., Roseboom, P. H., Weller, J. L., Bernard, M., Gastel, J. A., Zatz, M., LuVone, P. M., Rodriguez, I. R., Begay, V., Falcón, J., Cahill, G. M., Cassone, V. M. &

Baler, R. (1997). The melatonin rhythm-generating enzyme: molecular regulation of serotonin N-acetyltransferase in the pineal gland. *Recent Prog Horm Res*, *52*, 307-357.

Konturek, P. C., Brzozowski, T. & Konturek, S. J. (2011). Gut clock: implication of circadian rhythms in the gastrointestinal tract. *J Physiol Pharmacol*, *62*, 139-150.

Korf, H. W. & Oksche, A. (1986). The pineal organ. In: Pang PKT, Schreibman MP (Eds.) Vertebrate endocrinology. Fundamentals and biomedical implications. Vol. 1. Morphological considerations. Academic Press, New York. pp. 105-145.

Lerner, A. B., Case, J. D., Takahashi, Y., Lee, T. H. & Mori, N. (1958). Isolation of melatonin, pineal factor that lightens melanocytes. *J Am Chem Soc*, *80*, 2587.

Lethimonier, C., Madigou, T., Muñoz-Cueto, J. A., Lareyre, J. J. & Kah, O. (2004). Evolutionary aspects of GnRHs, GnRH neuronal systems and GnRH receptors in teleost fish. *Gen Comp Endocrinol*, *135*, 1-16.

Liu, T. & Borjigin, J. (2005). Reentrainment of the circadian pacemaker through 3 distinct stages. *J Biol Rhythm*, *20*, 1-10.

Liu, C., Fukuhara, C., Wessel, J. H., Iuvone, P. M. & Tosini, G. (2004). Localization of Aa-nat mRNA in the rat retina by fluorescence in situ hybridization and laser capture microdissection. *Cell Tissue Res*, *315*, 197-201.

Macias, M., Rodriguez-Cabezas, M. N., Reiter, R. J., Osuna, A. & Acuna-Castroviejo, D. (1999). Presence and effects of melatonin in *Trypanosoma cruzi*. *J Pineal Res*, *27*, 86-94.

Maitra, S. K. & Ray, A. K. (2000). Role of light in the mediation of acute effects of a single afternoon melatonin injection on steroidogenic activity of testis in the rat. *J Biosci*, *25*, 253-256.

Maitra, S. K. & Chattoraj, A. (2007). Role of photoperiod and melatonin in the regulation of ovarian functions in Indian carp *Catla catla*: basic information for future application. *Fish Physiol Biochem*, *33*, 367-382.

Maitra, S. K., Chattoraj, A. & Bhattacharyya, S. (2005). Implication of melatonin in oocyte maturation in Indian major carp *Catla catla*. *Fish Physiol Biochem*, *31*, 201-207.

Maitra, S. K., Chattoraj, A., Mukherjee, S. & Moniruzzaman, M. (2013). Melatonin: A potent candidate in the regulation of fish oocyte growth and maturation. *Gen Comp Endocrinol*, *181*, 215-222.

Martinez-Chavez, C. C., Al-Khamees, S., Campos-Mendoza, A., Penman, D. J. & Migaud, H. (2008). Clock controlled endogenous melatonin rhythms in Nile tilapia (*Oreochromis niloticus niloticus*) and African catfish (*Clarias gariepinus*). *Chronobiol Int*, *25*, 31-49.

Mazurais, D., Brierley, I. Anglade, I., Drew, J., Randall, C., Bromage, N., Michel, D., Kah, O. & Williams, L. M. (1999). Central melatonin receptors in the rainbow trout: comparative distribution of ligand binding and gene expression. *J Comp Neurol*, *409*, 313-324.

McCord, C. P. & Allen, F. B. (1917). Evidence associating pineal gland function with alterations in pigmentation. *J Exp Zool*, *23*, 207-224.

Menendez-Pelaez, A., Howes, K. A., Gonzalez-Brito, A. & Reiter, R. J. (1987). *N*-acetyltransferase activity, hydroxyindole-O-methyltransferase activity, and melatonin levels in the Harderian glands of the female Syrian hamster: changes during the light:dark cycle and the effect of 6-parachlorophenylalanine administration. *Biochem Biophy Res Commun*, *145*, 1213-1238.

Migaud, H., Davie, A., Martinez, Chavez, C. C. & Al-Khamees, S. (2007). Evidence for differential photic regulation of pineal melatonin synthesis in teleosts. *J Pineal Res*, *43*, 327-335.

Migaud, H., Davie, A. & Taylor, J. F. (2010). Current knowledge on the photoneuroendocrine regulation of reproduction in temperate fish species. *J Fish Biol*, *76*, 27-68.

Miller, L., Stier, M. & Lovenberg, W. (1980). Evidence for the presence of *N*-acetyl transferase in rat retina. *Comp Biochem Physiol C*, *66*, 213-216.

Moncaut, N., Somoza, G., Power, D. M. & Canario, A. V. (2005). Five gonadotrophin-releasing hormone receptors in a teleost fish: Isolation, tissue distribution and phylogenetic relationships. *J Mol Endocrinol*, *34*, 767-779.

Moniruzzaman, M. & Maitra, S. K. (2012). Influence of altered photoperiods on serum melatonin and its receptors (MT1 and MT2) in the brain, retina, and ovary in carp *Catla catla*. *Chonobiol Intl*, *29*, 175-188.

Mukherjee, S., Moniruzzaman, M. & Maitra, S. K. (2014a). Daily and seasonal profiles of gut melatonin and their temporal relationship with pineal and serum melatonin in carp *Catla catla* under natural photo-thermal conditions. *Biol Rhythm Res*, *45*, 301-315.

Mukherjee, S., Moniruzzaman, M. & Maitra, S. K. (2014b). Impact of artificial lighting conditions on the diurnal profiles of gut melatonin in a surface dwelling carp (*Catla catla*). *Biol Rhythm Res*, *45*, 831-848.

Nagahama, Y. (1994). Endocrine regulation of gametogenesis in fish. *Int. J. Dev. Biol.* *38*, 217–229.

Nagahama, Y. & Yamashita, M. (2008). Regulation of oocyte maturation in fish. *Dev Growth Differ*, *50*, 195-219.

Nakamura, Y., Tamura, H., Takayama, H. & Kato, H. (2003). Increased endogenous level of melatonin in preovulatory human follicles does not directly influence progesterone production. *Fertil Steril*, *80*, 1012-1016.

Okimoto, D. K. & Stetson, M. H. (1999). Presence of an intrapineal oscillator in the teleostean family Poecilidae. *Gen Comp Endocrinol*, *114*, 304-312.

Patiño, R. & Thomas, P. (1990). Effects of gonadotropin on ovarian intrafollicular processes during the development of oocyte maturational competence in a teleost, the Atlantic croaker: evidence for two distinct stages of gonadotropin control of final oocyte maturation. *Biol Reprod*, *43*, 818-827.

Patiño, R., Yoshizaki, G., Thomas, P. & Kagawa, H. (2001). Gonadotropin control of ovarian follicle maturation: the two-stage concept and its mechanism. *Comp Biochem Physiol B*, *129*, 427-439.

Poeggeler, B. & Hardeland, R. (1994). Detection and quantification of melatonin in a dinoflagellate, *Gonyaulax polyedra*: solutions to the problem of methoxyindole destruction in non-vertebrate material. *J Pineal Res*, *17*, 1-10.

Porter, M. J. R., Duncan, N. J., Mitchell, D. & Bromage, N. R. (1999). The use of cage lighting to reduce plasma melatonin in Atlantic salmon (*Salmo salar*) and its effects on the inhibition of grilsing. *Aquaculture*, *176*, 237-244.

Porter, M. J. R., Stefansson, SO., Nyhammer, G., Karlsen, O., Norberg, B.&Bromage, N. R. (2000). Environmental influences on melatonin secretion in Atlantic cod (*Gadus morhua* L) and their relevance to commercial culture. *Fish Physiol Biochem*, *23*, 191-200.

Raikhlin, N. T. & Kvetnoy, I. M. (1974). Lightening effect of the extract of human appendix mucosa on frog skin melanophores. *Bull Exp Biol Med*, *8*,114-116.

Randall, C. F., Bromage, N., Thorpe, J., Miles, M. & Muir, J. (1995). Melatonin rhythms in Atlantic salmon (*Salmo salar*) maintained under natural and out-of-phase seasonal photoperiods. *Gen Comp Endocrinol, 98*, 73-87.

Reiter, R. J. (1991). The pineal gland: reproductive interactions. In: Pang PKT, Schreibman MP (Eds.) Vertebrate endocrinology: Fundamentals and biomedical implications, Vol. 4, Part B, Academic Press, San Diego, California. pp. 269-310.

Reiter, R. J., Richardson, B. A., Matthews, S. A., Lane, S. J. & Ferguson, B. N. (1983). Rhythms in immunoreactive melatonin in the retina and Harderian gland of rats: persistence after pinealectomy. *Life Sci, 32*, 1229-1236.

Reiter, R. J., Tan, D. X., Acuna-Castroviejo, D., Burkhardt, S. & Karbownik, M. (2000). Melatonin: Mechanisms and actions as an antioxidant. *Curr Top Biophys, 24*, 171-183.

Reiter, R. J., Tan, D. X., Manchester, L. C., Paredes, S. D., Mayo, J. C. & Sainz, R. M. (2009). Melatonin and reproduction revisited. *Biol Reprod, 81*, 445-456.

Reiter, R. J., Rosales-Corral, S. A., Manchester, L. C. & Tan, D. X. (2013). Peripheral reproductive organ health and melatonin: Ready for prime time. *Int J Mol Sci, 14*, 7231-7272.

Reppert, S. M., Weaver, D. R. & Godson, C. (1996). Melatonin receptor step into light: cloning and classification of subtypes. *Trends Pharmacol Sci, 17*, 100-102.

Rikans, L. E. & Hornbrook, K. R. (1997). Lipid peroxidation, antioxidant protection and aging. *Biochim Biophys Acta, 1362*, 116-127.

Rodriguez, C., Mayo, J. C., Sainz, R. M., Antolin, I., Herrera, F., Martin, V. & Reiter, R. J. (2004). Regulation of antioxidant enzymes: A significant role for melatonin. *J Pineal Res, 36*, 1-9.

Ronnberg, L., Kauppila, A., Leppaluoto, J., Martikainen, H. &Vakkuri, O. (1990). Circadian and seasonal variation in human preovulatory fluid melatonin concentration. *J Clin Endocrinol Metab, 71*, 492-496.

Salhab, M., Dhorne-Pollets, S., Auclair, S., Guyader-Joly, C., Brisard, D., Dalbies-Tran, R., Dupont, J., Ponsart, G., Mermillod, P. & Uzbekova, S. (2013). *In vitro* maturation of oocytes alters gene expression and signaling pathways in bovine cumulus cells. *Mol Reprod Dev, 80*, 166-182.

Samiec, P. S., Drews-Botsch, C., Flagg, E. W., Kurtz, J. C., Sternberg, P., Reed, R. L. & Jones, D. P. (1998). Glutathione in human plasma: decline in association with aging, age-related macular degeneration, and diabetes. *Free Rad Biol Med, 24*, 699-704.

Saxena, P. K. & Anand, K. (1977). A comparison of ovarian recrudescence in the catfish, *Mystus tengara* (Ham.) exposed to short photoperiod, to long photoperiod and to melatonin. *Gen Comp Endocrinol, 33*, 506-511.

Servili, A., Lethimonier, C., Lareyre, J. J., López-Olmeda, J. F., Sánchez-Vázquez, F. J., Kah, O. & Muñoz-Cueto, J. A. (2010). The highly conserved gonadotropin-releasing hormone-2 form acts as a melatonin-releasing factor in the pineal of a teleost fish, the European sea bass *Dicentrarchuslabrax. Endocrinol, 151*, 2265-2275.

Servili, A., Herrera-Pérez, P., Rendón, M. C. & Muñoz-Cueto, J. A. (2013). Melatonin Inhibits GnRH-1, GnRH-3 and GnRH Receptor Expression in the Brain of the European Sea Bass, *Dicentrarchus labrax. Int J Mol Sci, 14*, 7603-7616.

Seth, M. & Maitra, S. K. (2010a). Photoreceptor proteins and melatonin rhythm generating AANAT in the carp pineal: Temporal organization and correlation with natural photo-thermal cues. *J Photochem Photobiol B, 99*, 21-28.

Seth, M. & Maitra, S. K. (2010b). Importance of light in temporal organization of photoreceptor proteins and melatonin-producing system in the pineal of carp *Catla catla*. *Chronobiol Int*, *27*, 463-486.

Singh, T. P. & Lal, P. (1994). Endocrine physiology of reproduction in Indian catfish. In: Singh HR. (Ed.) Advances in fish biology. Hindustan Publishing Corporation, Delhi, India. pp. 147-154.

Smith, J. A. (1981). The biochemistry and pharmacology of melatonin. In: Birau N, Schloot W (Eds.) Melatonin: Current status and perspectives. Pergamon Press, New York. pp. 135-146.

Soares, J. M., Manson, M. I., Ersahin, C. & Dubocovich, M. L. (2003). Functional melatonin receptors in rat ovaries at various stages of the oestrous cycle. *J Pharmacol Exp Ther*, *306*, 694-702.

Sundararaj, B. I. & Keshavanath, P. (1976). Effects of melatonin and prolactin treatment on the hypophysial-ovarian system in the catfish, *Heteropneustes fossilis* (Bloch). *Gen Comp Endocrinol*, *29*, 84-96.

Tamura, H., Takasaki, A., Miwa, I., Taniguchi, K., Maekawa, R., Asada, H., Taketani, T., Matsuoka, A., Yamagata, Y., Shimamura, K., Morioka, H., Ishikawa, H. & Reiter, R. J., Sugino, N. (2008). Oxidative stress impairs oocyte quality and melatonin protects oocytes from free radical damage and improves fertilization rate. *J Pineal Res*, *44*, 280-287.

Tamura, H., Nakamura, Y., Korkmaz, A., Manchester, L. C., Tan, D-X., Sugino, N. & Reiter, R. J. (2009). Melatonin and the ovary: physiological and pathophysiological implications. *Fertil Steril*, *92*, 328-343.

Tamura, H., Takasaki, A., Taketani, T., Tanabe, M., Kizuka, F., Lee, L., Tamura, I., Maekawa, R., Asada, H., Yamagata, Y. & Sugino, N. (2013). Melatonin as a free radical scavenger in the ovarian follicle. *Endocr J*, *60*, 1-13.

Tamura, H., Takasaki, A., Taketani, T., Tanabe, M., Lee, L., Tamura, I., Maekawa, R., Asada, H., Yamagata, Y. & Sugino, N. (2014). Melatonin and female reproduction. *J Obstet Gynaecol Res*, *40*, 1-11.

Tan, D. X., Chen, L. D., Poeggeler, B., Manchester, L. C. & Reiter, R. J. (1993). Melatonin: A potent, endogenous hydroxyl radical scavenger. *Endocrine J*, *1*, 57-60.

Tokumoto, T., Tokumoto, M., Horiguchi, R., Ishikawa, K. & Nagahama, Y. (2004). Diethylstilbestrol induces fish oocyte maturation. *Proc Natl Acad Sci USA*, *101*, 3686-3690.

Tosini, G. & Menaker, M. (1996). Circadian rhythms in cultured mammalian retina. *Sci*, *272*, 419-421.

Tosini, G. & Menaker, M. (1998). The clock in the mouse retina: melatonin synthesis and photoreceptor degeneration. *Brain Res*, *789*, 221-228.

Tosini, G., Baba, K., Hwang, C. K. & Iuvone, P. M. (2012). Melatonin: an underappreciated player in retinal physiology and pathophysiology. *Exp Eye Res*, *103*, 82-89.

Tsutsui, K. (2009). A new key hormone controlling reproduction, gonadotropin-inhibitory hormone (GnIH): Biosynthesis, mode of action and functional significance. *Progr Neurobiol*, *88*, 76-88.

Tsutsui, K., Bentley, G. E., Bedecarrats, G., Osugi, T., Ubuka, T. & Kriegsfled, L. J. (2010). Gonadotropin inhibitory hormone (GnIH) and its control of central and peripheral reproductive function. *Front Neuroendocrinol*, *31*, 284-295.

Tsutsui, K., Ubuka, T., Bentley, G. E. & Kriegsfeld, L. J. (2012). Gonadotropin-inhibitory hormone (GnIH): Discovery, progress and prospect. *Gen Comp Endocrinol, 177*, 305-314.

Ubuka, T., Bentley, G. E., Ukena, K., Wingfield, J. C. & Tsutsui, K. (2005). Melatonin induces the expression of gonadotropin-inhibitory hormone in avian brain. *Proc Natl Acad Sci USA, 120*, 3052-3057.

Ubuka, T., Kim, S., Huang, Y. C., Reid, J., Jiang, J., Osugi, T., Chowdhury, V. S., Tsutsui, K. & Bentley, G. E. (2008). Gonadotropin-inhibitory hormone neurons interact directly with gonadotropin-releasing hormone-I and II neurons in European starling brain. *Endocrinol, 149*, 268-278.

Vakkuri, O., Rintamaki, H. & Leppaluoto, J. (1985a). Presence of immunoreactive melatonin in different tissues of the pigeon. *Gen Comp Endocrinol, 58*, 69-75.

Vakkuri, O., Rintamaki, H. & Leppaluoto, J. (1985b). Plasma and tissue concentrations of melatonin after midnight light exposure and pinealectomy in the pigeon. *J Endocrinol, 105*, 263-268.

Vanecek, J. (1998). Cellular mechanisms of melatonin action. *Physiol Rev, 78*, 687-721.

Vaughan, M. K., Hubbard, G. B., Champney, G. M., Vaughan, G. M., Little, J. C. & Reiter, R. J. (1987). Splenic hypertrophy and extramedullary heamatopoiesis induced in male Syrian hamster by short photoperiod melatonin injections and eversed by melatonin pellets or pinealectomy. *Am J Anat, 179*, 131-136.

Velarde, E., Cerdá-Reverter, J. M., Alonso-Gómez, A. L., Sánchez, E., Isorna, E. & Delgado, M. J. (2010). Melatonin-synthesizing enzymes in pineal, retina, liver, and gut of the goldfish (*Carassius*): mRNA expression pattern and regulation of daily rhythms by lighting conditions. *Chronobiol Int, 27*, 1178-1201.

Vera, L. M., DeOliveira, C., Lopez-Olmeda, J. F., Ramos, J., Mananose, E., Madrid, J. E. & Sanchez-Vazquez, F. J. (2007). Seasonal and daily plasma melatonin rhythms and reproduction in Senegal sole kept under natural photoperiod and natural or controlled water temperature. *J Pineal Res, 43*, 50-55.

Vernadakis, A. J., Bemis, W. E. & Bittman, E. L. (1998). Localization and partial characterization of melatonin receptors in amphioxus, hagfish, lamprey, and skate. *Gen Comp Endocrinol, 110*, 67-78.

Vollrath L (1981) The pineal organ. In: Oksche A, Vollrath L (Eds.) Handbuch der mikroskopischen Anatomie des Menschen. Vol VI/7 Springer, Berlin, Heidelberg, New York. pp. 1-665.

Vuilleumier, R., Boeuf, G., Fuentès, M., Gehring, W. J. & Falcón, J. (2007). Cloning and early expression pattern of two melatonin biosynthesis enzymes in the turbot (*Scophthalmus maximus*). *Eur J Neurosci, 25*, 3047-3057.

Weil, C., Crim, L. W., Wilson, C. E. & Cauty, C. (1992). Evidence of GnRH receptors in cultured pituitary cells of the winter flounder (*Pseudopleuronectes americanus* W). *Gen Comp Endocrinol, 155*, 126-140.

Wiechmann, A. F. (1986). Melatonin: parallels in pineal gland and retina. *Exp Eye Res, 42*, 507-527.

Wiechmann, A. F. & Summers, J. A. (2008). Circadian rhythms in the eye: the physiological significance of melatonin receptors in ocular tissues. *Prog Retin Eye Res, 27*, 137-160.

Wojtowicz, M. & Jakiel, G. (2002). Melatonin and its role in human reproduction. *Ginekol Pol, 73*, 1231-1237.

Woo, M. M. M., Tai, C. J. & Kang, S. K. (2001). Direct action of melatonin in human granulosa-luteal cells. *J Clin Endocrinol Metab*, *86*, 4789-4797.

Yu, H. S., Pang, S. F. & Tang, P. L. (1981). Increase in the level of retinal melatonin and persistence of its diurnal rhythm in rats after pinealectomy. *J Endocrinol*, *91*, 477-481.

Zelko, I. N., Mariani, T. J. & Folz, R. J. (2002). Superoxide dismutase multigene family: a comparison of the CuZn-SOD (SOD1), Mn-SOD (SOD2), and EC-SOD (SOD3) gene structures, evolution, and expression. *Free Radic Biol Med*, *33*, 337-349.

Zmijewski, M. A., Sweatman, T. W. & Slominski, A. T. (2009). The melatonin- producing system is fully functional in retinal pigment epithelium (ARPE-19). *Mol Cell Endocrinol*, *307*, 211-216.

Zmora, N., González-Martínez, D., Muñoz-Cueto, J. A., Madigou, T., Mañanos-Sánchez, E., Doste, S. Z., Zohar, Y., Kah, O. & Elizur, A. (2002). The GnRH system in the European sea bass (*Dicentrarchus labrax*). *J Endocrinol*, *172*, 105-116.

Zohar, Y., Muñoz-Cueto, J. A., Elizur, A. & Kah, O. (2010). Neuroendocrinology of reproduction in teleost fish. *Gen Comp Endocrinol*, *165*, 438-455.

In: Indoleamines
Editor: Ángel Catalá

ISBN: 978-1-63482-097-4

Chapter 3

THE NEUROIMMUNE ROLES OF INDOLEAMINE PATHWAYS IN CANCER

Hui-Yen Chuang and Jeng-Jong Hwang*
Department of Biomedical Imaging and Radiological Sciences,
National Yang-Ming University, Taipei, Taiwan

ABSTRACT

A Latin aphorism "*Mens sana in corpore sano*" implies that the correlation between mental and physical health has been known for thousands of years. This phenomenon has been observed in several diseases, such as cardiovascular diseases, autoimmune diseases, and cancer. Currently, cancer kills more than six million people each year and has become a heavily economic burden caused by the treatments. Miovic and Block reported that 5-26% of advanced cancer patients have major depression, and are prescribed with antidepressants. A new medical field named "psycho-oncology" is launched based on these clinical observations. In the field of psyco-oncology discusses both the effects of cancer on a person's mental health as well as the behavioral factors that may affect the progress of cancer.

Serotonin is a major neurotransmitter contributing to the feeling of happiness, and there is a strong correlation between the serotonin level and depression. It has been shown that the serotonin level of cancer patients is significantly lower than that of healthy subjects. Serotonin is synthesized from tryptophan which is one of the essential amino acids for mammals. Dysregulation of tryptophan metabolic pathways could be responsible for developing depression symptoms in cancer patients. Tryptophan is metabolized through the kynurenine pathway by up-regulated indoleamine 2,3-dioxygenase (IDO) in cancers instead of the serotonin pathway used in neurons, resulted in the decrease of serotonin production. In addition, the intermediates and final product of the kynurenine pathway, such as 3-hydroxykynurenine (OHK) and quinolinic acid (QUIN), are potent neurotoxins. Increased levels of these neurotoxins may contribute to the apoptosis of astrocytes and other neurons, affect various networks in the brain, and result in depression symptoms or other mental disorders. The up-regulated kynurenine

* Corresponding author: Jeng-Jong Hwang, PhD, Professor, Department of Biomedical Imaging and Radiological Sciences, National Yang-Ming University, No. 155, Sec. 2, Li-Nong St, Beitou, Taipei 112, Taiwan, Tel: +886-2-28267064; Fax: +886-2-28201095; Email: jjhwang@ym.edu.tw.

pathway in cancer patients not only strikes the survival of neurons and regulations of several networks in the brain, but impacts the antitumor immunity of T cells. The shortage of tryptophan inhibits the proliferations of CD8+ T cells and CD4+ T-helper cells. In addition, kynurenine could cause these T cells to apoptosis. It also has been reported that numbers and activities of T cells are highly correlated with the prognosis and survival of patients. Other than decreasing the numbers and functions of T cells, kynurenine also turns tumor microenvironment toward more immunosuppressive.

Tryptophan is converted into 5-hydroxyl-L-tryptophan (5-HTP) by tryptophan hydroxylase (TPH) and its cofactor tetrahydrobiopterin (BH4) under the normal physical condition. However, the BH4 is also the cofactor of nitric oxide synthase (NOS) which could be induced by proinflammatory cytokines, such as interleukin-6 (IL-6), in the tumor microenvironment. Inducible nitric oxide synthase (iNOS) competes BH4 with TPH and causes the lower production of serotonin and other neurotransmitters, both worsen the depression symptoms in cancer patients. On the other hand, BH4 helps iNOS to convert arginine to nitric oxide (NO) that is further metabolized into reactive nitrogen species (RNS), such as peroxynitrite. These RNS cause post-translational modifications of proteins, and the RNS-modified cytokines would affect the recruitments of immune cells and antitumor immunity.

In overall, the metabolism of tryptophan significantly influences the synthesis of serotonin and generation of kynurenine, and is closely correlated to the cancer-induced depression and immunosuppression. Hence, the roles of indoleamines, such as tryptophan and serotonin, and their related pathways will be discussed in the following. Moreover, how these molecules modulate the immunity and mood in cancer patients will also be covered.

Clinical Observations of Cancer and Depression

As the Latin aphorism says "*Mens sana in corpore sano,*" the mental status mirrors the physical health of a person. According to WHO, depression will become the biggest health problem by 2030 and influence millions of people [1]. Disability resulted from depression and related medical expenses directly and indirectly make enormous costs to the society; it is estimated, about $90- 110billion in EU and US [2-4]. Being one of the leading death causes worldwide, cancer had to account for more than 8 million deaths in 2012 [5]. Furthermore, WHO projected that cancer deaths would increase to 11.5 million by 2030 [1]. Based on these statistics, both depression and cancer are critical issues needed to be solved. Interestingly, a strong correlation between cancer and depression has been observed in the clinic [6, 7], which again echoes the Latin aphorism "*Mens sana in corpore sano*".

Many research groups have analyzed the prevalence of depression in cancer patients in the last decades [7-10]. Varied results are found. Depression is not caused by single risk factor, and there are many confounding factors affecting the results in the studies including gender, ages, psychiatric history and medical interventions [11, 12]. However, depression seems to be more prevalent for the patients with advanced stages of cancer [9, 13, 14]. Studies showed that the depression could be a predictor of poor prognosis [15, 16], worse quality of life and longer hospitalization [14, 17-19] in advanced cancer. Various factors contribute to the development of depression. It might be related to the interactions among the neuro-endocrine-immune (NEI) system, which indicates that the neurotransmitters, hormones, and cytokines could regulate each other. For instance, patients with pancreatic [20, 21], head and neck [22, 23], and breast [24, 25] cancer have been reported that they tend to develop depression more than patients with other cancer due to cancer secreted cytokines or related complications.

As reported by Miovic and Block that 5-26% of advanced cancer patients have major depression, and are prescribed with antidepressants [26]. Many reports also found that patients behaved better after receiving antidepressant [27-29]. Nevertheless, Lloyd-Williams et al. commented that the introductions of antidepressants into cancer treatments are usually at inadequate doses and too late [30]. These studies implied that the depression is underdiagnosed and undertreated in cancer patients, especially in those with mild symptoms. Several scales are utilized in depression assessment in cancer patients. The most widely used one called the Hospital Anxiety and Depression Scale (HADS), which only focuses on the indicators related to anhedonia and shows positive results in patients receiving anticancer treatment [31]. However, HADS could not be used for screening of depression in cancer patients because its sensitivity and specificity [32]. The Edinburg Depression Scale (EDS) is adopted from postnatal depression evaluation and also exhibited some promising results [33]. Another scale used for depression evaluation called the Brief Edinburg Depression Scale (BEDS) was proposed in 2007 [34]. Compared with the original 10-item EDS, the BEDS has only 6-item but shows better ability to differentiate depression clearly in patients with advanced cancer. A newer scale with higher sensitivity and specificity is needed to improve the diagnosis of depression in cancer patients.

The attitude that patients face the cancer is positively correlated with the treatment outcomes and prognosis. Moorey and Greer reported that if patients viewed the cancer as a challenge and arose their fighting spirit, in this situation, the positive prognosis could be expected. In contrast, if the patients showed helplessness-hopelessness attitude, and prognosis would be relatively worse [35]. These findings, again, indicate that mental health status interferes physical health status.

It is known that many patients with autoimmune diseases will develop depression because the imbalance between Th1 and Th2 cytokines [36]. Moreover, patients receiving IFN-α therapy, which applied to types of cancer and hepatitis B and C, have been reported consequence of depression [37, 38]. Accumulating evidence shows antitumor therapies enhancing levels of pro-inflammatory cytokines would induce or worsen depression in patients [39, 40]. Sickness behavior that is similar to major depression is observed in people and animal with elevated levels of TNF-α, IL-1 and IL-6 [41-43]. Cytokines in circulation enter brain through leaky regions of the blood-brain barrier (BBB) or by active transport because the large sizes. Various mechanisms that activated immune cells (e.g. macrophages, monocytes, and CD4+T cells) can migrate into the brain freely and release cytokines are proposed [44-47]. Proinflammatory cytokines result in depression by increasing neuron damages, expressions of serotonin and dopamine transporters and both level and activity of IDO; on the other hand, decreasing serotonin availability and neurogenesis [48, 49]. These observations show that the regulations between mental and physical health is not unidirectional, but complex interactions, which imply that both mental and physical medications should be introduced for cancer treatment.

ABNORMAL METABOLISM OF TRYPTOPHAN IN CANCER

Tryptophan is one of the essential amino acids (EAA) and is indispensable for protein synthesis and normal physiology. Except for protein synthesis, most of tryptophan is metabolized through the kynurenine pathway to generate the coenzyme, nicotinamide adenine dinucleotide (NAD). As the precursor of serotonin, a known neurotransmitter related to depression, tryptophan is also metabolized *via* serotonin pathways [50]. The kynurenine pathway begins with indoleamine 2,3-dioxygenase (IDO) or tryptophan 2,3-dioxygenase (TDO) expressed mainly in the liver. On the other hand, the serotonin pathway starts from tryptophan hydroxylase (TPH) and further becomes serotonin or melatonin.

In normal physiology, only 1% of dietary tryptophan is used for serotonin synthesis in the brain, and more than 90% of dietary tryptophan is metabolized through the kynurenine pathway. However, many reports showed that IDO expression is up-regulated in cancer, deprives the free tryptophan in circulation, and causes reductions of serotonin and melatonin. IDO could not only catalyze tryptophan, but 5-HTP, serotonin or the final product of the serotonin pathway, melatonin. Several kynurenines are formed in the kynurenine pathway, and function contradictorily as neuroprotectants and neurotoxins [51], which are intimately related to depression and other neuronal diseases. Up-regulated IDO expression and activity impairs both cell- and humoral-mediated antitumor immunity by shifting cytokine expression patterns from Th1 to Th2 [36] and depleting of tryptophan [52] in the tumor microenvironment. .

NEROPROTECTIVE METABOLITES IN KYNURENINE PATHWAY

It has been hypothesized that major depression is due to the balance between neuroprotectants and neurotoxins is lost [53]. The neuroprotectants formed in the kynurenine pathways include kynurenic acid (KYNA), picolinic acid and NAD [51]. In normal physiology, kynurenine is mainly converted into KYNA by kynurenine aminotransferase (KATs; KATII is the most important enzyme for KYNA synthesis) instead of catabolized into 3-hydroxykynurenine (OHK) in astrocytes because lack of kynurenine monooxygenase (KMO). In the brain, KYNA functions as an endogenous N-methyl-D-aspartic acid (NMDA) receptor antagonist at the nanomolar concentration. However, the concentration of KYNA is reduced due to up-regulated KMO expression resulted from increased pro-inflammatory cytokine; consequently, the metabolism would be shifted to quinolinic pathway and reduces KYNA production. KYNA blocks the excitotoxicity caused by NMDA receptor agonists at their glycine site to protect neurons from damages [54]; and it also has been reported that it may be beneficial to Alzheimer's disease, Parkinson disease and Huntington disease [55-58]. However, it seems unimaginable that KYNA only targets to NMDA receptors and being a neuroprotectant. Klein et al. proposed an NMDA-independent neuroprotection mechanism using human neuroblastoma SH-SY5Y cells and mouse cortical neurons. They found that KYNA could enhance viabilities of both SH-SY5Y cells and mouse cortical neurons by inducing the expression of neprilysin (NEP) gene and protein at low micromolar concentration [59]. Downregulation of NEP is found related to amyloid beta (Aβ) deposition and neural deaths usually found in Alzheimer's disease [60]. Effects of KYNA on neurons depend on its concentration, KYNA tends to enhanceα-amino-3-hydroxy-5-methyl-4-isoxazolepropionic acid (AMPA) receptor at nano-to micromolar concentration. However, higher concentration of KYNA interferes with glutamatergic transmission and shows neuroinhibitory effect [61]. High concentration of KYNA has been reported which is related to impaired cognition in schizophrenia by inducing antagonistic effect on α7 nicotinic acetylcholine receptor (α7nAchR) [62]. On the other hand, improved cognition status was found when KAT is inhibited by PF-04859989, a brain-penetrable inhibitor in a dose-dependent manner in rats [63].

Picolinic acid, the isomer of nitric acid (also known as vitamin B3), is another kynurenine that produced along the kynurenine pathway [51]. Picolinic acid has been reported that has antitumor, antimicrobial and antiviral abilities [64-66], and is one of the endogenous antagonists of QUIN as KYNA. Unlike KYNA, picolinic acid blocks neurotoxic instead of the neuroexcitatory effects caused by QUIN. Though less potent than KYNA, less cholinergic neuron loss in the brain was found when co-injection of picolinic acid with QUIN was performed [67]. Picolinic acid is also known as iron [68] and zinc chelator [69], and this might be another reason that picolinic acid is a neuroprotectant. As a redox-metal, iron is also known to induce neuron damages as copper. Zinc has been reported which causes neuron injury through various mechanisms. The neurotoxicity induced by NMDA receptor agonist like QUIN is related to unbound zinc co-released with glutamates and following enormous calcium influx. Several iron or zinc-chelators have been developed for treating ischemia-related neuron injury and degenerative neuron damages [70, 71]. It is worth to mention that the enzyme responsible for picolinic acid synthesis, aminocarboxymuconate semialdehyde decarboxylase (ACMSD), lacks in SK-N-SH neuroblastoma but not the neural cells [72].

Nicotinamide adenine dinucleotide (NAD) is the last neuroprotectant produced in the kynurenine pathway. *De novo* NAD production uses tryptophan as a precursor in animals. Thus, it has been reported that IDO is one of the factors controls the abundance of NAD. Energy transfer during the redox reaction between NAD and NADH, its reduced counterpart, makes NAD becoming a critical coenzyme for ATP generation for survival of cells including neurons. Attenuations of axonal degeneration have been found in the culture with additional extracellular NAD [73]. Moreover, axonal degeneration is regulated by apoptosis pathways and NAD sensitive pathway [74], which are related to changes in mitochondrial permeability when cell are under stress. Acute or chronic stresses of neurons resulted from ischemia/reperfusion, inflammation or trauma, showing elevated poly(ADP-ribosyl) polymerase (PARP)-mediated depletions of intracellular NAD [75]. An interesting finding proposed by Ying et al. that astrocytes are more resistant to oxidants because the higher concentration of NAD is present in astrocytes [76]. NAD is also the precursor of cyclic ADP-ribose, a secondary messenger involving in the calcium signaling in many cells including neurons [77].

NEUROTOXIC METABOLITES IN KYNURENINE PATHWAY

3-hydroxykynurenine (OHK), 3-hydroxyanthranilic acid (HAA) and quinolinic acid (QUIN) are the neurotoxic metabolites generated in the kynurenine pathway. Increased level of OHK is found in several neuronal degeneration diseases such as Huntington disease [78] and Alzheimer's disease [79]. OHK functions as an oxidative stress generator, and it has been shown that antioxidants could block apoptosis, chromatin fragmentation induced by OHK-related neurotoxicity. Additionally, the neurotoxicity of OHK seems to be region-selective, and the cortical and striatal neurons are the most sensitive and cerebral neurons are resistant to OHK [80]. It has been found that the ratio of KYNA/OHK could be reversed by treating patients with antipsychotics in schizophrenia [81]. In the study, authors suggested that the higher ratio of KYNA/OHK exhibits better clinical outcomes meaning that the neural damages are slowing down. Kocki et al. also reported a similar findings, and they further discovered that antidepressants enhanced the expressions of KATs (KATI and KATII) genes and reduced the expression of KMO gene in a time-dependent manner. Thus, the kynurenine pathway is shifted to KYNA pathway other than QUIN pathway [82]. Interestingly, high concentration of OHK has dual functions. High concentration of OHK serves as a reducing agent and a radical scavenger when short exposure but induces neurotoxicity when the exposure is prolonged [83].

HAA is catabolized from OHK by kynureninase (KYNU) and is a potential generator of superoxide and hydrogen peroxide when the copper is present [84]. A study showed that HAA would inhibit the proliferation of activated CD8+ T cells stimulated by IL-2, IL-7 and IL-15 [85]. These activated CD8+ T cells are important in cell-mediated antitumor immunity, which causes apoptosis of cancer cells. Moreover, HAA selectively induces apoptosis in Th1 but not Th2 cells in mice, and arrests cell cycle of CD4+ T cells at G1/S phase [86]. As a result, cytokine pattern in cancer patients shifts to Th2 pattern, which usually causes poor prognosis [87].

Being an NMDA receptor agonist, QUIN elicits both neurotoxicity and neuroexcitatory, and then causes neural damages. In normal condition, astrocytes uptake QUIN generated by infiltrating macrophages, microglia, and dendritic cells (DCs), and convert it into NAD to protect neurons. However, this conversion of NAD from QUIN is easily saturated because excess QUIN is produced under inflammation conditions such as Alzheimer's disease; thus, causing the accumulation of QUIN [88]. QUIN could be an ROS generator when the transition metals are present and results in lipid peroxidation that is highly related to neuronal degeneration diseases. QUIN (<350 nM) induces astrogliosis, which is associated with scar formation and immune regulation in the central nerve system (CNS). When the concentration of QUIN is higher than 350 nM, the apoptosis of astrocytes will be seen [89, 90]. Astrocytes are involved in neuron repair, nutrient supply and maintenance of BBB integrity. Disruption of BBB might be found in QUIN accumulated brain followed by increased numbers of infiltrating macrophages and lymphocytes. These activated macrophages further create inflammation conditions, secrete more pro-inflammatory cytokines and recruit more macrophages [91]. Elevated neuronal nitric oxide synthase (nNOS) and inducible nitric oxide synthase (iNOS) are found in QUIN accumulated neurons and astrocytes. Nitric oxide is a critical molecule for development of depression [92]. Inhibition of NOS has been thought as a way to treat depression [93-95]. Both enhanced extracellular serotonin and dopamine levels in the hippocampus, an important region of the brain related to development of depression, are detected after NOS inhibition. *Via* the above mechanisms, QUIN serves as a neurotoxin and results in neuronal damages.

ANTITUMOR IMMUNITY AFFECTED BY ABNORMAL METABOLISM OF TRYPTOPHAN IN CANCER

IDO is a well-known immunosuppressive factor, which hinders the antitumor effects *via* reducing the cell-mediated immunity and disturbing the Th1/ Th2 balance of the humoral immunity [52]. Up-regulated IDO depletes available tryptophan, which is critical for T cells proliferation. Furthermore, kynurenine is toxic to T cells, and results in the apoptosis of T cells. Cytotoxic T cells (CTLs or CD8+ T cells) eliminate cancer cells through both apoptosis and necrosis pathways. After recognizing the specific antigens presented by cancer cells, activated CD8+ T cells release granules containing perforin and granzymes, punching holes in the cell membrane of cancer cells. Thus, the contents of cancer cells leak out, resulting in necrosis of cancer cells. On the other hand, activated CD8+ T cells trigger the apoptosis of cancer cells either through Fas-FasL pathway or the mitochondria-dependent apoptosis start from the specific recognition [96]. Kynurenine produced in the kynurenine pathway affects not only neurons, but also immune cells. Receptors for kynurenine, aryl hydrocarbon receptor (AHR), is expressed by helper T cells (Th or CD4+ T cells), DCs and cancer cells. Binding of kynurenine to AHR facilitates the conversion of CD4+ T cells into regulatory T cells (CD4+CD25+FOXP3+, Tregs) [97], which inhibit the functions of activated CD8+ T cells; thus, the antitumor effects will be reduced. Nguyen et al. reported that AHR is required for expressing IDO in DCs. DCs is one of the professional antigen-presenting cells (APCs) as macrophages. They added exogenous kynurenine to naïve T cells co-cultured with lipopolysaccharide (LPS)-/ CpG oligodeoxynucleotide-stimulated bone marrow-derived dendritic cells (BMDCs) expressing AHR, and found these CD4+ T cells preferred to convert into Tregs rather than Th17 cells. Moreover, the proliferations of naïve T cells were also found [98]. The IDO-AHR interaction in brain tumor has been shown that is associated with tumor progression and poor prognosis. In the study, exogenous kynurenine enhanced migration and proliferations of glioma cells through activating AHR, and changed the autocrine and paracrine signaling of glioma cells [99].

IDO could be activated by interferons, especially, the IFN-α and IFN-γ. IFN-γ is a critical antitumor cytokine secreted by activated CD8+ T cells. IFN-γ activates IDO activity through JAK/STATs pathway in plasmacytoid dendritic cells (pDCs) and results in the depletion of tryptophan in microenvironment. IDO could also be activated by IFN-α/β, which are used as antiproliferation cytokines in cancer treatment through noncanonical NF-□B pathway, and enhances expressions of IDO, TGF-β, and IFN-α/β [100]. Other than up-regulating the IDO expression, TGF-β is also involved in the conversion of Tregs from naïve CD4+ T cells [100]. A "bystander" effect of IDO-expressing dendritic cells in the tumor has been proposed to have widely suppressive effects on T cells and impair the cell-mediated antitumor immunity [101]. Constitutive IDO expression in cancer cells triggers the metastasis of cancer by creating an immunosuppressive microenvironment, which is associated with IDO-AHR-IL-6-STAT3 signaling loop [102].

In our previous studies, we found that pretreated tumor-bearing mice with immunomodulating agents, such as curcumin and sorafenib, to repress IDO expression could significantly augment the therapeutic efficacy of adoptive T cell therapy (ACT) through increased activations and numbers of tumor-infiltrating CD8+ T cells [103, 104]. Fewer Tregs and myeloid-derived suppressor cells (MDSCs) were detected in mice treated with serial low doses of sorafenib because the inhibition of STAT3 pathway [104]. TGF-β receptor inhibitor and IDO inhibitor used for the combination therapy were also shown with better tumor responses than ACT alone, suggested that IDO inhibition might be a novel strategy for cancer therapy [103].

The balance of Th1/Th2 cytokines is also disturbed by up-regulation of IDO. The Th1/Th2 ratio will be lowered when IDO is overexpressed. Higher levels of Th2 cytokines, such as IL-10, TGF-β and IL-6, result in angiogenesis and hypoxia. The changes of cytokines and chemokines will recruit Tregs and MDSCs to the tumor lesions, and polarize the tumor-associated macrophages (TAMs) into M2-subtype [105, 106]. Furthermore, Tregs, MDSCs and M2-TAMs all weaken the antitumor functions of CD8+ T cells and natural killer cells (NK cells) by direct inhibition or cytokine secretions [107].

IDO seems to be an ideal target for cancer treatment because up-regulation of IDO shifts humoral immunity from Th1 to Th2 and dampens the antitumor abilities of CD8+ T cells and NK cells (Figure 1). Several IDO inhibitors have been developed, and promising outcomes were found in preclinical studies [108-110]. Currently, three IDO inhibitors including 1-methyl-D-tryptophan (1-MT; indoximod), INCB024360, and NLG-919 have entered the clinical trials for patients with melanoma or other solid malignancies, and are combined with other therapies including immunotherapy [111-112].

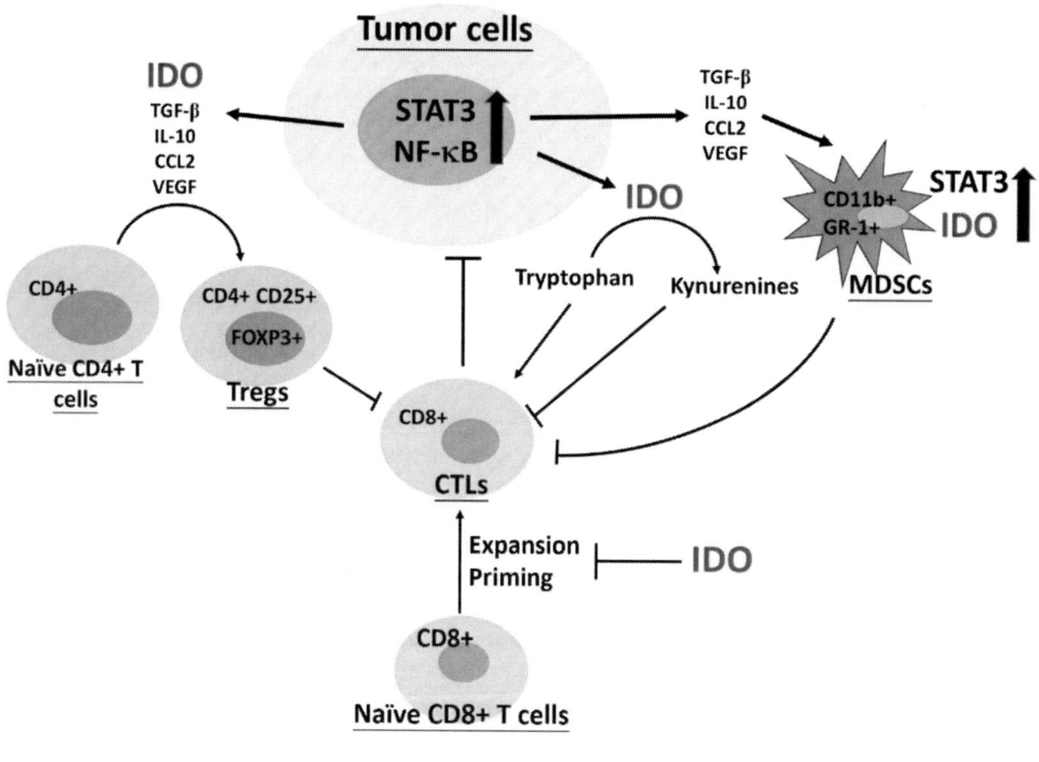

Figure 1. Antitumor immunity affected by tumor cells expressing IDO and other immunosuppressive factors.

EFFECTS OF SEROTONIN

Serum levels of serotonin and melatonin are significantly reduced in patients with cancer due to up-regulation of IDO and kynurenine pathway [113, 114]. Though the antitumor effects of serotonin have been reported [115, 116], the role of serotonin in cancer progression is still debating. Angiogenesis and tumor progression promoted by serotonin *via* angiostatin pathway has been observed in colon cancer by down -regulating MMP-12 expression [117]. Other studies show that the serotonin activates serotonin receptors expressed by cancer cells and endothelial cells to cause tumor progression. Moreover, the tumor progression could be blocked by treating with serotonin receptor antagonist or TPH inhibitor in several types of cancer including cholangiocarcinoma [118], lung cancer [119], and pancreatic cancer [120]. Serotonin is also important for chemotaxis, cytokine secretion and proliferation of immune cells [121]; however, serotonin could not be synthesized in most of immune cells. Up-regulation of serotonin transporter (SERT) is found in activated DCs [122], and B cells [123] to uptake serotonin. It is speculated that T cells uptake serotonin *via* dopamine transporter (DAT) because the identification of SERT in T cells was failed. However, the expression of TPH was found increased in activated T cells [122, 124].

Up to present, how serotonin affects these immune cells in cancer is still obscure. Effects of serotonin on antigen presentation by DCs were found inconsistently. The antigen presentation by DCs to CD8+ T cells was found to be attenuated by serotonin in chronic inflammation [125]. A contradictory finding was reported that serotonin enhanced the productions of both innate and adaptive cytokines, and augmented the stimulation of naïve T cells *via* NF-κB pathway in gut inflammation model [126]. Enhanced migration of immature DCs and expression of Th2 cytokine in DCs in allergic diseases were reported in many studies [127, 128]. A subpopulation of APCs bearing-Ia-antigen is required for antigen-specific T cell proliferation [129]. It has been reported that serotonin could suppress the expression of IFN-γ induced Ia-antigen on macrophage, and an IFN-γ concentration dependent serotonin modulation of IFN-γ induced phagocytosis on macrophage [130]. Serotonin enhances phagocytosis when low levels of IFN-γ co-exist in the medium; conversely, phagocytosis is inhibited by serotonin when the concentration of IFN-γ is high [131]. Recently, Casas-Engel et al. suggest that serotonin skews the polarization of macrophages to M2 subtype through up-regulations of serotonin receptors types 2B and 7 (5-HT_{2B} and 5-HT_7) [132]. According to these findings, it seems that serotonin impairs antigen presentation of DCs and macrophages and also switches the cytokine expressions to Th2 profile, which can further reduce the antitumor immunity.

Nevertheless, *via* 5-HT_{1A} receptor, serotonin enhances IL-2 stimulated T cells proliferation and activation, which are associated with antitumor immunity substantially [133, 134]. In addition to 5-HT_{1A} receptor, 5-HT_7 receptor was also found up-regulated during T cell activation. The authors additionally indicated that the 5-HT_7 receptor was a critical cofactor for T cells proliferation and activation, which was validated by pre-treating T cells with 5-HT_7 antagonist [124]. Contradictory effects of serotonin on T cells were also found. Higher IFN-γ and IL-17 produced by T cells in $TPH1^{-/-}$ mice were found with better prognosis of skin tumor [135]. Another study showed that serotonin- activated CD8+ T cells could secret IL-16 and recruit CD4-expressing cells (e.g. CD4+ T cell, monocyte, and DCs) [136]. Furthermore, the level of IL-16 is tightly correlated to cancer progression [137, 138].

In summary, the effects of serotonin on immune cells and cancer are not fully understood because the complex interactions in the immune regulation. Further studies are needed to perform to figure out the role of serotonin in the immune regulation in cancer under different conditions.

EFFECTS OF MELATONIN

The final product of the serotonin pathway, melatonin, is mainly synthesized in the pineal gland and is related to the regulation of the circadian cycle. Melatonin is a powerful free radical scavenger and antioxidant [139]. It is known that free radicals could be generated in ischemia/reperfusion process damage normal cells including neurons, myocardium and hepatocytes [140-143]. Melatonin converts hydroxyl radicals to cyclic 3-hydroxymelatonin (C3HOM), which is excreted through the urinary system [144]. C3OHM is also a free radical scavenger and antioxidant, and is more potent than melatonin. The antioxidant mechanism of C3HOM is different from melatonin, prevents apoptosis by blocking release of cytochrome c from mitochondria. It has been suggested that partial antioxidative ability of melatonin might be related to C3HOM [145]. Melatonin showed better scavenging activity of peroxyl radicals produced in lipid peroxidation than vitamin C and E [146]. Moreover, Morishima et al. reported that melatonin prevented the cardiomyopathy [147], renal injury [148] and hepatic injury [149] induced by widely used chemotherapeutic drugs *via* inhibiting lipid peroxidation. Though only few reports suggested that melatonin could interact with NO directly, body of evidence supported that melatonin inhibits the expression of iNOS and reduce the formation of NO in various models [150-152].

The relationship between immune regulation, season, and light/dark cycle has been reported [153], implies that melatonin has impact on the immunity. Weight losses of the spleen or thymus after pinealectomy [154] shows that melatonin level is associated with proliferations of immune cells. Besides, the level of melatonin in the pineal gland and circulation decreases with aging [154], which is associated with higher susceptibility to infection, autoimmune disease and cancer in elders [155, 156]. Reduced Th1 cytokines productions by melatonin were found in contact hypersensitivity [157], inflammatory bowel disease (IBD) [158], islet graft [159] and ovarian graft [160]. However, melatonin increased Th1 cytokine expression in aging-related immunosuppression in mice [161]. Elevated IL-12 production was found in monocytes stimulated with low concentrations of LPS and IFN-γ after treated with melatonin, and the authors suggested that increased IL-12 would lead differentiation of T cells into Th1 subtype [162, 163].

The immune cells are strongly affected by free radical, NO, and peroxynitrite formed by superoxide and nitric oxide. NO and reactive oxygen species (ROS) help the host to eliminate foreign pathogens under normal physiology; however, the excess NO showing in the acidic microenvironment, i.e., the tumor microenvironment, triggers the microenvironment becoming immunosuppressive [164]. Antigens and cytokines could be nitration by peroxynitrite generated in the tumor microenvironment, and nitrotyrosinylation of antigens and cytokines impairs the antigen recognition abilities and prevents the infiltration of antitumor CD8+ T cells [165, 166]. Melatonin reduces levels of NO and other free radicals preventing the nitration of antigens or cytokines; furthermore, it has been shown that melatonin enhanced the antigen presentation of macrophages and T cells proliferation, and was consistent with increased MHC-II expression [167]. Except for being a professional APC, macrophage is also critical for generations of superoxide anion and NO, which may induce DNA damage. Higher expressions of iNOS and COX-2 in macrophages are often observed in cancer, which could be reduced by melatonin through inhibition of NF-κB pathway [168, 169]. Macrophages are derived from monocytes in bone marrow, and both two cells types could eliminate foreign pathogens by phagocytosis [170]. Melatonin administration increased numbers of monocytes in bone marrow and spleen from mice [171], and augmented the phagocytosis of testicular macrophage correlated with the concentration of intracellular free calcium [172]. Moreover, higher production of the antitumor cytokine, IL-12, was found in both macrophages and human myeloid monocytic cell line after melatonin treatment [173]. To date, no direct evidence shows that melatonin has effects on DCs. Melatonin administration increased numbers of NK cells significantly in cancer patients treated with or without IL-2 [174, 175]. NK cells exhibited higher activity and antibody-dependent cell cytotoxicity (ADCC) after melatonin treatment [176]. There is no evidence showing that melatonin affects numbers of functions of CD8+ T cells directly. However, it has been reported that conversion of Tregs from CD4+ T cells were inhibited by melatonin treatment [177]. In addition, melatonin administration shifts the differentiation of CD4+ T cells to Th1 subtype due to the changes of cytokine profiles.

Till now, melatonin is only used as an adjuvant therapy in cancer treatment because underlying mechanisms are still unclear. Combinational therapies of melatonin and chemotherapy or radiotherapy exhibit better tumor control and less normal tissue damages. Seely et al. found that combined melatonin and chemotherapy improved therapeutic outcomes and decreased 1-year mortality through a systemic review and meta-analysis. Moreover, the adverse side-effects from chemotherapy were also ameliorated [178]. Similar results showed that melatonin significantly better therapeutic efficacy and reduced side-effects caused by chemo-/radiotherapy [179]. As a free radical scavenger and antioxidant, melatonin showed radioprotection ability from *in vitro* to *in vivo*. Vijayalaxmi et al. showed that pre-treated human lymphocytes with melatonin for 20 minutes could reduce genetic damages caused by 150 cGy γ-radiation [180]. The survival of whole-body irradiation of mice was increased if the mice received melatonin prior to 815 cGy irradiation, and this protective effect was dependent on the dose of melatonin [181]. Also reported by Vijayalaxmi et al., both the incidence of micronuclei and the percentages of polychromatic erythrocytes were decreased in melatonin treated mice, compared with irradiated ones [182]. Jang et al. reported that mice pretreated with 200 mg melatonin showed fewer radiation-injuries caused by 12 Gy lung irradiation through down-regulating oxidative stress and productions of TGF-β and TNF-α [183]. Recently, the mechanism that melatonin reduces radiation-induced oral mucositis, a frequent side-effect seen in patients receiving radiotherapy, was proposed by Ortiz et al. The author indicated that melatonin reduced generations of ROS by inhibiting the NK-κB/NLRP3 (NLR-related protein 3 nucleotide-binding domain leucine-rich repeat containing receptors-related protein 3) mediated-inflammation [184]. In addition to the radioprotection, the better therapeutic outcome of melatonin combined with radiotherapy was also reported in patients with glioblastoma. Patients received melatonin and radiotherapy exhibited higher 1-year survival rate and less radiation-induced toxicities [185]. According to previous findings, several articles have suggested that melatonin could be used in radiotherapy to reduce normal tissue damages and enhance therapeutic outcomes. However, the optimal dosage and administration scheme of melatonin still need further investigation. [186, 187]. Improved outcomes of chemotherapy and protection by melatonin were also discovered [188, 189].

Accumulating evidence have shown that melatonin exerts antitumor effect without combined with other treatments in types of cancer. Apoptosis and necrosis seem to be increased by melatonin by altering the balance of Bcl-2/ Bax in pancreatic cancer cells [190]. Melatonin could be a novel SIRT1 inhibitor, inhibited in vitro and in vivo proliferation of TRAMP prostate cancer model [191]. Another study also indicated that melatonin functioned as an SIRT1 inhibitor in breast cancer, downregulated MDM2 and enhanced p53 in breast cancer cells [192].

Downregulation of angiogenesis by melatonin was shown in breast cancer [193] and liver cancer [194] through downregulating expression of VEGF-R2 and VEGF, respectively. Increased apoptosis was observed in lung cancer by increasing lipid content and reducing intensity of nucleic acids and secondary structure of proteins [195]; reversing the anti-apoptosis induced by hypoxia [196]. Inhibition of migration and metastasis caused by melatonin was also found in oral cancer through inhibiting HIF-1α and ROCK expressions [197]. Decreased migration and invasiveness found in liver cancer after melatonin treatment was caused by inhibitions of MMP-9 and NK-κB [198]. Inhibition of cancer-initiating cells is an alternative way to control the tumor growth, melatonin also exhibited the ability to block the proliferation of glioma-initiating cells by inducing autophagy [199].

OTHER PATHWAYS RELATED TO ABNORMAL TRYPTOPHAN METABOLISM IN CANCER

Other than up-regulating IDO, the tryptophan metabolism is also interfered by other pathways. Highly proliferative cancer cells usually exhibit higher levels and activities of NF-κB and STAT3 induced by pro-inflammatory cytokines to stimulate the expression of iNOS. iNOS converts L-arginine into NO with the cofactor, tetrahydrobiopterin (BH4), which is also the cofactor involved in serotonin synthesis. NO interacts with excessive superoxide anion (O_2^-) produced in the tumor microenvironment, and forms the peroxynitrite ($ONOO^-$), which may result dysfunction of mitochondria [200], DNA damages [201], lipid peroxidation [202], and nitrotyrosinylation of proteins [203]. All of these could further impair the functions of APCs including macrophages and DCs. Accordingly, the antitumor immunity will be diminished. Furthermore, these damages also trigger the apoptosis of neurons and are related to the development of depression [200, 204-206].

ANTIDEPRESSANTS AND CANCER

Currently, antidepressants are used as an adjuvant therapy in cancer treatment to relieve the nausea, fatigue and/or other depressive symptoms in cancer patients to improve the quality of life of cancer patients [207]. Antidepressants are shown with several antitumor mechanisms even contradictory results, which may be due to the differences in chemical structures and functional targets of each antidepressant.

The following reports showed some effects on cancer inhibition or progression caused by antidepressant treatment. Both rat glioma C6 cells and human glioblastoma U87MG cells treated with imipramine, a tricyclic antidepressant (TCA), were found with significant cell death but not the rat astrocytes. The authors suggested that the cell death caused by imipramine was *via* necrosis rather than apoptosis, which was mediated by PI3K/AKT/mTOR signaling pathway [208]. Amitriptyline, another TCA, also displayed antitumor capability on human uterine leiomyosarcoma HTB114 cells, which shows both radioresistance and chemoresistance. It is known that the proliferation of uterine leiomyosarcoma is related to neurotrophins (NTs) and an autocrine loop triggered by NTs. The apoptosis of HTB114 cells caused by amitriptyline seemed to be associated with up-regulating $p75^{NTR}$, the neurotrophin receptor (NTR), through inhibition of AKT pathways [209]. The antitumor effect of amitriptyline was found *via* inhibition of cyclin-D2 transactivation and histone deacetylases (HDACs) activity in myeloma cells [210]. A comprehensive study of TCAs on small cell lung cancer (SCLC) and pancreatic cancer was proposed by Jahchan et al. in 2013. Like leiomyosarcoma, both SCLC and pancreatic cancer are belonged to neuroendocrine tumors. The authors concluded that TCAs could be used as a second-line therapy in SCLC patients becoming refractory to cisplatin/etoposide, but not combined with other chemotherapies because higher toxicities might appear [211].

In clinic, TCAs are replaced by the new generations of antidepressants of selective serotonin reuptake inhibitors (SSRIs) [211]. Fluoxetine, one of the SSRIs, has been shown to function as an antitumor agent in oral cancer and colon cancer and melanoma. Lin et al. showed that fluoxetine evoked the intracellular [Ca^{2+}] by inducing phospholipase C (PLC)-dependent [Ca^{2+}] release from smooth endoplasmic reticulum (SER), and resulted in [Ca^{2+}] dependent and independent apoptosis in oral cancer [212]. Stopper et al. found that fluoxetine arrested colon cancer cells at G1 phase by altering the glucose metabolism through inhibition of NF-κB pathway [213].

In our previous study, we have shown that mirtazapine, a noradrenergic and specific serotonergic antidepressant (NaSSA) had antitumor effects on colon cancer *in vivo*. Though no obvious cell death and apoptosis found in *in vitro* study, significantly improved tumor controls were found in mirtazapine-treated mice. The better tumor control was associated with improved tumor microenvironment including elevated IFN-γ, IL-12 and much more tumor-infiltrating CD4+ and CD8+ T cells. In addition, we also found that recovery of serotonin level in the serotonergic system using [^{123}I]-ADAM *ex vivo* autoradiography. [214]. Recently, mirtazapine was also found to prevent metastasis of invasive oral cancer by up-regulating HTR2 expression, which led to the restoration of Lin-7c/β-catenin pathway [215].

Stimulatory effects of antidepressants on cancer progression were also found. Kubera et al. discovered that pretreatment of fluoxetine in high-active mice would increase metastasis, which might be related to decreased levels of IL-12 and IFN-γ, and antitumor cytotoxicity of spleen-derived NK cells [216]. Szpunar et al. showed that treated 4T1 tumor-bearing mice with desipramine increases tumor growth but not metastasis with reduced VEGF, IL-6, and expressions of pro-metastatic cytokines. Furthermore, treated mice with a highly selective α2 -adrenergic receptor agonist, dexmedetomidine, increased both tumor growth and metastasis. The authors suggested that the dexmedetomidine activated α2- adrenergic receptors, altered extracellular matrix, and promoted progression and metastasis of cancer [217]. Thus, the usage of antidepressants needs detailed investigations and be used with caution.

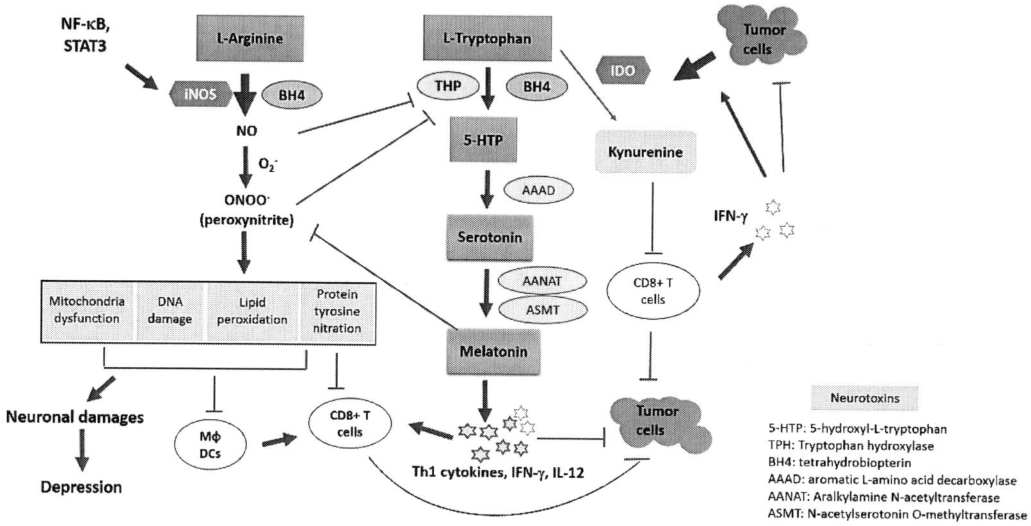

Figure 2. The proposed interactions between antitumor immunity and depression in cancer.

CONCLUSION

To summarize how indoleamines interplay with cancer and depression, we depict a correlation among the interactions of IDOs, cancer and depression as shown in Figure 2. Briefly, the serotonin synthesis pathway is down-regulated in the cancer due to up-regulated IDO expression. The serotonin level is highly associated with the development of depression. Additionally, shortages of serotonin, melatonin and increased levels of neurotoxic metabolites generated in kynurenine pathway (i.e., OHK, HAA and QUIN) directly affect the neuronal system. Free radicals are tightly correlated with the occurrence of many neuronal degenerative diseases as described above. Both melatonin and neuroprotective metabolites generated in the kynurenine pathway (i.e., KYNA, picolinic acid and NAD), whichare powerful free radical scavengers and antioxidants, are all reduced in the cancer due to overexpression of IDO.

Kynurenine is not only toxic to neurons, but to immune cells as well. Reduced numbers and activities of antitumor CD8+ T cells, higher percentages of Tregs and MDSCs, and shifted the balance of Th1/Th2 cytokine profile are caused by the up-regulated kynurenine pathway. In addition, melatonin, a pro-Th1 agent, elicits productions of antitumor cytokines IL-12 and IFN-γ, removes excessive free radicals and prevents the formation of peroxynitrite. As a result, antitumor immunity is depleted because altered Th1/Th2 balance and impaired antigen presentation by macrophages or DCs, which are related to the reduction of melatonin in cancer.

The interactions between cancer and depression still need further investigations as well as the roles of antidepressants in cancer. Based on the previous researches, using antidepressants to treat cancer rather than cancer-induced depression seems to be possible. However, it must be careful when prescribe antidepressants to patients with cancer, the optimal dosages and treatment schedules have to be studied thoroughly because some adverse effects have been reported.

ACKNOWLEDGMENTS

This work was supported by grants NSC100-2314-B-010-016, NSC100-2321-B-010-013, NSC101-2321-B-010-006 and NSC102-2321-B-010-005 from Ministry of Science and Technology, Taiwan.

REFERENCES

[1] Mathers, CD; Loncar, D. Projections of global mortality and burden of disease from 2002 to 2030. *PLoS Med*, 2006, 3, e442.
[2] Gustavsson, A; Svensson, M; Jacobi, F; Allgulander, C; et al. Cost of disorders of the brain in Europe 2010. *Eur Neuropsychopharmacol*, 2011, 21, 718-779.
[3] lesen, J; Gustavsson, A; Svensson, M; et al. The economic cost of brain disorders in Europe. *Eur J Neurol*, 2012, 19, 155-162.
[4] Mrazek, DA; Hornberger, JC; Altar, CA; et al. A review of the clinical, economic, and societal burden of treatment-resistant depression: 1996-2013. *Psychiatr Serv*, 2014, 65, 977-987.
[5] Ferlay, J; Soerjomataram, I; Ervik, M; et al. GLOBOCAN 2012 v1.0, Cancer Incidence and Mortality Worldwide: IARC CancerBase No. 11
[6] Pinquart, M; Duberstein, PR. Depression and cancer mortality: a meta-analysis. *Psychol Med*, 2010, 40, 1797-1810.
[7] Linden, W; Vodermaier, A; Mackenzie, R; et al. Anxiety and depression after cancer diagnosis: prevalence rates by cancer type, gender, and age. *J Affect Disord*, 2012, 141, 343-351.
[8] Raison, CL; Miller, AH. Depression in cancer: new developments regarding diagnosis and treatment. *Biol Psychiatry*, 2003, 54, 283-294.
[9] Massie, MJ. Prevalence of depression in patients with cancer. *J Natl Cancer Inst Monogr*, 2004, 32, 57-71.
[10] Li, M; Fitzgerald, P; Rodin, G. Evidence-based treatment of depression in patients with cancer. *J Clin Oncol*, 2012, 30, 1187-1196.
[11] Wedding, U; Pientka, L; Höffken, K. Quality-of-life in eldérly patients with cancer: a short review. *Eur J Cancer*, 2007, 43, 2203-2210.
[12] Linden, W; Girgis, A. Psychological treatment outcomes for cancer patients: what do meta-analyses tell us about distress reduction? *Psychooncology*, 2012, 21, 343-350.
[13] Breitbart, W; Bruera, E; Chochinov, H; et al. Neuropsychiatric syndromes and psychological symptoms in patients with advanced cancer. *J Pain Symptom Manage*, 1995, 10, 131-141.
[14] Hopwood, P; Stephens, RJ. Depression in patients with lung cancer: prevalence and risk factors derived from quality-of-life data. *J Clin Oncol*, 2000, 18, 893-903.
[15] Sotelo, JL; Musselman, D; Nemeroff, C. The biology of depression in cancer and the relationship between depression and cancer progression. *Int Rev Psychiatry*, 2014, 26, 16-30.
[16] Meyer, F; Fletcher, K; Prigerson, HG; et al. Advanced cancer as a risk for major depressive episodes. *Psychooncology*, 2014 doi: 10.1002/pon.3722.

[17] Reyes-Gibby, CC; Anderson, KO; Morrow, PK; et al. Depressive symptoms and health-related quality of life in breast cancer survivors. *J Womens Health (Larchmt)*, 2012, 21, 311-318.

[18] Que, JC; Sy, Ortin, TT; Anderson, KO; et al. Depressive symptoms among cancer patients in a Philippine tertiary hospital: prevalence, factors, and influence on health-related quality of life. *J Palliat Med*, 2013, 16, 1280-1284.

[19] Henderson, VP; Clemow, L; Massion, AO; et al. The effects of mindfulness-based stress reduction on psychosocial outcomes and quality of life in early-stage breast cancer patients: a randomized trial. *Breast Cancer Res Treat*, 2012, 131, 99-109.

[20] Turaga, KK; Malafa, MP; Jacobsen, PB; et al. Suicide in patients with pancreatic cancer. *Cancer*, 2011, 117, 642-647.

[21] Boyd, AD; Riba, M. Depression and pancreatic cancer. *J Natl Compr Canc Netw*, 2007, 5, 113-116.

[22] De, Boer, MF; McCormick, LK; Pruyn, JF; et al. Physical and psychosocial correlates of head and neck cancer: a review of the literature. *Otolaryngol Head Neck Surg*, 1999, 120, 427-436.

[23] Archer, J; Hutchison, I; Korszun, A. Mood and malignancy: head and neck cancer and depression. *J Oral Pathol Med*, 2008, 37, 255-270.

[24] Reich, M; Lesur, A; Perdrizet-Chevallier, C. Depression, quality of life and breast cancer: a review of the literature. *Breast Cancer Res Treat*, 2008, 110, 9-17.

[25] Chen, X; Zheng, Y; Zheng, W; et al. Prevalence of depression and its related factors among Chinese women with breast cancer. *Acta Oncol*, 2009, 48, 1128-1136.

[26] Miovic, M; Block, S. Psychiatric disorders in advanced cancer. *Cancer*, 2007, 110, 1665-1676.

[27] Rodríguez, Vega, B; Orgaz, Barnier, P; Bayón, C; et al. Differences in depressed oncologic patients' narratives after receiving two different therapeutic interventions for depression: a qualitative study. *Psychooncology*, 2012, 21, 1292-1298.

[28] Rayner, L; Price, A; Evans, A; et al. Antidepressants for the treatment of depression in palliative care: systematic review and meta-analysis. *Palliat Med*, 2011, 25, 36-51.

[29] Laoutidis, ZG; Mathiak, K. Antidepressants in the treatment of depression/depressive symptoms in cancer patients: a systematic review and meta-analysis. *BMC Psychiatry*, 2013, 13, 140.

[30] Lloyd-Williams, M; Friedman, T; Rudd, N. A survey of antidepressant prescribing in the terminally ill. *Palliat Med*, 1999, 13, 243-248.

[31] Zigmond, AS; Snaith, RP. The hospital anxiety and depression scale. *Acta Psychiatr Scand*, 1983, 67, 361-370.

[32] Ibbotson, T; Maguire, P; Selby, P; et al. Screening for anxiety and depression in cancer patients: the effects of disease and treatment. *Eur J Cancer.*, 1994, 30A, 37-40.

[33] Cox, JL; Holden, JM; Sagovsky, R. Detection of postnatal depression. Development of the 10-item Edinburgh Postnatal Depression Scale. *Br J Psychiatry*, 1987, 150, 782-786.

[34] Lloyd-Williams, M; Shiels, C; Dowrick, C. The development of the Brief Edinburgh Depression Scale (BEDS) to screen for depression in patients with advanced cancer. *J Affect Disord*, 2007, 99, 259-264.

[35] Moorey, S; Greer, S. Psychological therapy for patients with cancer: a new approach. American psychiatric press, Washington.

[36] Martino, M; Rocchi, G; Escelsior, A; et al. Immunomodulation Mechanism of Antidepressants: Interactions between Serotonin/Norepinephrine Balance and Th1/Th2 Balance. *Curr Neuropharmacol*, 2012, 10, 97-123.

[37] Wichers, M; Maes, M. The psychoneuroimmuno-pathophysiology of cytokine-induced depression in humans. *Int J Neuropsychopharmacol*, 2002, 5, 375-388.

[38] Asnis, GM; De, La, Garza, R, 2nd. Interferon-induced depression in chronic hepatitis C: a review of its prevalence, risk factors, biology, and treatment approaches. *J Clin Gastroenterol*, 2006, 40, 322-335.

[39] Young, JJ; Bruno, D; Pomara, N. A review of the relationship between proinflammatory cytokines and major depressive disorder. *J Affect Disord*, 2014, 169, 15-20.

[40] Slavich, GM; Irwin, MR. From stress to inflammation and major depressive disorder: a social signal transduction theory of depression. *Psychol Bull*, 2014, 140, 774-815.

[41] Ghia, JE; Blennerhassett, P; Collins, SM. Impaired parasympathetic function increases susceptibility to inflammatory bowel disease in a mouse model of depression. *J Clin Invest*, 2008, 118, 2209-2218.

[42] Hocaoglu, C; Kural, B; Aliyazıcıoglu, R; et al. IL-1β, IL-6, IL-8, IL-10, IFN-γ, TNF-α and its relationship with lipid parameters in patients with major depression. *Metab Brain Dis*, 2012, 27, 425-430.

[43] Bufalino, C; Hepgul, N; Aguglia, E; et al. The role of immune genes in the association between depression and inflammation: a review of recent clinical studies. *Brain Behav Immun*, 2013, 31, 31-47.

[44] Plotkin, SR; Banks, WA; Kastin, AJ. Comparison of saturable transport and extracellular pathways in the passage of interleukin-1 alpha across the blood-brain barrier. *J Neuroimmunol*, 1996, 67, 41–47.

[45] Rivest, S; Lacroix, S; Vallieres, L; et al. How the blood talks to the brain parenchyma and the paraventricular nucleus of the hypothalamus during systemic inflammatory and infectious stimuli. *Proc Soc Exp Biol Med*, 2000, 223, 22-38.

[46] Watkins, LR; Goehler, LE; Relton, JK; et al. Blockade of interleukin-1 induced hyperthermia by subdiaphragmatic vagotomy: evidence for vagal mediation of immune-brain communication. *Neurosci Lett*, 1995, 183, 27-31.

[47] D'Mello, C; Le, T; Swain, MG. Cerebral microglia recruit monocytes into the brain in response to tumor necrosis factor-alpha signaling during peripheral organ inflammation. *J Neurosci*, 2009, 29, 2089-2102.

[48] Wichers, MC; Koek, GH; Robaeys, G; et al. IDO and interferon-alpha-induced depressive symptoms: a shift in hypothesis from tryptophan depletion to neurotoxicity. *Mol Psychiatry*, 2005, 10, 538-544.

[49] Myint, AM; Kim, YK. Cytokine-serotonin interaction through IDO: a neurodegeneration hypothesis of depression. *Med Hypotheses*, 2003, 61, 519-525.

[50] Oxenkrug, GF. Tryptophan kynurenine metabolism as a common mediator of genetic and environmental impacts in major depressive disorder: the serotonin hypothesis revisited 40 years later. *Isr J Psychiatry Relat Sci*, 2010, 47, 56-63.

[51] Chen, Y; Guillemin, GJ. Kynurenine Pathway Metabolites in Humans: Disease and Healthy States. *Int J Tryptophan Res*, 2009, 2, 1-19.

[52] Prendergast, GC. Immune escape as a fundamental trait of cancer: focus on IDO. *Oncogene*, 2008, 27, 3889-3900.

[53] Licinio, J; Wong, ML. The role of inflammatory mediators in the biology of major depression: central nervous system cytokines modulate the biological substrate of depressive symptoms, regulate stress-responsive systems, and contribute to neurotoxicity and neuroprotection. *Mol Psychiatry*, 1999, 4, 317-327.

[54] Leeson, PD; Baker, R; Carling, RW; et al. Kynurenic acid derivatives. Structure-activity relationships for excitatory amino acid antagonism and identification of potent and selective antagonists at the glycine site on the N-methyl-D-aspartate receptor. *J Med Chem*, 1991, 34, 1243-1252.

[55] Hartai, Z; Juhász, A; Rimanóczy, A; et al. Decreased serum and red blood cell kynurenic acid levels in Alzheimer's disease. *Neurochem Int*, 2007, 50, 308-313.

[56] Hartai, Z; Klivenyi, P; Janaky, T; et al. Kynurenine metabolism in plasma and in red blood cells in Parkinson's disease. *J Neurol Sci*, 2005, 239, 31-35.

[57] Beal, MF; Matson, WR; Swartz, KJ; et al. Kynurenine pathway measurements in Huntington's disease striatum: evidence for reduced formation of kynurenic acid. *J Neurochem*, 1990, 55, 1327-1339.

[58] Tan, L; Yu, JT; Tan, L. The kynurenine pathway in neurodegenerative diseases: mechanistic and therapeutic considerations. *J Neurol Sci*, 2012, 323, 1-8.

[59] Klein, C; Patte-Mensah, C; Taleb, O; et al. The neuroprotector kynurenic acid increases neuronal cell survival through neprilysin induction *Neuropharmacology*, 2013, 70, 254-260.

[60] Iwata, N; Tsubuki, S; Takaki, Y; et al. Metabolic regulation of brain Abeta by neprilysin. *Science*, 2001, 292, 1550-1552.

[61] Prescott, C; Weeks, AM; Staley, KJ; et al. Kynurenic acid has a dual action on AMPA receptor responses. *Neurosci Lett*, 2006, 402, 108-112.

[62] Alexander, KS; Pocivavsek, A; Wu, HQ; et al. Early developmental elevations of brain kynurenic acid impair cognitive flexibility in adults: reversal with galantamine. *Neuroscience*, 2013, 238, 19-28.

[63] Dounay, AB; Anderson, M; Bechle, BM; et al. Discovery of brain-penetrant, irreversible kynurenine aminotransferase II inhibitors for schizophrenia. *ACS Medicinal Chemistry Letters*, 2012, 3, 187-192.

[64] Leuthauser, SW; Oberley, LW; Oberley, TD. Antitumor activity of picolinic acid in CBA/J mice. *J Natl Cancer Inst*, 1982, 68, 123-126.

[65] Cai, S; Sato, K; Shimizu, T; et al. Antimicrobial activity of picolinic acid against extracellular and intracellular Mycobacterium avium complex and its combined activity with clarithromycin, rifampicin and fluoroquinolones. *J Antimicrob Chemother*, 2006, 57, 85-93.

[66] Fernandez-Pol, JA; Klos, DJ; Hamilton, PD. Antiviral, cytotoxic and apoptotic activities of picolinic acid on human immunodeficiency virus-1 and human herpes simplex virus-2 infected cells. *Anticancer Res*, 2001, 21, 3773-3776.

[67] Cockhill, J; Jhamandas, K; Boegman, RJ; et al. Action of picolinic acid and structurally related pyridine carboxylic acids on quinolinic acid-induced cortical cholinergic damage. *Brain Res*, 1992, 599, 57-63.

[68] Mikogami, T; Marianne, T; Spik, G. Effect of intracellular iron depletion by picolinic acid on expression of the lactoferrin receptor in the human colon carcinoma cell subclone HT29-18-C1. *Biochem J*, 1995, 308, 391-397.

[69] Jhamandas, KH; Boegman, RJ; Beninger, RJ; et al. Role of zinc in blockade of excitotoxic action of quinolinic acid by picolinic acid. *Amino Acids*, 1998, 14, 257-261.

[70] Papazisis, G; Pourzitaki, C; Sardeli, C; et al. Deferoxamine decreases the excitatory amino acid levels and improves the histological outcome in the hippocampus of neonatal rats after hypoxia-ischemia. *Pharmacol Res*, 2008, 57, 73-78.

[71] Frederickson, CJ; Suh, SW; Koh, JY; et al. Depletion of intracellular zinc from neurons by use of an extracellular chelator in vivo and in vitro. *J Histochem Cytochem*, 2002, 50, 1659-1662.

[72] Guillemin, GJ; Cullen, KM; Lim, CK; et al. Characterization of the kynurenine pathway in human neurons. *J Neurosci*, 2007, 27, 12884-12892.

[73] Kaneko, S; Wang, J; Kaneko, M; et al. Protecting axonal degeneration by increasing nicotinamide adenine dinucleotide levels in experimental autoimmune encephalomyelitis models. *J Neurosci*, 2006, 26, 9794-9804.

[74] Schoenmann, Z; Assa-Kunik, E; Tiomny, S; et al. Axonal degeneration is regulated by the apoptotic machinery or a NAD+-sensitive pathway in insects and mammals. *J Neurosci*, 2010, 30, 6375-6386.

[75] Ha, HC; Snyder, SH. Poly(ADP-ribose) polymerase is a mediator of necrotic cell death by ATP depletion. *Proc Natl Acad Sci U S A*, 1999, 96, 13978-13982.

[76] Ying, W; Sevigny, MB; Chen, Y; et al. Poly(ADP-ribose) glycohydrolase mediates oxidative and excitotoxic neuronal death. *Proc Natl Acad Sci U S A*, 2001, 98, 12227-12232.

[77] Lee, HC. Physiological functions of cyclic ADP-ribose and NAADP as calcium messengers. *Annu Rev Pharmacol Toxicol*, 2001, 41, 317-345.

[78] Pearson, SJ; Reynolds, GP. Increased brain concentrations of a neurotoxin, 3-hydroxykynurenine, in Huntington's disease. *Neurosci Lett*, 1992, 144, 199-201.

[79] Bonda, DJ; Mailankot, M; Stone, JG; et al. Indoleamine 2,3-dioxygenase and 3-hydroxykynurenine modifications are found in the neuropathology of Alzheimer's disease. *Redox Rep*, 2010, 15, 161-168.

[80] Wichers, MC; Maes, M. The role of indoleamine 2,3-dioxygenase (IDO) in the pathophysiology of interferon-α-induced depression. *J Psychiatry Neurosci*, 2004, 29, 11-17.

[81] Myint, AM; Schwarz, MJ; Verkerk, R; et al. Reversal of imbalance between kynurenic acid and 3-hydroxykynurenine by antipsychotics in medication-naïve and medication-free schizophrenic patients. *Brain Behav Immun*, 2011, 25, 1576-1581.

[82] Kocki, T; Wnuk, S; Kloc, R; et al. New insight into the antidepressants action: modulation of kynurenine pathway by increasing the kynurenic acid/3-hydroxykynurenine ratio. *J Neural Transm*, 2012, 119, 235-243.

[83] Colín-González, AL; Maldonado, PD; Santamaría, A. 3-Hydroxykynurenine: an intriguing molecule exerting dual actions in the central nervous system. *Neurotoxicology*, 2013, 34, 189-204.

[84] Goldstein, LE; Leopold, MC; Huang, X; et al. 3-Hydroxykynurenine and 3-hydroxyanthranilic acid generate hydrogen peroxide and promote alpha-crystallin cross-linking by metal ion reduction. *Biochemistry*, 2000, 39, 7266-7275.

[85] Weber, WP; Feder-Mengus, C; Chiarugi, A; et al. Differential effects of the tryptophan metabolite 3-hydroxyanthranilic acid on the proliferation of human CD8+ T cells

induced by TCR triggering or homeostatic cytokines. *Eur J Immunol*, 2006, 36, 296-304.

[86] Platten, M; Ho, PP; Youssef, S; et al. Treatment of autoimmune neuroinflammation with a synthetic tryptophan metabolite. *Science*, 2005, 310, 850-855.

[87] Seruga, B; Zhang, H; Bernstein, LJ; et al. Cytokines and their relationship to the symptoms and outcome of cancer. *Nat Rev Cancer*, 2008, 8, 887-899.

[88] Guillemin, GJ; Brew, BJ; Noonan, CE; et al. Indoleamine 2,3 dioxygenase and quinolinic acid immunoreactivity in Alzheimer's disease hippocampus. *Neuropathol Appl Neurobiol*, 2005, 31, 395-404.

[89] Guillemin, GJ; Wang, L; Brew, BJ. Quinolinic acid selectively induces apoptosis of human astrocytes: potential role in AIDS dementia complex. *J Neuroinflammation*, 2005, 2, 16.

[90] Ting, KK; Brew, BJ; Guillemin, GJ. Effect of quinolinic acid on human astrocytes morphology and functions: implications in Alzheimer's disease. *J Neuroinflammation*, 2009; 6: 36.

[91] Mantovani, A; Sozzani, S; Locati, M; et al. Macrophage polarization: tumor-associated macrophages as a paradigm for polarized M2 mononuclear phagocytes. *Trends Immunol*, 2002, 23, 549-555.

[92] Dhir, A; Kulkarni, SK. Nitric oxide and major depression. *Nitric Oxide*, 2011, 24, 125-131.

[93] Volke, V; Wegener, G; Bourin, M; et al. Antidepressant- and anxiolytic-like effects of selective neuronal NOS inhibitor 1-(2-trifluoromethylphenyl)-imidazole in mice. *Behav Brain Res*, 2003, 140, 141-147.

[94] Ergün, Y; Ergün, UG; Orhan, FO; et al. Co-administration of a nitric oxide synthase inhibitor and melatonin exerts an additive antidepressant-like effect in the mouse forced swim test. *Med Sci Monit*, 2006, 12, BR307-312.

[95] Joca, SR; Guimarães, FS. Inhibition of neuronal nitric oxide synthase in the rat hippocampus induces antidepressant-like effects. *Psychopharmacology (Berl)*, 2006, 185, 298-305.

[96] Ju, ST; Panka, DJ; Cui, H; et al. Fas(CD95)/FasL interactions required for programmed cell death after T-cell activation. *Nature*, 1995, 373, 444-448.

[97] Mezrich, JD; Fechner, JH; Zhang, X; et al. An interaction between kynurenine and the aryl hydrocarbon receptor can generate regulatory T cells. *J Immunol*, 2010, 185, 3190-3198.

[98] Nguyen, NT; Kimura, A; Nakahama, T; et al. Aryl hydrocarbon receptor negatively regulates dendritic cell immunogenicity via a kynurenine-dependent mechanism. *Proc Natl Acad Sci U S A*, 2010, 107, 19961-19966.

[99] Opitz, CA; Litzenburger, UM; Sahm, F; et al. An endogenous tumour-promoting ligand of the human aryl hydrocarbon receptor. *Nature*, 2011, 478, 197-203.

[100] Chen, WJ. IDO: more than an enzyme. *Nat Immuno*, 2011, 12, 809-811.

[101] Mellor, AL; Munn, DH. IDO expression by dendritic cells: tolerance and tryptophan catabolism. *Nat Rev Immunol*, 2004, 4, 762-774.

[102] Litzenburger, UM; Opitz, CA; Sahm, F; et al. Constitutive IDO expression in human cancer is sustained by an autocrine signaling loop involving IL-6, STAT3 and the AHR. *Oncotarget*, 2014, 5, 1038-1051.

[103] Chang, YF; Chuang, HY; Hsu, CH; et al. Immunomodulation of curcumin on adoptive therapy with T cell functional imaging in mice. *Cancer Prev Res (Phila)*, 2012, 5, 444-452.

[104] Chuang, HY; Chang, YF; Liu, RS; et al. Serial low doses of sorafenib enhance therapeutic efficacy of adoptive T cell therapy in a murine model by improving tumor microenvironment. *PLoS One*, 2014, 9, e109992.

[105] Liu, G; Ma, H; Qiu, L; et al. Phenotypic and functional switch of macrophages induced by regulatory CD4+CD25+ T cells in mice. *Immunol Cell Biol*, 2011, 89, 130-142.

[106] Hao, NB; Lü, MH; Fan, YH; et al. Macrophages in tumor microenvironments and the progression of tumors. *Clin Dev Immunol*, 2012, 2012, 948098.

[107] Sica, A; Bronte, V. Altered macrophage differentiation and immune dysfunction in tumor development. *J Clin Invest*, 2007, 117, 1155-1166.

[108] Koblish, HK; Hansbury, MJ; Bowman, KJ; et al. Hydroxyamidine inhibitors of indoleamine-2,3-dioxygenase potently suppress systemic tryptophan catabolism and the growth of IDO-expressing tumors. *Mol Cancer Ther*, 2010, 9, 489-498.

[109] Liu, X; Shin, N; Koblish, HK; et al. Selective inhibition of IDO1 effectively regulates mediators of antitumor immunity. *Blood*, 2010, 115, 3520-3530.

[110] Li, M; Bolduc, AR; Hoda, MN; et al. The indoleamine 2,3-dioxygenase pathway ontrols complement-dependent enhancement of chemo-radiation therapy against murine glioblastoma. *J Immunother Cancer*, 2014, 2, 21.

[111] Nayak, A; Hao, Z; Sadek, R; et al. A Phase I study of NLG919 for adult patients with recurrent advanced solid tumors. *J Immunother Cancer*, 2014, 2(Suppl 3), P250.

[112] Kennedy, E; Rossi, GR; Vahanian, NN; et al. Phase 1/2 trial of the indoleamine 2,3-dioxygenase pathway (IDO) inhibitor indoximod plus ipilimumab for the treatment of unresectable stage 3 or 4 melanoma. *J Clin Oncol*, 2014, 32(15 Suppl).

[113] Capuron, L; Ravaud, A; Neveu, PJ; et al. Association between decreased serum tryptophan concentrations and depressive symptoms in cancer patients undergoing cytokine therapy. *Mol Psychiatry*, 2002, 7, 468-473.

[114] Bartsch, C; Bartsch, H; Bellmann, O; et al. Depression of serum melatonin in patients with primary breast cancer is not due to an increased peripheral metabolism. *Cancer*, 1991, 67, 1681-1684.

[115] Burtin, C; Scheinmann, P; Salomon, JC; et al. Decrease in tumour growth by injections of histamine or serotonin in fibrosarcoma-bearing mice: influence of H1 and H2 histamine receptors. *Br J Cancer*, 1982, 45, 54-60.

[116] Manda, T; Nishigaki, F; Mori, J; et al. Important role of serotonin in the antitumor effects of recombinant human tumor necrosis factor-alpha in mice. *Cancer Res*, 1988, 48, 4250-4255.

[117] Nocito, A; Dahm, F; Jochum, W; et al. Serotonin regulates macrophage-mediated angiogenesis in a mouse model of colon cancer allografts. *Cancer Res*, 2008, 68, 5152-5158.

[118] Alpini, G; Invernizzi, P; Gaudio, E; et al. Serotonin metabolism is dysregulated in cholangiocarcinoma, which has implications for tumor growth. *Cancer Res*, 2008, 68, 9184-9193.

[119] Asada, M; Ebihara, S; Yamanda, S; et al. Depletion of serotonin and selective inhibition of 2B receptor suppressed tumor angiogenesis by inhibiting endothelial nitric oxide

synthase and extracellular signal-regulated kinase 1/2 phosphorylation. *Neoplasia*, 2009, 11, 408-417.

[120] Gurbuz, N; Ashour, AA; Alpay, SN; et al. Down-regulation of 5-HT1B and 5-HT1D receptors inhibits proliferation, clonogenicity and invasion of human pancreatic cancer cells. *PLoS One*, 2014, 9, e105245.

[121] Ahern, GP. 5-HT and the immune system. *Curr Opin Pharmacol.*, 2011, 11, 29-33.

[122] O'Connell, PJ; Wang, X; Leon-Ponte, M; et al. A novel form of immune signaling revealed by transmission of the inflammatory mediator serotonin between dendritic cells and T cells. *Blood*, 2006, 107, 1010-1017.

[123] Meredith, EJ; Holder, MJ; Chamba, A; et al. The serotonin transporter (SLC6A4) is present in B-cell clones of diverse malignant origin: probing a potential anti-tumor target for psychotropics. *FASEB J*, 2005, 19, 1187-1189.

[124] León-Ponte, M; Ahern, GP; O'Connell, PJ. Serotonin provides an accessory signal to enhance T-cell activation by signaling through the 5-HT7 receptor. *Blood*, 2007, 109, 3139-3146.

[125] Branco-de-Almeida, LS; Kajiya, M; Cardoso, CR; et al. Selective serotonin reuptake inhibitors attenuate the antigen presentation from dendritic cells to effector T lymphocytes. *FEMS Immunol Med Microbiol*, 2011, 62, 283-294.

[126] Li, N; Ghia, JE; Wang, H; et al. Serotonin activates dendritic cell function in the context of gut inflammation. *Am J Pathol*, 2011, 178, 662-671.

[127] Katoh, N; Soga, F; Nara, T; et al. Effect of serotonin on the differentiation of human monocytes into dendritic cells. *Clin Exp Immunol*, 2006, 146, 354-361.

[128] Müller, T; Dürk, T; Blumenthal, B; et al. 5-hydroxytryptamine modulates migration, cytokine and chemokine release and T-cell priming capacity of dendritic cells in vitro and in vivo. *PLoS One*, 2009, 4, e6453.

[129] Cowing, C; Pincus, SH; Sachs, DH; et al. A subpopulation of adherent accessory cells bearing both I-A and I-E or C subregion antigens is required for antigen-specific murine T lymphocyte proliferation. *J Immunol* 1978, 121, 1680-1686.

[130] Sternberg, EM; Trial, J; Parker, CW. Effect of serotonin on murine macrophages: suppression of Ia expression by serotonin and its reversal by 5-HT2 serotonergic receptor antagonists. *J Immunol*, 1986, 137, 276-282.

[131] Sternberg, EM; Wedner, HJ; Leung, MK; et al. Effect of serotonin (5-HT) and other monoamines on murine macrophages: modulation of interferon-gamma induced phagocytosis. *J Immunol*, 1987, 138, 4360-4365.

[132] de, las, Casas-Engel, M; Domínguez-Soto, A; Sierra-Filardi, E; et al. Serotonin skews human macrophage polarization through HTR2B and HTR7. *J Immunol*, 2013, 190, 2301-2310.

[133] Aune, TM; Golden, HW; McGrath, KM. Inhibitors of serotonin synthesis and antagonists of serotonin 1A receptors inhibit T lymphocyte function in vitro and cell-mediated immunity in vivo. *J Immunol*, 1994, 153, 489-498.

[134] Young, MR; Matthews, JP. Serotonin regulation of T-cell subpopulations and of macrophage accessory function. *Immunology*, 1995, 84, 148-152.

[135] Nowak, EC; de, Vries, VC; Wasiuk, A; et al. Tryptophan hydroxylase-1 regulates immune tolerance and inflammation. *J Exp Med*, 2012, 209, 2127-2135.

[136] Laberge, S; Cruikshank, WW; Beer, DJ; et al. Secretion of IL-16 (lymphocyte chemoattractant factor) from serotonin-stimulated CD8+ T cells in vitro. *J Immunol*, 1996, 156, 310-315.
[137] Compérat, E; Rouprêt, M; Drouin, SJ; et al. Tissue expression of IL16 in prostate cancer and its association with recurrence after radical prostatectomy. *Prostate*, 2010, 70, 1622-1627.
[138] Kovacs, E. The serum levels of IL-12 and IL-16 in cancer patients. Relation to the tumour stage and previous therapy. *Biomed Pharmacother*, 2001, 55, 111-116.
[139] Reiter, RJ; Tan, DX; Manchester, LC; et al. Biochemical reactivity of melatonin with reactive oxygen and nitrogen species: a review of the evidence. *Cell Biochem Biophys*, 2001, 34, 237-256.
[140] Hess, ML; Manson, NH. Molecular oxygen: friend and foe. The role of the oxygen free radical system in the calcium paradox, the oxygen paradox and ischemia/ reperfusion injury. *J Mol Cell Cardiol*, 1984, 16, 969-985.
[141] Cao, W; Carney, JM; Duchon, A; et al. Oxygen free radical involvement in ischemia and reperfusion injury to brain. *Neurosci Lett*, 1988, 88, 233-238.
[142] Lucchesi, BR. Myocardial ischemia, reperfusion and free radical injury. *Am J Cardiol*, 1990, 65, 14I-23I.
[143] Nauta, RJ; Tsimoyiannis, E; Uribe, M; et al. Oxygen-derived free radicals in hepatic ischemia and reperfusion injury in the rat. *Surg Gynecol Obstet*, 1990, 171, 120-125.
[144] Tan, DX; Manchester, LC; Reiter, RJ; et al. A novel melatonin metabolite, cyclic 3-hydroxymelatonin: a biomarker of in vivo hydroxyl radical generation. *Biochem Biophys Res Commun*, 1998, 253, 614-620.
[145] Tan, DX; Hardeland, R; Manchester, LC; et al. Cyclic-3-hydroxymelatonin (C3HOM), a potent antioxidant, scavenges free radicals and suppresses oxidative reactions. *Curr Med Chem*, 2014, 21, 1557-1565.
[146] Pieri, C; Moroni, F; Marra, M; et al. Melatonin is an efficient antioxidant. *Arch Gerontol Geriatr*, 1995, 20, 159-165.
[147] Morishima, I; Matsui, H; Mukawa, H; et al. Melatonin, a pineal hormone with antioxidant property, protects against adriamycin cardiomyopathy in rats. *Life Sci*, 1998, 63, 511-521.
[148] Parlakpinar, H; Sahna, E; Ozer, MK; et al. Physiological and pharmacological concentrations of melatonin protect against cisplatin-induced acute renal injury. *J Pineal Res*, 2002, 33, 161-166.
[149] Catalá, A; Zvara, A; Puskás, LG; et al. Melatonin-induced gene expression changes and its preventive effects on adriamycin-induced lipid peroxidation in rat liver. *J Pineal Res*, 2007, 42, 43-49.
[150] Dong, WG; Mei, Q; Yu, JP; et al. Effects of melatonin on the expression of iNOS and COX-2 in rat models of colitis. *World J Gastroenterol*, 2003, 9, 1307-1311.
[151] Kilic, E; Kilic, U; Reiter, RJ; et al. Tissue-plasminogen activator-induced ischemic brain injury is reversed by melatonin: role of iNOS and Akt. *J Pineal Res*, 2005, 39, 151-155.
[152] Ortiz, F; García, JA; Acuña-Castroviejo, D; et al. The beneficial effects of melatonin against heart mitochondrial impairment during sepsis: inhibition of iNOS and preservation of nNOS. *J Pineal Res*, 2014, 56, 71-81.

[153] Carrillo-Vico, A; Lardone, PJ; Alvarez-Sánchez, N; et al. Melatonin: buffering the immune system. *Int J Mol Sci*, 2013, 14, 8638-8683.

[154] Fabris, N; Mocchegiani, E; Provinciali, M. Plasticity of neuroendocrine-thymus interactions during aging. *Exp Gerontol*, 1997, 32, 415-429.

[155] Pawelec, G; Akbar, A; Caruso, C; et al. Is immunosenescence infectious? *Trends Immunol*, 2004, 25, 406-410.

[156] Hakim, FT; Flomerfelt, FA; Boyiadzis, M; et al. Aging, immunity and cancer. *Curr Opin Immunol*, 2004, 16, 151-156.

[157] Majewska, M; Zajac, K; Zemelka, M; et al. Influence of melatonin and its precursor L-tryptophan on Th1 dependent contact hypersensitivity. *J Physiol Pharmacol*, 2007, 58 Suppl 6, 125-132.

[158] Mei, Q; Yu, JP; Xu, JM; et al. Melatonin reduces colon immunological injury in rats by regulating activity of macrophages. *Acta Pharmacol Sin*, 2002, 23, 882-886.

[159] Lin, GJ; Huang, SH; Chen, YW; et al. Melatonin prolongs islet graft survival in diabetic NOD mice. *J Pineal Res*, 2009, 47, 284-292.

[160] Hemadi, M; Abolhassani, F; Akbari, M; et al. Melatonin promotes the cumulus-oocyte complexes quality of vitrified-thawed murine ovaries; with increased mean number of follicles survival and ovary size following heterotopic transplantation. *Eur J Pharmacol*, 2009, 618, 84-90.

[161] Inserra, P; Zhang, Z; Ardestani, SK; et al. Modulation of cytokine production by dehydroepiandrosterone (DHEA) plus melatonin (MLT) supplementation of old mice. *Proc Soc Exp Biol Med*, 1998, 218, 76-82.

[162] Garcia-Mauriño, S; Gonzalez-Haba, MG; Calvo, JR; et al. Melatonin enhances IL-2, IL-6, and IFN-gamma production by human circulating CD4+ cells: a possible nuclear receptor-mediated mechanism involving T helper type 1 lymphocytes and monocytes. *J Immunol*, 1997, 159, 574-581.

[163] Garcia-Mauriño, S; Gonzalez-Haba, MG; Calvo, JR; et al. Involvement of nuclear binding sites for melatonin in the regulation of IL-2 and IL-6 production by human blood mononuclear cells. *J Neuroimmunol*, 1998, 92, 76-84.

[164] Wink, DA; Hines, HB; Cheng, RY; et al. Nitric oxide and redox mechanisms in the immune response. *J Leukoc Biol*, 2011, 89, 873-891.

[165] Nagaraj, S; Gupta, K; Pisarev, V; et al. Altered recognition of antigen is a novel mechanism of CD8+ T cell tolerance in cancer. *Nat Med*, 2007, 13, 828-835.

[166] Molon, B; Ugel, S; Del, Pozzo, F; et al. Chemokine nitration prevents intratumoral infiltration of antigen-specific T cells. *J Exp Med*, 2011, 208, 1949-1962.

[167] Pioli, C; Caroleo, MC; Nistico, G; et al. Melatonin increases antigen presentation and amplifies specific and non specific signals for T-cell proliferation. *Int J Immunopharmacol*, 1993, 15, 463-468.

[168] Gilad, E; Wong, HR; Zingarelli, B; et al. Melatonin inhibits expression of the inducible isoform of nitric oxide synthase in murine macrophages: role of inhibition of NFkappaB activation. *FASEB J*, 1998, 12, 685-693.

[169] Deng, WG; Tang, ST; Tseng, HP; et al. Melatonin suppresses macrophage cyclooxygenase-2 and inducible nitric oxide synthase expression by inhibiting p52 acetylation and binding. *Blood*, 2006, 108, 518-524.

[170] Dale, DC; Boxer, L; Liles, WC. The phagocytes: neutrophils and monocytes. *Blood*, 2008, 112, 935-945.

[171] Conti, A; Conconi, S; Hertens, E; et al. Evidence for melatonin synthesis in mouse and human bone marrow cells. *J Pineal Res*, 2000, 28, 193-202.
[172] Pawlak, J; Singh, J; Lea, RW; et al. Effect of melatonin on phagocytic activity and intracellular free calcium concentration in testicular macrophages from normal and streptozotocin-induced diabetic rats. *Mol Cell Biochem*, 2005, 275, 207-213.
[173] Cutolo, M; Villaggio, B; Candido, F; et al. Melatonin influences interleukin-12 and nitric oxide production by primary cultures of rheumatoid synovial macrophages and THP-1 cells. *Ann N Y Acad Sci*, 1999, 876, 246-254.
[174] Lissoni, P; Tisi, E; Barni, S; et al. Biological and clinical results of a neuroimmunotherapy with interleukin-2 and the pineal hormone melatonin as a first line treatment in advanced non-small cell lung cancer. *Br J Cancer*, 1992, 66, 155-158.
[175] Miller, SC; Pandi-Perumal, SR; Esquifino, AI; et al. The role of melatonin in immuno-enhancement: potential application in cancer. *Int J Exp Pathol*, 2006, 87, 81-87.
[176] Giordano, M; Palermo, MS. Melatonin-induced enhancement of antibody-dependent cellular cytotoxicity. *J Pineal Res*, 1991, 10, 117-121.
[177] Liu, H; Xu, L; Wei, JE; et al. Role of CD4+ CD25+ regulatory T cells in melatonin-mediated inhibition of murine gastric cancer cell growth in vivo and in vitro. *Anat Rec (Hoboken)*, 2011, 294, 781-788.
[178] Seely, D; Wu, P; Fritz, H; et al. Melatonin as adjuvant cancer care with and without chemotherapy: a systematic review and meta-analysis of randomized trials. *Integr Cancer Ther*, 2012, 11, 293-303.
[179] Wang, YM; Jin, BZ; Ai, F; et al. The efficacy and safety of melatonin in concurrent chemotherapy or radiotherapy for solid tumors: a meta-analysis of randomized controlled trials. *Cancer Chemother Pharmacol*, 2012, 69, 1213-1220.
[180] Vijayalaxmi; Reiter, RJ; Sewerynek, E; et al. Marked reduction of radiation-induced micronuclei in human *blood* lymphocytes pretreated with melatonin. *Radiat Res*, 1995, 143, 102-106.
[181] Vijayalaxmi; Meltz, ML; Reiter, RJ; et al. Melatonin and protection from whole-body irradiation: survival studies in mice. *Mutat Res*, 1999, 425, 21-27.
[182] Vijayalaxmi; Meltz, ML; Reiter, RJ; et al. Melatonin and protection from genetic damage in *blood* and bone marrow: whole-body irradiation studies in mice. *J Pineal Res*, 1999, 27, 221-225.
[183] Jang, SS; Kim, HG; Lee, JS; et al. Melatonin reduces X-ray radiation-induced lung injury in mice by modulating oxidative stress and cytokine expression. *Int J Radiat Biol*, 2013, 89, 97-105.
[184] Ortiz, F; Acuña-Castroviejo, D; Doerrier, C; et al. Melatonin blunts the mitochondrial/NLRP3 connection and protects against radiation-induced oral mucositis. *J Pineal Res*, 2014 doi:10.1111/jpi.12191.
[185] Lissoni, P; Meregalli, S; Nosetto, L; et al. Increased survival time in brain glioblastomas by a radioneuroendocrine strategy with radiotherapy plus melatonin compared to radiotherapy alone. *Oncology*, 1996, 53, 43-46.
[186] Shirazi, A; Ghobadi, G; Ghazi-Khansari, M. A radiobiological review on melatonin: a novel radioprotector. *J Radiat Res*, 2007, 48, 263-272.
[187] Mihandoost, E; Shirazi, A; Mahdavi, SR; et al. Can Melatonin Help Us in Radiation Oncology Treatments? *Biomed Res Int*, 2014, 2014, 578137.

[188] Lissoni, P; Paolorossi, F; Ardizzoia, A; et al. A randomized study of chemotherapy with cisplatin plus etoposide versus chemoendocrine therapy with cisplatin, etoposide and the pineal hormone melatonin as a first-line treatment of advanced non-small cell lung cancer patients in a poor clinical state. *J Pineal Res*, 1997, 23, 15-19.
[189] Sanchez-Barcelo, EJ; Mediavilla, MD; Alonso-Gonzalez, C; et al. Melatonin uses in *oncology*: breast cancer prevention and reduction of the side effects of chemotherapy and radiation. *Expert Opin Investig Drugs*, 2012, 21, 819-831.
[190] Xu, C; Wu, A; Zhu, H; et al. Melatonin is involved in the apoptosis and necrosis of pancreatic cancer cell line SW-1990 via modulating of Bcl-2/Bax balance. *Biomed Pharmacother*, 2013, 67, 133-139.
[191] Jung-Hynes, B; Schmit, TL; Reagan-Shaw, SR; et al. Melatonin, a novel Sirt1 inhibitor, imparts antiproliferative effects against prostate cancer in vitro in culture and in vivo in TRAMP model. *J Pineal Res*, 2011, 50, 140-149.
[192] Proietti, S; Cucina, A; Dobrowolny, G; et al. Melatonin down-regulates MDM2 gene expression and enhances p53 acetylation in MCF-7 cells. *J Pineal Res*, 2014, 57, 120-129.
[193] Jardim-Perassi, BV; Arbab, AS; Ferreira, LC; et al. Effect of melatonin on tumor growth and angiogenesis in xenograft model of breast cancer. *PLoS One*, 2014, 9, e85311.
[194] Carbajo-Pescador, S; Ordoñez, R; Benet, M; et al. Inhibition of VEGF expression through blockade of Hif1α and STAT3 signalling mediates the anti-angiogenic effect of melatonin in HepG2 liver cancer cells. *Br J Cancer*, 2013, 109, 83-91.
[195] Plaimee, P; Weerapreeyakul, N; Thumanu, K; et al. Melatonin induces apoptosis through biomolecular changes, in SK-LU-1 human lung adenocarcinoma cells. *Cell Prolif*, 2014, 47, 564-577.
[196] Lee, YJ; Lee, JH2; Moon, JH; et al. Overcoming hypoxic-resistance of tumor cells to TRAIL-induced apoptosis through melatonin. *Int J Mol Sci*, 2014, 15, 11941-11956.
[197] Goncalves, Ndo N; Rodrigues, RV; Jardim-Perassi, BV; et al. Molecular markers of angiogenesis and metastasis in lines of oral carcinoma after treatment with melatonin. *Anticancer Agents Med Chem*, 2014, 14, 1302-1311.
[198] Ordoñez, R; Carbajo-Pescador, S; Prieto-Dominguez, N; et al. Inhibition of matrix metalloproteinase-9 and nuclear factor kappa B contribute to melatonin prevention of motility and invasiveness in HepG2 liver cancer cells. *J Pineal Res*, 2014, 56, 20-30.
[199] Martín, V; Sanchez-Sanchez, AM; Puente-Moncada; N; et al. Involvement of autophagy in melatonin-induced cytotoxicity in glioma-initiating cells. *J Pineal Res*, 2014, 57, 308-316.
[200] Stewart, VC; Heales, SJ. Nitric oxide-induced mitochondrial dysfunction: implications for neurodegeneration. *Free Radic Biol Med*, 2003, 34, 287-303.
[201] Nguyen, T; Brunson, D; Crespi, CL; et al. DNA damage and mutation in human cells exposed to nitric oxide in vitro. *Proc Natl Acad Sci U S A*, 1992, 89, 3030-3034.
[202] Radi, R; Beckman, JS; Bush, KM; et al. Peroxynitrite-induced membrane lipid peroxidation: the cytotoxic potential of superoxide and nitric oxide. *Arch Biochem Biophys*, 1991, 288, 481-487.
[203] Stamler, JS; Simon, DI; Osborne, JA; et al. S-nitrosylation of proteins with nitric oxide: synthesis and characterization of biologically active compounds. *Proc Natl Acad Sci U S A*, 1992, 89, 444-448.

[204] Irie, M; Miyata, M; Kasai, H. Depression and possible cancer risk due to oxidative DNA damage. *J Psychiatr Res*, 2005, 39, 553-560.

[205] Bilici, M; Efe, H; Köroğlu, MA; et al. Antioxidative enzyme activities and lipid peroxidation in major depression: alterations by antidepressant treatments. *J Affect Disord*, 2001, 64, 43-51.

[206] Nakamura, T; Lipton, SA. Redox modulation by S-nitrosylation contributes to protein misfolding, mitochondrial dynamics, and neuronal synaptic damage in neurodegenerative diseases. *Cell Death Differ*, 2011, 18, 1478-1486.

[207] Reich, M; Lesur, A; Perdrizet-Chevallier, C. Depression, quality of life and breast cancer: a review of the literature. *Breast Cancer Res Treat*, 2008, 110, 9-17.

[208] Jeon, SH; Kim, SH; Kim, Y; et al. The tricyclic antidepressant imipramine induces autophagic cell death in U-87MG glioma cells. *Biochem Biophys Res Commun*, 2011, 413, 311-317.

[209] Pula, G; Pistilli, A; Montagnoli, C; et al. The tricyclic antidepressant amitriptyline is cytotoxic to HTB114 human leiomyosarcoma and induces p75(NTR)-dependent apoptosis. *Anticancer Drugs*, 2013, 24, 899-910.

[210] Mao, X; Hou, T; Cao, B; et al. The tricyclic antidepressant amitriptyline inhibits D-cyclin transactivation and induces myeloma cell apoptosis by inhibiting histone deacetylases: in vitro and in silico evidence. *Mol Pharmacol*, 2011, 79, 672-680.

[211] Jahchan, NS; Dudley, JT; Mazur, PK; et al. A drug repositioning approach identifies tricyclic antidepressants as inhibitors of small cell lung cancer and other neuroendocrine tumors. *Cancer Discov*, 2013, 3, 1364-1377.

[212] Lin, KL; Chou, CT; Cheng, JS; et al. Effect of fluoxetine on [Ca^{2+}] and cell viability in OC2 human oral cancer cells. *Chin J Physiol*, 2014, 57, 256-264.

[213] Stopper, H; Garcia, SB; Waaga-Gasser, AM; et al. Antidepressant fluoxetine and its potential against colon tumors. *World J Gastrointest Oncol*, 2014, 6, 11-21.

[214] Fang, CK; Chen, HW; Chiang, IT; et al. Mirtazapine inhibits tumor growth via immune response and serotonergic system. *PLoS One.*, 2012, 7(7), e38886.

[215] Uzawa, K; Kasamatsu, A2; Shimizu, T; et al. Suppression of metastasis by mirtazapine via restoration of the Lin-7C/β-catenin pathway in human cancer cells. *Sci Rep*, 2014, 4, 5433.

[216] Kubera, M; Grygier, B; Wrona, D; et al. Stimulatory effect of antidepressant drug pretreatment on progression of B16F10 melanoma in high-active male and female C57BL/6J mice. *J Neuroimmunol*, 2011, 240-241, 34-44.

[217] Szpunar, MJ; Burke, KA; Dawes, RP; et al. The antidepressant desipramine and α2-adrenergic receptor activation promote breast tumor progression in association with altered collagen structure. *Cancer Prev Res (Phila)*, 2013, 6, 1262-1272.

In: Indoleamines
Editor: Ángel Catalá

ISBN: 978-1-63482-097-4

Chapter 4

THE TRYPTOPHAN "DAMAGE" SIGNAL AND INDOLEAMINE 2, 3-DIOXYGENASE

J. W. Earl[1, 2*]***, C. E. Nath***[1]***, M. G. Hanlon***[3]***, W. C. Leach***[4]***, R. C. Dale***[2, 5]***, G. J. Logan***[6]***, I. E. Alexander***[6]*** and G. A. Bishop***[7]

[1]Biochemistry Department, Children's Hospital Westmead, Westmead, NSW, Australia
[2]Department of Paediatrics and Child Health, University of Sydney, Sydney, NSW, Australia
[3]Department of Neurosurgery, Royal Prince Alfred Hospital, Sydney, NSW, Australia
[4]Microbiology Department, Children's Hospital Westmead, Westmead, NSW, Australia
[5]Neurology Department, Children's Hospital Westmead, Westmead, NSW, Australia
[6]Gene Therapy Research Unit, Children's Medical Research Institute, Westmead, NSW, Australia
[7]Transplant Laboratory, Blackburn Building, University of Sydney, Sydney, NSW, Australia

ABSTRACT

Tryptophan (TRP) has been linked to the immune response mainly through the immunosuppressive activities of indoleamine 2, 3-dioxygenase (IDO). Here we show evidence that TRP acts as a biochemical signal from damaged cells that attracts and enables immune responses. TRP levels above 2-3μmol/L are essential for T-Lymphocyte proliferation, but it has never been previously established whether normal physiological levels of TRP are sufficiently low to limit T-Lymphocyte proliferation. Although blood plasma TRP in humans varies from 30 to 65μmol/L, most of this TRP is bound to protein so free-TRP levels are much lower. Interstitial fluid levels of drugs and metabolites are generally regarded as being similar to plasma free-levels. Human cerebrospinal fluid (CSF) TRP levels vary from 0.9 to 3.3μmol/L in non-inflammatory conditions, but albumin is also present at levels of 1.5 to 5.2μmol/L and able to bind TRP in a 1 : 1 ratio. Using ultrafiltration, we measured free TRP levels in CSF from 10 non-inflammatory patients and found these were 22% lower than total TRP in these same patients. This

* Corresponding author: A/Prof John Earl, Biochemistry Department, The Children's Hospital at Westmead, Locked Bag 4001 Westmead NSW 2145 Australia, john.earl@sydney.edu.au.

indicated a normal reference interval of 0.5 to 2.9μmol/L for CSF free-TRP which is either below or bordering upon the threshold for T-Lymphocyte proliferation previously determined in tissue culture medium. We also measured free TRP in 4 patients with central nervous system inflammatory disease and found free-TRP levels of 0.5+/-0.3 which were well below the threshold.

Activated T-lymphocytes in inflammatory conditions release γ-interferon, which induces IDO in a wide range of different cells including epithelial cells, fibroblasts, dendritic cells, neurones, astrocytes, microglia, trophoblasts and macrophages. IDO is able to lower TRP levels below the threshold for T-lymphocyte proliferation, thus acting as a TRP noise suppression system and feedback control mechanism. Here we show that neutrophils exhibit strong TRP chemotaxis, migrating towards higher TRP levels, which raises the possibility that other myeloid and lymphoid cells might also migrate in a TRP gradient. We observed that freeze-thaw damaged liver tissue slices or fresh slices incubated with *Staphylococcus aureus* or *Escherichia coli* produce very large quantities of TRP along with other amino acids. We propose that the TRP signal arises from damaged tissue, raising TRP levels well above the threshold for T-lymphocyte proliferation, that IDO dampens signal noise by removing excess TRP, highlighting inflammatory foci, maintaining a TRP gradient for inward chemotaxis of neutrophils and possibly other immune cells.

INTRODUCTION

Cognate immune responses depend on T cells, which have specific receptors for antigen. Recognition of foreign (non-self) antigen by clones of T cells bearing the receptors for that antigen leads to their expansion, resulting in elimination of the foreign antigen. This clonal selection theory was first proposed over fifty years ago by Burnet to explain the specific antibody response of B cells [1] and has since been extended to T cells. T cells drive the immune response to antigen by helping B cell and antibody responses as well as by becoming cytotoxic effector cells. For their initial stimulation, they must interact with specialised professional antigen-presenting cells (APCs) which provide co-stimulatory signals [2-4]. APCs have pattern recognition receptors for conserved molecular motifs of pathogens. Binding of a pathogen-derived molecular motif such as bacterial lipopolysaccharide to its receptor on an APC activates the APC to express co-stimulatory signals and enables it to be a more efficient stimulator of T cells [5]. In this way, the limited repertoire of pattern recognition receptors for pathogens on APCs allows their rapid but non-specific response that subsequently leads to efficient stimulation of the particular clones of T cells with specific receptors for the pathogen. This two-step process ensures that the resulting immune response is both rapid and specific.

There remain some apparent exceptions to the above process, including the lack of an immune response to foreign antigens such as commensal gut bacteria or paternal antigens on the developing foetus as well as the existence of auto-antibodies to a wide range of self-antigens [6]. To address these concerns Matzinger has suggested that "danger" signals arising from damaged cells can stimulate an immune response and that in the absence of these signals the response does not develop [6]. Consequently, it has been proposed that self-antigen can stimulate a response in the presence of danger signals while foreign antigen does not elicit a response in its absence. As it is difficult to precisely define what constitutes "danger", more recently the concept that "damage" signals arising from necrotic cells stimulate immunity has gained support [7,8]. A number of such damage signals have been described, including high mobility group protein 1 (HMGB1) [9] hydrophobicity [10] heat shock proteins, uric acid and cholesterol crystals [8]. Central to the damage model is that components of the innate immune system, including macrophages, dendritic cells, polymorphonuclear leucocytes, complement proteins and coagulation factors can respond to self-molecules that have been liberated as a result of cellular damage [11].

The key characteristics of a Biological Signal include the transmitter or source of the signal, the signal strength, signal to noise characteristics, the specificity of the signal, and the receiver and its transduction pathways which implement the signal's message. TRP has not been recognised in the past as a "damage" signal since previous research has been primarily focussed upon the enzyme Indoleamine 2, 3-Dioxygenase (IDO) which metabolises TRP to kynurenine [12-16], the depletion of TRP by IDO [17] and the role of the Kynurenine Pathway metabolites in immune regulation [18]. More recently the essential role of TRP as a specific signal controlling T-cell proliferation has emerged [19, 20]. The primary source of TRP in animals is dietary and the liver maintains TRP homeostasis by taking up TRP from blood, converting it into various proteins or metabolising it using the enzyme Tryptophan 2, 3-dioxygenase. Liver homeostasis generates a normal reference interval for blood total TRP (protein-bound and free) of 30 – 65μmol/L [21-24]. However immune signalling takes place across interstitial fluid, lymphatic fluid or cerebrospinal fluid so the level of total TRP in blood is not necessarily relevant to the immune signalling role of TRP, especially as the blood free tryptophan level is about one tenth this level (*vide infra*). In order to understand the role of TRP in the immune response we need to better understand the sources and signal strength of TRP in these extracellular fluids.

SOURCE OF THE TRYPTOPHAN SIGNAL

In preliminary analytical studies we measured TRP in homogenised commercial chicken liver and found 87 ± 18 mg/Kg TRP (n = 5) whereas in fresh homogenised rat liver we found much lower levels of 8 ± 4 mg/Kg TRP (n = 5). This suggested that tissue autolysis may be raising TRP levels in poorly preserved samples. To investigate this, fresh rat liver tissue slices were incubated at 35°C in saline and compared with freeze/thawed slices, monitoring the rate of TRP release into the medium. We found that over a period of ten hours at 35° C, TRP was released linearly over time from the freeze/thawed rat liver slices 40 times faster than for freshly prepared rat liver slices (Figure 1). TRP was released from freeze/thawed slices at a rate of 60 mg/Kg/hr and after ten hours the released TRP was equivalent to a tissue content of 600 mg/Kg. As the total TRP content of liver is around 2100 – 2300 mg/Kg [25] this represented more than one quarter of the total TRP content of the liver being released in ten hours. We found that 1, 10-phenanthroline, a metalloproteinase inhibitor, was very effective in blocking the production of TRP (Figure 2) which confirmed that the released TRP was being generated by catabolism of protein. These studies are consistent with the proposed role of TRP as a "damage" signal.

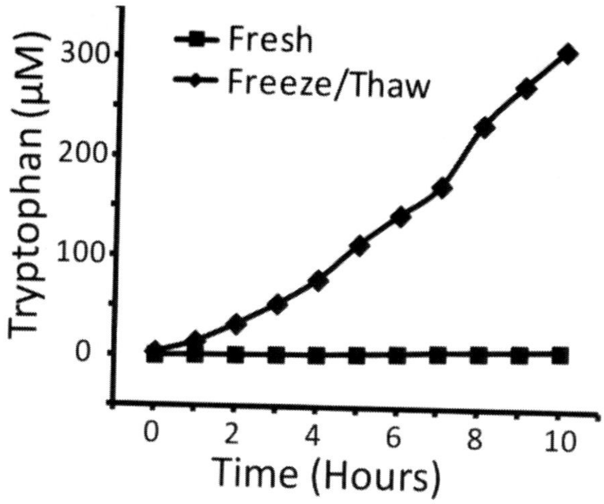

Figure 1. Fresh rat liver was harvested, stored on ice for 2 hours then cut into 0.6 g slices which were placed in sterile screw cap bottles with 4 ml normal saline in an incubator at 35°C. Each hour 20 μl of saline was removed, treated with 20 μl 0.65M perchloric acid and the supernatant frozen for analysis of TRP by HPLC with electrochemical detection [60]. Rat liver which had been previously harvested and stored frozen at -40°C was thawed and treated in an identical manner.

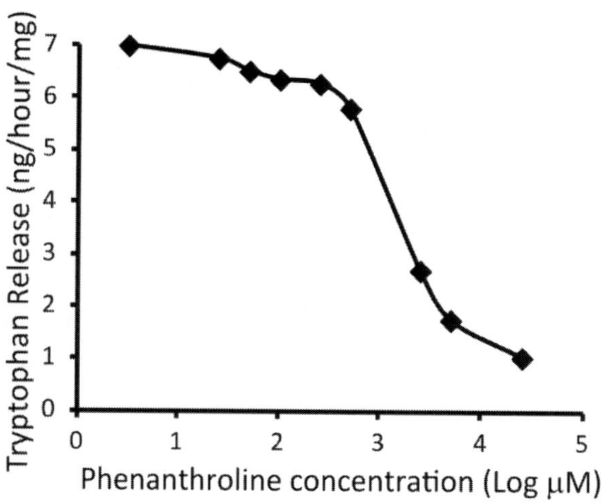

Figure 2. Varying amounts of 1, 10-phenanthroline in methanol were added to a number of glass vials and the methanol evaporated. Freshly prepared rat liver (2.8 g) was homogenised in 20 ml phosphate buffered saline and, 200 μl added to each vial, stoppered and incubated for 18 hours at 35°C. Samples (100 μl) were removed, treated with 100 μl acetonitrile, centrifuged and the supernatant frozen for analysis of TRP by HPLC with electrochemical detection [60].

To examine the effect of bacteria upon TRP release we incubated freshly prepared rat liver tissue with saline containing Escherichia Coli (E-Coli) or Staphylococcus Aureus (S-Aureus). Figure 3A shows the rate of TRP release into the media from fresh liver slices at 35° during the first ten hours. Initially TRP levels in the media rose slowly, then the rate of TRP release increased markedly between ten and 28 hours (Figure 3B). S-Aureus increased the overall rate of TRP release by two fold while E-Coli increased it by four fold compared with the controls. This again demonstrates that TRP has the characteristics of a "damage" signal, being released by the action of pathogens. Interestingly during the first 9 hours of incubation (Figure 3A) there was little difference between controls and S-Aureus treatment but the TRP level was lower for E-Coli suggesting E-Coli may have some ability to remove extracellular TRP and delay the appearance of the TRP "damage" signal.

THE STRENGTH OF THE TRP SIGNAL

TRP is a pro-inflammatory signal as T cell proliferation is prevented by TRP depletion. When varying amounts of TRP were added back into the depleted media containing co-cultured human monocytes and T cells it was shown that maximal T cell proliferation occurred above 2-3 μmol/L TRP and the half-maximal TRP concentration was 0.5 - 1.0 μmol/L [19]. Subsequently a higher T cell proliferation threshold of 10 μmol/L was determined for a monocyte and lymphocyte co-culture [20]. As low levels of TRP inhibit T-cell proliferation it is important to know what physiological levels of TRP are present in microenvironments where T-cell proliferation occurs.

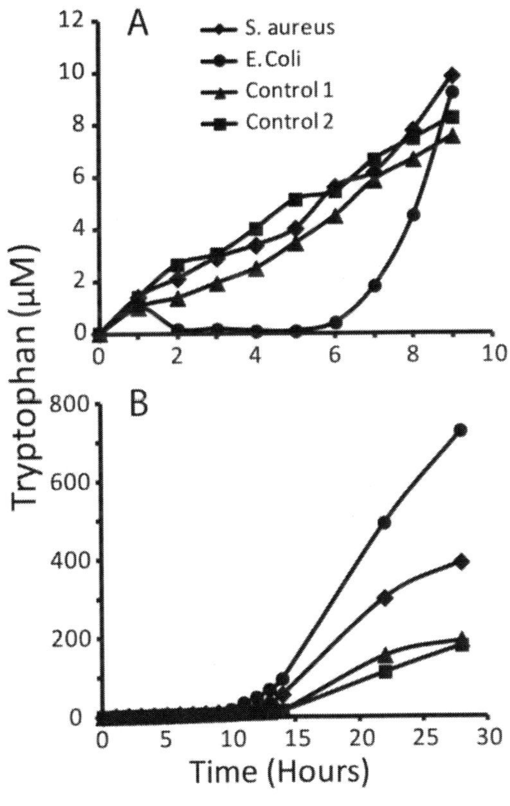

Figure 3. Freshly prepared rat liver slices were incubated with 4 ml saline or 4 ml saline spiked with 10^8 E-Coli or 10^8 S-Aureus for 28 hours at 35°C. Saline aliquots (20 μl) were removed at regular intervals, treated with 20 μl 0.65M perchloric acid and the supernatant frozen for analysis of TRP by HPLC with electrochemical detection [60]. A. Time course during the first ten hours. B. Time course up to 30 hours.

The reference interval for blood plasma TRP in humans is around 30 to 65 μmol/L [21, 22, 23, 24] however around 90% of this TRP is bound to protein whilst approximately 10% remains unbound as "free-TRP". Eynard et al. [26] found TRP levels of 61.5 ± 5.0 μmoles/L and free-TRP levels of 6.1 ± 0.7 μmoles/L in 7 healthy volunteers. Badawy et al. [27] found TRP levels of 49.9 ± 8.1 μmoles/L (10.19 ± 1.66 μg/ml) and free-TRP levels of 4.7 ± 0.8 μmoles/L (0.97 ± 0.17 μg/ml) in 18 healthy control patients. Free levels of drugs [28, 29], steroids [30] and Vitamin D [31] determine partitioning into cerebrospinal fluid, interstitial fluid or lymphatic fluid and govern efficacy and toxicity. This is also very likely to be true for TRP so free TRP levels and the levels of TRP in physiological fluids are fundamentally important to our understanding of the role of TRP in immune signalling.

Figure 4 shows the normal distribution curve for CSF TRP in 1087 children with neurological problems investigated during the period 1987 – 1997 using CSF amino acid analysis by HPLC with fluorescence detection [32]. We found 121 patients (11%) had CSF TRP below our normal reference interval lower limit of 1.4 μmol/L and 241 (21%) had TRP above the upper limit of 3.2 μmol/L. This illustrates that about half this group of children with a variety of neurological disease had CSF TRP below the 2-3 μmol/L threshold [19] for T cell proliferation. On the other hand some patients had very high levels of CSF TRP where T-cell proliferation might be enabled. These included children with bacterial meningitis and some babies with incomplete Blood Brain Barrier closure.

We previously reported a mean CSF TRP of 2.3 ± 0.5 μmol/L in 22 children who had a prior lumbar puncture for myelography but subsequently showed no evidence of intercurrent illness or neurological disease [32]. This corresponds to a normal reference interval of 1.4 - 3.2 μmol/L for CSF TRP in children based upon a 95% confidence interval. Great Ormond Street Hospital for Children uses a somewhat similar CSF TRP reference interval of 1 - 3 μmol/L. [33] This normal reference interval for CSF TRP is well below the reported threshold of 10 μmol/L by Kudo [20] for T cell proliferation and close to the reported threshold of 2-3μmol/L by Munn [19] indicating that T cell proliferation may be limited in CSF under normal physiological conditions. CSF contains 100 – 350 mg/L (1.5- 5.2 μmol/L) of albumin [34] which can bind TRP in a 1 : 1 ratio [35] so it is possible that protein binding of TRP may also occur in CSF. We measured the levels of CSF free-TRP in ten children. As CSF Neopterin is a reliable marker for Central Nervous System (CNS) inflammatory disease [36], we screened patients to ensure they had no evidence for CNS inflammation, selecting those with normal levels of Neopterin (16.0 ± 2.9) and normal neurotransmitter metabolites. CSF TRP in this group was 2.1 ± 0.3 μmol/L and free-TRP was found to be 1.7 ± 0.3 μmol/L (Table 1) which was on average 22% lower than total TRP. These results indicate a reference interval for CSF free-TRP in the absence of inflammatory disease of 0.5 - 2.9 μmol/L which indicates, even more strongly, that the level of free-TRP in CSF under normal conditions is unlikely to support T-Cell proliferation. Further work needs to be done to establish the TRP threshold in physiological fluids because there may be considerable differences between physiological fluids and tissue culture media.

We also measured CSF TRP and free-TRP in 4 patients with severe CNS inflammatory disease (CSF Neopterin: 909.0 ± 532.5 nmol/L) and found TRP levels of 0.8 ± 0.4 μmol/L and free-TRP levels of 0.5 ± 0.3 μmol/L. Interferon-γ (IFN-γ) from activated T-cells induces the production of GTP-cyclohydrolase-1 which produces Neopterin and IDO which metabolises TRP, accounting for the much lower TRP in this group. It is interesting that similar amounts of TRP were bound to protein in both the normal group (0.46 ± 0.13 µmol/L) and the inflammatory group (0.38 ± 0.09 µmol/L). This bound TRP may allow facilitative transport to tissue cells even when IDO levels are highly elevated. The dynamics of TRP diffusion, TRP protein binding and transport of TRP through interstitial fluids are important considerations for understanding immune regulation.

Table 1A.

	Age	Neopterin	Biopterin	5-HIAA	HVA	Total TRP	Free TRP	Bound TRP	% Binding
	1.5	16.3	30.5	0.15	0.45	2.53	2.01	0.52	21 %
	1.7	21.9	29.0	0.16	0.71	1.58	1.17	0.41	26 %
	2.4	17.4	27.7	0.18	0.55	2.61	2.04	0.57	22 %
	2.7	13.1	32.7	0.13	0.51	2.37	1.77	0.60	25 %
	3.0	14.7	29.8	0.16	0.58	2.26	1.96	0.30	13 %
	3.8	14.1	40.1	0.15	0.53	1.99	1.50	0.49	25 %
	4.0	11.9	25.9	0.17	0.60	2.15	1.50	0.65	30 %
	6.7	17.5	29.2	0.31	0.43	1.74	1.31	0.43	25 %
	8.0	18.3	26.5	0.12	0.47	1.85	1.47	0.38	21 %
	8.3	14.4	25.4	0.15	0.51	2.29	2.04	0.25	11 %
Mean	4.2	16.0	29.7	0.17	0.53	2.14	1.68	0.46	22 %
STD DEV	2.5	2.9	4.3	0.05	0.08	0.34	0.33	0.13	

Table 1B.

	Age	Neopterin	Biopterin	5-HIAA	HVA	Total TRP	Free TRP	Bound TRP	% Binding
	3.0	1335.0	132.6	0.06	0.35	0.79	0.49	0.30	38 %
	3.5	387.3	45.6	0.21	0.79	1.36	0.87	0.49	36 %
	7.5	1399.7	175.0	0.04	0.21	0.60	0.30	0.30	50 %
	8.0	513.9	29.6	0.12	0.43	0.60	0.19	0.41	68 %
Mean	5.5	909.0	95.7	0.11	0.45	0.84	0.46	0.38	48 %
STD DEV	2.6	532.5	65.6	0.08	0.25	0.36	0.30	0.09	

Samples from children collected and previously analysed for Amino Acids, Neurotransmitter and Pterin analysis were further analysed to develop a reference interval for free-TRP levels. A. Ten patients were selected with normal levels of Neopterin and normal neurotransmitter metabolites demonstrating no evidence for CNS inflammatory disease. B. CSF from four patients with severe CNS inflammatory disease, very high levels of CSF Neopterin and low TRP levels were analysed for free-TRP levels. CSF (0.2ml) was placed in the upper chamber of an Amicon Ultra (0.5ml) Centrifugal filter (10,000 Daltons Molecular Weight), sealed and centrifuged for 2 min at 12,000×g in an Eppendopf Minispin Centrifuge. Ultrafitrate in the lower centrifuge tube was frozen for later analysis of TRP by HPLC with electrochemical detection [60].

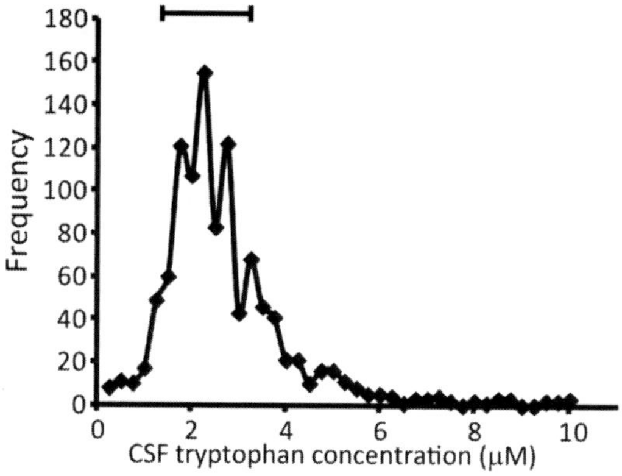

Figure 4. Normal distribution curve for CSF TRP in 1087 children with neurological problems investigated during the period 1987 – 1997 using CSF amino acid analysis by HPLC with fluorescence detection [31]. The reference interval for CSF TRP of 1.4 - 3.2 μmol/L derived from 22 children found after investigation to have a normal myelogram is shown by the bar.

We propose that T cell activation and TRP sufficiency act as a combination lock upon inflammatory processes. Consequently *in-vivo* free-TRP levels have a role in either preventing or supporting cognate immunity. The local concentration of TRP is dynamic, depending upon various physiological processes which provide or remove TRP, including protein binding, rate of TRP transport from blood vessels, rates of cellular uptake or release and fluid flow dynamics. However for an environmentally generated "damage" signal to be coherent we might expect these physiological effects to be insignificant compared with the much larger quantities of TRP released from damaged tissue, the rate of diffusion of TRP towards or away from the signalling site driven by the TRP concentration gradient, the timing of IDO induction in the local environment in relation to the onset of the "damage" signal and the eventual depletion of extracellular TRP by induced IDO.

As there are undoubtedly differences between physiological fluids and tissue culture media, T-cell proliferation studies really need to be performed directly in CSF, blood plasma and ideally also in interstitial fluid and lymphatic fluid, before we can fully ascertain whether T-cell proliferation is normally enabled or disabled under non-inflammatory physiological conditions.

TRYPTOPHAN SIGNAL REGULATION AND SIGNAL NOISE

The immune signalling role of TRP takes place across interstitial fluid, lymphatic fluid or cerebrospinal fluid. Protein binding of TRP in blood results in free-TRP levels much lower than total TRP [26, 27] and these free levels drive TRP transport into body fluids. Excessive TRP is removed by Tryptophan 2, 3-Dioxygenase in the liver [37] or by Indolamine 2, 3-Dioxygenase (IDO) which is constitutively expressed in a variety of tissues including brain, lung, spleen, alimentary tract, prostate, uterus [12] and placenta [38]. IDO from intestinal cells has a broad substrate specificity catalysing oxidative ring cleavage of the D and L Isomers of TRP as well as D and L 5-hydroxytryptophan (5-HTP), tryptamine and serotonin [12]. IFN-γ produced by lymphocytes and macrophages induces IDO in a variety of tissues including lung, spleen, small intestine, colon, kidney, pancreas, uterus, heart and brain [39] and in a number of cell types including mononuclear phagocytes [40], macrophages [15] lung fibroblasts [13] and various tumour cells [41]. The IDO enzyme induced in human fibroblasts has higher specificity for TRP with a K_m of 3 μM for L-Tryptophan and shows little activity with D-TRP or 5-HTP [14]. More recently two forms of IDO have been identified in humans encoded by the IDO1 and IDO2 genes both located together on Chromosome 8 [42]. IDO2 is preferentially inhibited by D 1-Methyltryptophan and is not as widely expressed as IDO1 [43]. IDO1 is highly induced by IFN-γ and also by IFN-α, IFN-β and lipopolysaccharide [44]. Guanosine triphosphate cyclohydrolase 1 (GTP-CH1), which produces pterins [45, 46] and Nitric Oxide Synthase (INOS) which produces nitric oxide (NO) [47] are also induced by IFN-γ.

We examined the induction of IDO and GTP-CH1 in HeLa cell clones after stimulation with IFN-γ [48]. HeLa cells not stimulated with IFN-γ showed no ability to metabolise TRP or to produce pterins. There was a delay time of around 12 hours for induction of IDO, after which time TRP levels in the media of one clone fell rapidly from 57 μmol/L to below 4 μmol/L in 3 days (Figure 5). There was minimal production of kynurenine in the first 12 hours and then media levels of kynurenine climbed to a peak of 40 mmol/L after 3 days. Interestingly there was a much longer delay time for the induction of GTP-CH1 as the pterins, Neopterin and 6-Hydroxymethylpterin, did not appear in the media until after 3 days. A second clone showed similar delay characteristics but was somewhat less active in both removing TRP and producing pterins.

We have shown that a small number of Hela cells (40,000) were very effective in removing high levels of extracellular TRP after IDO induction by IFN-γ. This removal of excessive amounts of the TRP signal by IDO is typical of many other biological signal noise reduction processes. There was a delay time for TRP removal and a further delay time for the appearance of pterins. In cells generating NO we might expect an additional delay for NO production as INOS activity depends upon the availability of Tetrahydrobiopterin [49]. Diffusion rates depend upon the molecular weight of the signal molecule so smaller molecules such as TRP, pterins and NO are much faster signals than IFN-γ and other cytokines and chemokines. These timing delay sequences, signal speed differences, signal strength differences, noise control and differences in the extent of signal transduction extending outwards from where "damage" signals originate may regulate the localisation, duration and amplitude of the inflammatory process across the local microenvironment.

TRYPTOPHAN SIGNAL TRANSDUCTION

T-Cell proliferation is regulated by TRP availability and the TRP signal is transduced by metabolism resulting in TRP metabolites with further signalling and immunoregulatory roles [43,50]. IDO converts TRP into N-formyl-kynurenine which is the first step in the Kynurenine Pathway leading to the formation of 3-hydroxykynurenine, kynurenine, 3-hydroxyanthranilic acid and subsequently picolinic acid, quinolinic acid and nicotinic acid [51]. Kynurenine, 3-hydroxykynurenine and 3-hydroxyanthranilic suppress T-cell proliferation by promoting apoptosis, acting as feedback regulators of TRP's pro-inflammatory role. However the levels required are relatively high with I_{50} values of 170 μmol/L for kynurenine, 71 μmol/L for 3-hydroxyanthranilic acid and 17 μmol/L for 3-hydroxykynurenine [52]. It seems unlikely IDO could produce such high metabolite levels by acting upon the normal physiological fluid levels of TRP so T-cell apoptosis may require an environment where induced IDO has acted upon a large TRP "damage" signal to produce high metabolite levels. TRP transduction has a further feedback regulatory role as low levels of kynurenine (0.05 μmol/L) act upon the Aryl Hydrocarbon (Aryl HC) receptor leading to T-cell differentiation [53]. Higher levels of kynurenine (50 – 100 μmol/L) also promote mast cell activation through the Aryl HC receptor [54].

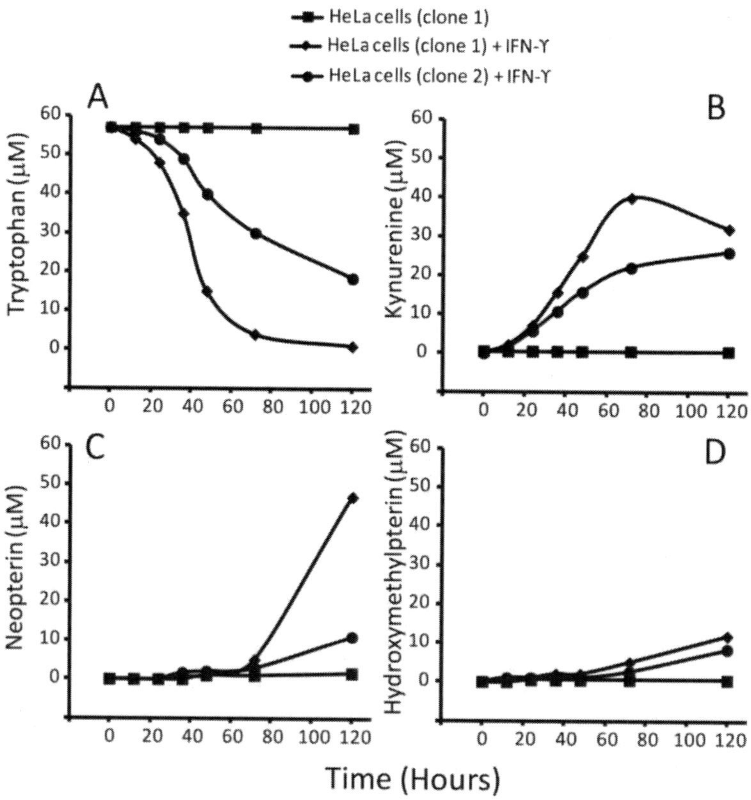

Figure 5. HeLa cells (40,000 cells) co-cultured with 20,000 peripheral blood lymphocytes were incubated for up to five days as previously described [48] . Culture media was harvested after various times periods for TRP and pterin analysis, centrifuged at 500×g for 5 minutes then stored at -80° C until analysed. Samples (100 μl) were removed, treated with 100 μl acetonitrile, centrifuged and the supernatant frozen for analysis of TRP and kynurenine by HPLC with electrochemical detection [60]. Samples (100 μl) were also removed, treated with 100 μl 1% Iodine and 2% Potassium Iodide in 0.1M Hydrochloric Acid solution, for 20 min, and analysed for pterins using HPLC with fluorescence detection [61].

TRP is synthesised by the Shikimic Acid pathway in plants, fungi and microorganisms. Metazoans lacking this pathway require TRP as an essential nutrient. As 30% of host-associated bacteria do not have a complete Shikimic Acid pathway [55] the depletion of TRP by IDO is an effective strategy for limiting the growth of many pathogens [18, 40, 56-58]. TRP metabolites 3-hydroxykynurenine and α-picolinic acid have antibacterial activity against methicillin-resistant Staphylococcus Aureus, Staphylococcus epidermidis, E-Coli and Pseudomonas Aeruginosa [59].

The removal of TRP by IDO inevitably sets up TRP concentration gradients as TRP will diffuse out of areas where it has been elevated by tissue damage into more depleted areas. This would prevent the overall build-up of TRP, act as a noise reduction process and highlight inflammatory foci. To test whether a TRP gradient might play a role in cell movement we adapted a standard clinical assay for neutrophil chemotaxis using modified Boyden chambers. The lower chamber contained 0.5ml of Phosphate Buffered Saline (PBS) which contained 80 μmol/L TRP and the upper chamber contained 0.1 ml of PBS with no detectable TRP. To calibrate the diffusion rate, TRP levels were measured in the chambers over time and it was observed that TRP steadily diffused into the upper chamber reaching a concentration of 7.5μmol/L after 30min. We investigated 3 human donors (3 replicates) for Neutrophil Chemotaxis using TRP as chemotactic stimulant. The distance travelled by the neutrophils "leading front" and the numbers of neutrophils at 30μm through the membrane are standard indicators of the effectiveness of the chemotactic response. For every replicate in all three donors, neutrophils demonstrated a chemotactic response to TRP equal to or greater than the chemotactic response to casein, the standard chemotactic stimulant. Further work needs to be carried out to see whether other myeloid and lymphoid cells respond to TRP as a chemotactic stimulant.

CONCLUSION

It is clear for immune signalling that we need to know much more about spatial and temporal effects determining the sequence of events, the speed of signalling and rate of signal molecule movement, the local concentrations of signal molecules and the processes which affect concentration and signalling including protein binding and turnover rates. We propose that the TRP signal arises from damaged tissue, raising TRP levels well above the threshold for T-lymphocyte proliferation, that IDO dampens signal noise by removing excess TRP, highlighting inflammatory foci, maintaining a TRP gradient for inward chemotaxis of neutrophils and possibly other myeloid and lymphoid cells.

REFERENCES

[1] A modification of Jerne's theory of antibody production using the concept of clonal selection. Burnet FM. *Aust J Sci.*, 1957, 20(2), 67–9.

[2] A theory of self-nonself discrimination. Bretscher P, Cohn M. *Science*, 1970, 169 (3950), 1042-9

[3] Activation of transplant immunity: effect of donor leukocytes on thyroid allograft rejection. Talmage DW, Dart G, Radovich J, Lafferty KJ. *Science.*, 1976, 191 (4225), 385-8.

[4] Dendritic cells and the control of immunity. Banchereau J, Steinman RM. *Nature.*, 1998, 392 (6673), 245-52.

[5] The immune system evolved to discriminate infectious nonself from noninfectious self. Janeway CA, Jr. *Immunol Today.*, 1992, 13 (1), 11-16.

[6] Tolerance, danger, and the extended family. Matzinger P. *Annu Rev Immunol.*, 1994,12, 991-1045

[7] The inflammatory response to cell death. Rock KL, Kono H. *Annu Rev Pathol*, 2008, 3, 99-126

[8] The danger theory: 20 years later. Pradeu T, Cooper EL. *Front Immunol*, 2012, 3(287), 1-9

[9] Release of chromatin protein HMGB1 by necrotic cells triggers inflammation. Scaffidi P, Misteli T, Bianchi ME. *Nature.*, 2002, 418 (6894), 191-5

[10] Hydrophobicity: an ancient damage-associated molecular pattern that initiates innate mmune responses. Seong SY, Matzinger P. *Nat Rev Immunol.*, 2004, 4 (6), 469-78

[11] An innate sense of danger. Matzinger P. *Semin Immunol.*, 1998, 10 (5), 399-415

[12] Indoleamine 2,3-dioxygenase. Purification and some properties. Shimizu T, Nomiyama S, Hirata F, Haaishi O. *J Biol Chem*, 1978, 253, 4700-6

[13] Induction of Tryptophan Degradation *In Vitro* and *In Vivo*: A γ-interferon stimulated activity. Byrne GI, Lehmann LK, Kirschbaum JG, Borden EC, Lee CM, Brown RB. *J Interferon Res.*, 1986, 6, 389-96

[14] Characterization of an Indoleamine 2, 3-Dioxygenase Induced by Gamma-Interferon in Cultured Human Fibroblasts. Pfefferkorn ER, Rebhun S, Eckel M. *J Interferon Res*, 1986, 6, 267-79

[15] Human macrophages degrade tryptophan upon induction by interferon-γ. Werner ER, Bitterlich B, Fuchs D, Hausen A, Reibnegger G, Szabo G, Dierich MP, Wachter H. *Life Sci*, 1987, 41, 273-80

[16] Human indolylamine 2,3-dioxygenase. Its tissue distribution, and characterization of the placental enzyme. Yamazaki F, Kuriowa T, Takikawa O, Kido R. *Biochem J*, 1985, 230, 635-8

[17] Mechanism of interferon-gamma action. Characterization of indoleamine 2,3-dioxygenase in cultured human cells induced by interferon-gamma and evaluation of the enzyme-mediated tryptophan degradation in its anticellular activity. Takikawa O, Kuroiwa T, Yamazaki F, Kido R. *J Biol Chem*, 1988, 263, 2041-8

[18] Antiparasitic and antiproliferative effects of indoleamine 2, 3-dioxygenase enzyme expression in human fibroblasts. Gupta SL, Carlin M, Pyati P, Dai W, Pfefferkorn ER and Murphy MJ. *Infect Immun*, 1994, 62, 2277-84

[19] Inhibition of T Cell Proliferation by Macrophage Tryptophan Catabolism. Munn DH, Shafizadeh E, Attwood JT, Bondarev I, Pashine A, Mellor AL. *J Exp Med*, 1999, 189(9), 1363-72

[20] Tryptophan degradation by human placental indolamine 2,3-dioxygenase regulates lymphocyte proliferation. Kudo Y, Boyd CAR, Sargent IL, Redman CWG. *J Physiol*, 2001, 535, 207-15

[21] Free and bound tryptophan in serum of untreated patients with chronic renal failure. Walser M, Hill SB. *Kidney Int*, 1993, 44(6), 1366-71

[22] Rapid and sensitive high-performance liquid chromatographic method for the analysis of tryptophan, tyrosine and phenylalanine in biological samples. Carducci C, Moretti F. *J Chromatog*, 1991, 553, 149-54

[23] Determination of tryptophan and its metabolites in human plasma and serum by high performance liquid chromatography with automated sample clean-up system. Morita I. *J Chromatog*, 1990, 526, 367–74

[24] Plasma tryptophan, kynurenine and 3-hydroxykynurenine measurement using automated on line solid phase extraction HPLC-tandem mass spectrometry. de Jong WHA, Smit R. Bakker SJL, de Vries EGE, Kema IP. *J Chromatog B*, 2009, 877, 603-9

[25] United States Department of Agriculture: National Nutrient Database for Standard Reference. http://ndb.nal.usda.gov/ndb/

[26] Platelet serotonin content and free and total plasma tryptophan in healthy volunteers during 24 hours. Eynard N, Flachaire E, Lestra C, Broyer M, Zaidan R, Claustrat B, Quincy C. *Clin Chem*, 1993, 39(11), 2337-40

[27] Heterogeneity of serum tryptophan concentration and availability to the brain in patients with the chronic fatigue syndrome. Badawy AA, Morgan CJ, Llewelyn MB, Albuquerque SR, Farmer A. *J Psychopharmacol*, 2005, 19(4), 385-91

[28] Importance of protein binding for the interpretation of serum or plasma drug concentrations. Greenblatt DJ, Sellers EM, Kock-Weser J. *J Clin Pharmacol*, 1982, 22(5-6), 259-63

[29] The determination and interpretation of the therapeutic index in drug development. Muller PY, Milton MN. *Nat Rev Drug Disc*, 2012, 11, 751-61

[30] Steroid hormone-binding proteins in Blood Plasma. Burton RM, Westphal U. *Metabolism*, 1972, 21(3), 253–76

[31] Serum protein binding of 1,25-dihydroxyvitamin D: A reevaluation by direct measurement of free metabolite levels. Bikle DD, Siiteri PK, Ryzen E, Haddad JG, Gee E. *J Clin Endocrin Metab*, 1985, 61(5), 969-75

[32] Determination of γ-Aminobutyric Acid and Other Amino Acids in Cerebrospinal Fluid of Pediatric Patients by Reversed-Phase Liquid Chromatography. Goldsmith RF, Earl JW, Cunningham AM. *Clin Chem*, 1987, 33(10), 1736-40

[33] Great Ormond Street Hospital for Children. Assay: CSF Amino Acids http://www.labs.gosh.nhs.uk/media/186743/CPRef%20range%20CSF%20amino%20acids.pdf

[34] Sydney South West Pathology Service: Handbook of Laboratory Tests: Cerebrospinal Fluid Albumin. http://www.swslhd.nsw.gov.au/sswps/default_hb.htm

[35] The specific binding of L-tryptophan to serum albumin. McMenamy RH, Oncley JL. *J Biol Chem*, 1958, 233, 1436-47

[36] CSF Neopterin in paediatric neurology. A marker of active CNS inflammation. Dale RC, Brilot F, Fagan E, Earl J. *Develop Med Child Neurol*, 2009, 51(4), 317-23

[37] The functions and regulation of Tryptophan Pyrrolase. Badawy AA. *Life Sci*, 1977, 21(6), 755-68

[38] Indoleamine 2, 3-dioxygenase tissue distribution and cellular localization in mice: Implications for its biological functions. Dai X, Zhu BT. *J Histochem Cytochem*, 2010, 58, 17-28

[39] Inhibition of experimental asthma by Indoleamine 2, 3-Dioxygenase. Hayashi T, Beck L, Rossetto C, Gong X, Takikawa O, Takabayashi K, Broide DH, Carson DA, Raz E. *J Clin Invest*, 2004, 114, 270-79

[40] Interferon-induced Indoleamine 2, 3-dioxygenase activity in human mononuclear phagocytes. Carlin JM, Borden EC, Sondel PM, Byrne GI. *J Leukocyte Biol*, 1989, 45, 29-34

[41] Relationship between interferon-g, indoleamine 2, 3-dioxygenase, and tryptophan catabolism. Tayor MW, Feng G. *FASEB J*, 1991, 5(11), 2516-22

[42] Characterization of an indoleamine 2,3 –dioxygenase-like protein found in humans and mice. Ball HJ, Sanchez-Perez A, Weiser S, Austin CJD, Astelbauer F, Miu J, McQuillan JA, Stocker R, Jermin LS, Hunt NH. *Gene*, 2007, 396, 203-13

[43] Novel Tryptophan Catabolic Enzyme IDO2 is the preferred Biochemical Target of the antitumour Indoleamine 2, 3-Dioxygenase Inhibitory Compound d-1-Methyl-Tryptophan. Metz R, DuHadaway JB, Kamasani U, Laury-Kleintop L, Muller AJ, Prendergast GC. *Cancer Res*, 2007, 67, 7082-7

[44] Molecules in focus: Indoleamine 2, 3-dioxygenase. King NJC, Thomas SR. *Int J Biochem Cell Biol*, 2007, 39, 2167-72

[45] Interferon-g enhances biosynthesis of pterins in peripheral blood mononuclear cells by induction of GTPcyclohydrolase 1 activity. Schroedon G, Troppmair J, Adolf G, Huber C, Niederwieser A. *J Interferon Res*, 1986, 6, 697-703

[46] Biosynthesis and metabolism of pterins in peripheral blood mononuclear cells and leukemia lines of man and mouse. Schroedon G, Troppmair J, Fontana A, Huber C, Curtius H-C, Niederwieser A. *Eur J Biochem*, 1987, 166(2), 303-12

[47] Induction of Nitric Oxide Synthase by cytokines in vascular smooth muscle cells. Busse R, Mulsch A. *FEBS Lett*, 1990, 275(1-2), 87– 90

[48] HeLa cells cocultured with peripheral blood lymphocytes acquire an immuno-inhibitory phenotype through up-regulation of indoleamine 2, 3-dioxygenase activity. Logan GJ, Smyth CM, Earl JW, Zaikina I, Rowe PB, Smythe JA, Alexander IE. *Immunology.*, 2002, 105(4), 478-87.

[49] Regulation of biosynthesis of nitric oxide. Nathan C, Xie Q-W. *J Biol Chem*, 1994, 269, 25722-9

[50] Indoleamine 2, 3-dioxygenase and regulation of T-cell immunity. Mellor A. *Biochem Biophys Res Commun*, 2005, 338, 20-4

[51] Regulation of enzymes involved in the conversion of Tryptophan into Nicotinamide Adenine Dinucleotide in a colourless strain. Brown AT, Wagner C. *J Bacteriol*, 1970, 102(2), 456–63

[52] Studying the immunosuppressive role of indoleamine 2, 3-dioxygenase; tryptophan metabolites supress rat allogenic T-cell responses in vitro and in vivo. Bauer TM, Jiga LP, Chuang J-J, Randazzo M, Opelz G, Terness P. *Transplant Int*, 2005, 18, 95-100

[53] An interaction between kynurenine and the aryl hydrocarbon receptor can generate regulatory T cells. Mezrich JD, Fechner JH, Zhang X, Johnson BP, Burlingham WJ, Bradfield CA. *J Immunol*, 2010, 185(6), 3190-8

[54] A tryptophan metabolite, kynurenine, promotes mast cell activation through the aryl hydrocarbon receptor. Kawasaki H, Chang HW, Tseng HC, Hsu SC, Yang SJ, Hung CH, Zhou Y, Huang SK. *Allergy*, 2014, 69(4), 445-52

[55] Global genome analysis of the shikimic acid pathway reveals greater gene loss in host associated than in free-living bacteria. Zucko J, Dunap WC, Shick JM, Cullum J, Cercelet F, Amin B, Hamman L, Lau T, Williams J, Hranueli D, Long PF. *BMC Genomics*, 2010, 11, 628

[56] Interferon-g blocks the growth of Toxoplasma gondii in human fibroblasts by inducing the host cells to degrade tryptophan. Pfefferkorn ER. *Proc Natl Acad Sci USA*, 1984, 81, 908-12

[57] Severe tryptophan starvation blocks onset of conventional persistence and reduces reactivation of *chlamydia trachomatis*. Leonhardt RM, Lee S-J, Kavathas PB, Cresswell P. *Infect Immun*, 2007, 75, 5105-7
[58] Indoleamine 2, 3-dioxygenase and metabolic control of immune responses. Munn H, Mellor AL. *Trends Immunol*, 2013, 34(3), 137-43
[59] Anti-infectious activity of tryptophan metabolite in the L-tryptophan-L-kynurenine pathway. Narui K, Noguchi N, SaitoA, Kakimu K, Motomura N, Kubo K, Takamoto S, Sasatsu M. *Biol Pharm Bull*, 2009, 32(1), 41-4
[60] Determination of tryptophan and metabolites in rat brain and pineal tissue by reversed-phase high-performance liquid chromatography with electrochemical detection. Meford IN, Barchas JD. *J Chromatog*, 1980, 181, 187-93
[61] Improved diagnosis of classical vs atypical phenylketonuria by liquid chromatography. Matsubara Y, Heininger JA, Lin YY. *Clin Chem*, 1984, 30(20), 278-80

In: Indoleamines
Editor: Ángel Catalá

ISBN: 978-1-63482-097-4

Chapter 5

HEALTH EFFECTS OF INDOLEAMINES IN THE PREVENTION OF COGNITIVE DECLINE ASSOCIATED WITH BRAIN AGING AND AGE-RELATED DISEASES

M. R. Ramis*[1*]*, F. Sarubbo*[1†]*, S. Aparicio*[1‡]*, D. Moranta*[1,2#]*,
***A. Miralles*[‖1] *and S. Esteban*[1Π]**

[1]Laboratorio de Neurofisiología, Departamento de Biología, Universitat de les Illes Balears, Mallorca, Spain

[2]Laboratorio de Infección e Inmunidad. Fundación de Investigación Sanitaria de las Islas Baleares (FISIB)-Instituto de Investigación Sanitaria de Palma (IdISPa), Centro de Investigación Biomédica en Red Enfermedades Respiratorias (CIBERES). Hospital Joan March, Mallorca, Spain

ABSTRACT

New treatments for several diseases of young and middle age have increased life expectancy significantly, but as a consequence a variety of diseases are emerging, especially disorders associated with brain dysfunction. Even in the absence of brain disorders, aging is associated with cognitive deficits, such as memory and learning impairment as well as motor disabilities that compromise the quality of life. It is generally accepted that damage resulting from reactive oxygen species, or oxidative stress, plays a role in the aging process, and that the brain is especially vulnerable to its effects. In this context, a main objective of investigation is the search for treatments or strategies that delay or reduce the effects of brain aging, i.e., the neuroprotective effects of antioxidants.

---

[*] M.R. Ramis, PhD. student, margarita.ramis@uib.es.
[†] F. Sarubbo, PhD. student, fioresarubbo@hotmail.com.
[‡] S. Aparicio, PhD., sara.aparicio@alvarez.es.
[#] D. Moranta, PhD., david.moranta@uib.es.
[‖] A. Miralles, PhD., amiralles@uib.es.
[Π] Corresponding author: Laboratorio de Neurofisiología, Departamento de Biología, Universitat de les Illes Balears, Ctra. Valldemossa Km 7,5 E-07122 Palma de Mallorca, Spain. Tel.: +3497173145; fax: +34 971 173184; S. Esteban, PhD., *e-mail address*: susana.esteban@uib.es.

The indoleamines serotonin and melatonin are biosynthesized from the essential amino acid tryptophan. The rate-limiting enzyme in the synthesis of serotonin, tryptophan hydroxylase (TPH) determines the effective concentration of serotonin *in vivo*. At night, serotonin is used as a substrate for melatonin synthesis. Thus, there is a balance between serotonin's use as a neurotransmitter and its availability as precursor for melatonin synthesis, which is deeply dependent on the environmental light. Serotonin is one of the major neurotransmitters influencing hippocampal neurotransmission, which is involved in many functions throughout the brain such as learning and memory; while melatonin, besides its known role in the circadian control, has an important role in the aging process since it acts as a powerful and direct toxic free radicals and reactive intermediates scavenger. The synthesis of serotonin and melatonin decreases with age, partly as a consequence of reduced plasma tryptophan by its degradation to kynurenic derivatives, and this may explain some physiological alterations during brain aging. In addition, TPH has been shown to be susceptible to oxidative damage by reactive oxygen species, and consequently its reduced activity could explain the decline of indoleamines in aged brain. Some studies suggest that treatments aimed at modulating the activity of TPH enzyme, which could be accompanied by restorative effect of the cognitive functions, were worsened in aging. In this manner, the synthesis of serotonin, and consequently that of melatonin can be modulated with potent antioxidants, and thus the age-related decrease of serotonin and melatonin could be prevented, and perhaps some complaints of aging could also be delayed. In summary, dietary intake of antioxidants could be used as a therapy to mitigate the cognitive decline associate with aging and it might aid to delay some aging-related diseases emerging as the greatest health threats of the Twenty First century.

Keywords: Indoleamines; aging; brain; serotonin; melatonin; tryptophan; TPH; antioxidants

INTRODUCTION

1. Aging and Cognitive Decline

New treatments for several diseases of young and middle age have increased life expectancy significantly. Between 2000 and 2050, the proportion of the world's population over 60 years will double from about 11% to 22%. The absolute number of people aged 60 years and over is expected to increase from 605 million to 2 billion over the same period, according to the World Health Organization (WHO). However, a variety of diseases have emerged, especially disorders associated with brain dysfunction [1, 2].

Aging is a universal, complex and multifactorial process characterized as degenerative in nature, which causes progressive loss of function and an increased risk of death [3, 4]. This process is accompanied by stereotypical structural and neurophysiological changes in the brain [2]. Related with this, even in the absence of brain disorders, aging is associated with cognitive deficits, such as memory and learning impairment, and alteration in motor functions [5-8]. Specifically, the working memory (WHO refers to a brain capacity to encode, maintain, and manipulate information for a relatively short time), declines in late adulthood [9-12] and is considered to be one of the main contributing factors of cognitive impairment in old age [13]. Thus, it is not surprising that attempts to improve working memory have been far explored and a large number of studies have been investigating the trainability of working memory across the lifespan [14-17]. Moreover, there is a less-coordinated activation with aging, suggesting also a global loss of integrative function between systems of brain areas, as well as decreases in brain structure size and white matter integrity, in addition to less neural activity localized in some brain regions. These modifications in the synaptic physiology of neurons may contribute to altered connectivity and higher-order integration in aging [5, 18]. Furthermore, these neuronal level changes involve deficits in cognitive processes and motor behaviour, closely related to changes in serotonergic neurotransmission in the brain as well as noradrenergic and dopaminergic systems [6, 7, 19-32].

2. Indoleamines: Serotonin and Melatonin Synthesis

The indoleamines serotonin (5-HT, 5-hidroxytryptamine) and melatonin (N-acetyl-5-methoxytryptamine) are biosynthesized from the essential amino acid L-tryptophan. Serotonin is synthesized in two reactions beginning with the hydroxylation of tryptophan to 5-hydroxytryptophan (5-HTP), which is then decarboxylated to serotonin. The rate-limiting enzyme in the synthesis of serotonin, tryptophan hydroxylase (TPH) determines the effective concentration of serotonin in vivo. This enzyme requires for its activity the presence of tetrahydrobiopterine, oxygen, $NADPH_2$ and a metal, iron or copper. Minor changes in TPH activity can cause drastic changes in serotonin content and serotonergic functions [33]. TPH is known to exist in two isoforms: TPH-1, mainly expressed in the pineal gland and in gut enterochromaffin cells where it plays a fundamental role in melatonin synthesis [34], and TPH-2, preferentially expressed in the brain where it plays a fundamental role in serotonin synthesis [35].

Serotonin is involved in many functions throughout the brain, including cognitive processes [26, 31]. It is one of the major neurotransmitters influencing hippocampal neurotransmission, with an important role in learning and memory. It has been reported that the increase in brain serotonin improves cognitive performance [19]. In addition, in the dark phase serotonin is used as a substrate for melatonin synthesis by the enzyme TPH-1. Although melatonin is secreted mainly by the pineal gland or epiphysis, the enzymes that produce it are present in numerous other tissues including retina, cochlea, skin, immune system, gastrointestinal tract and reproductive tract [36-42]. Melatonin is synthesized from tryptophan by a consecutive action of four enzymes (Figure 1); TPH firstly converts tryptophan to 5-hydroxytryptophan, then the L-aromatic amino acid decarboxylase leads to serotonin, that is transformed to N-acetyl-serotonin by N-acetyltransferase (NAT), followed by its methylation by acetylserotonin methyltransferase (also known as hydroxyindole-O-methyltransferase, HIOMT) to obtain melatonin or acetyl-5-methoxytryptamine [43, 44]. As a consequence, there is a balance between the use of serotonin as a neurotransmitter or as precursor for melatonin synthesis, which is deeply dependent on the environmental light. The concentration of melatonin in the pineal gland follows circadian variations (increases at night and decreases during the day), depending of the degree of NAT activity, thus light inhibits melatonin biosynthesis [43, 45]. The pineal is the only endocrine gland that is directly influenced by the external environment via the retina and has been considered the photoneuroendocrine transducer that synthesizes and secretes melatonin at night in response to an endogenous internal clock trainable by environmental photoperiodic stimuli [45-47].

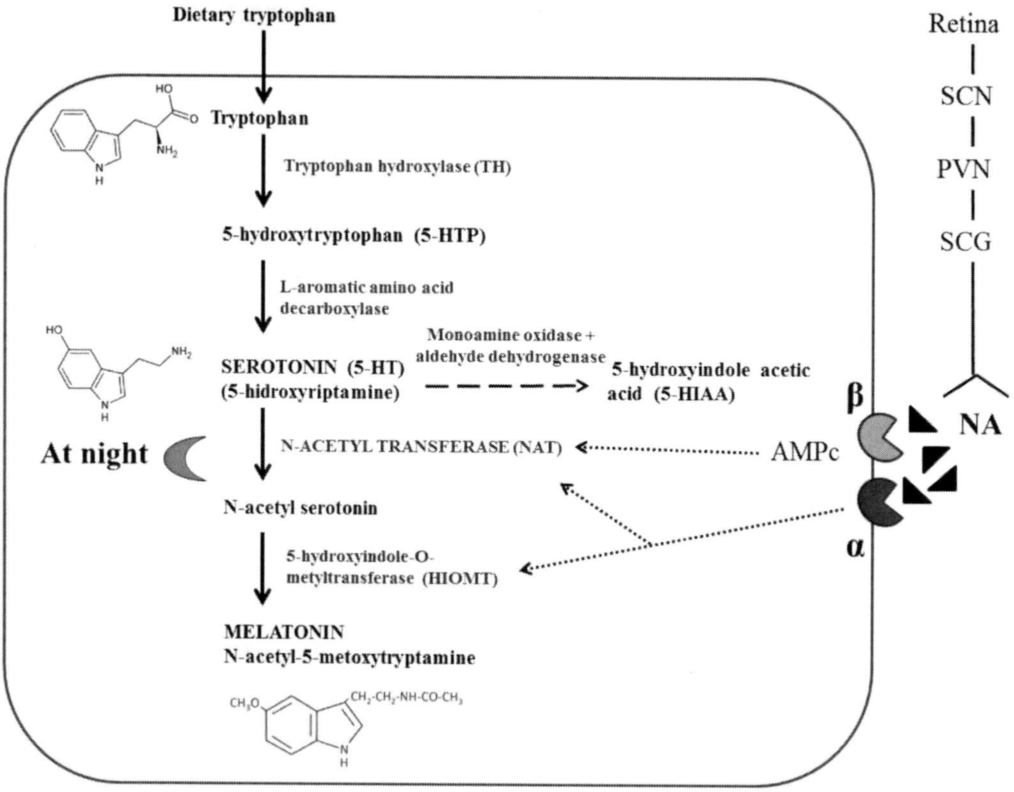

Figure 1. Biosynthesis of serotonin and melatonin from the essential amino acid tryptophan in the pinealocyte. NA: noradrenalin; SCN: suprachiasmatic nucleus; PVN: paraventricular nucleus; SCG: superior cervical ganglion.

The information is transmitted from the photoreceptors of the retina to suprachiasmatic nucleus (SCN), and then to paraventricular nucleus (PVN) that reach the superior cervical ganglion (SCG) where the noradrenergic fibers extend their terminals to the pineal gland [48]. Stimulation of adrenergic receptors on pinealocyte synergistically increases intracellular levels of AMPc and GMPc which enhance the activity of NAT to converting serotonin to N-acetyl serotonin, and finally to melatonin by the enzyme HIOMT. Therefore, NAT activity is under the control of the retino-pineal route and represents the limiting factor for melatonin synthesis, through norepinephrine release (Figure 1).

Both the synthesis of serotonin and melatonin depend on the availability of L-tryptophan that must be ingested in sufficient quantities. The quantity of tryptophan recommended daily is about 250 - 350 mg, of which only a small part is converted into serotonin. Indeed, in addition to the metabolic pathway leading to serotonin, tryptophan is used in protein synthesis and it is transformed by hepatic tryptophan pyrrolase (tryptophan 2-3-dioxygenase) and by indolaleamine-2,3-dioxygenase, into N-formylkynurenine and then into kynurenine, precursor of xanthurenic acid and nicotinic acid. The activation of these 2 enzymes of the kynurenine pathway could reduce the amount of tryptophan available for serotonin biosynthesis [49] and melatonin production. Activity of tryptophan pyrrolase is increased by cortisol [50] and ethanol consumption [51]. Moreover, the activity of indoleamine-2,3-dioxygenase increases during stimulation of the immune system and the selective inhibition of the enzymatic activity may have a role in the inhibition of immune responses to tumors [52, 53].

3. Negative Effects of Oxidative Stress on Aging Brain

Since neurodegenerative changes and dementia are major and debilitating features of aging, the curtailing actions of free radicals and other reactants in the brain are potentially of great importance in to alleviating or to forestalling the cognitive and mental decline in older individuals. In this way, the free radical theory of aging postulated by Harman [54] predicts that the production of intracellular reactive oxygen species (ROS) is the major determinant of lifespan. Moreover, natural human antioxidant defenses are not always sufficient to maintain the proper ROS balance [55]. The classical free radical theory of aging expects that minimizing deleterious free radical reactions can increase the healthy lifespan. Data available indicates that this can be done by keeping body weight down, at a level compatible with a sense of well-being, while ingesting diets adequate in essential nutrients but designed to minimize random free radical reactions in the body. Such diets would contain minimal amounts of components prone to enhance free radical reactions [56] and increased amounts of substances capable of decreasing free radical reaction damage, as antioxidants [57].

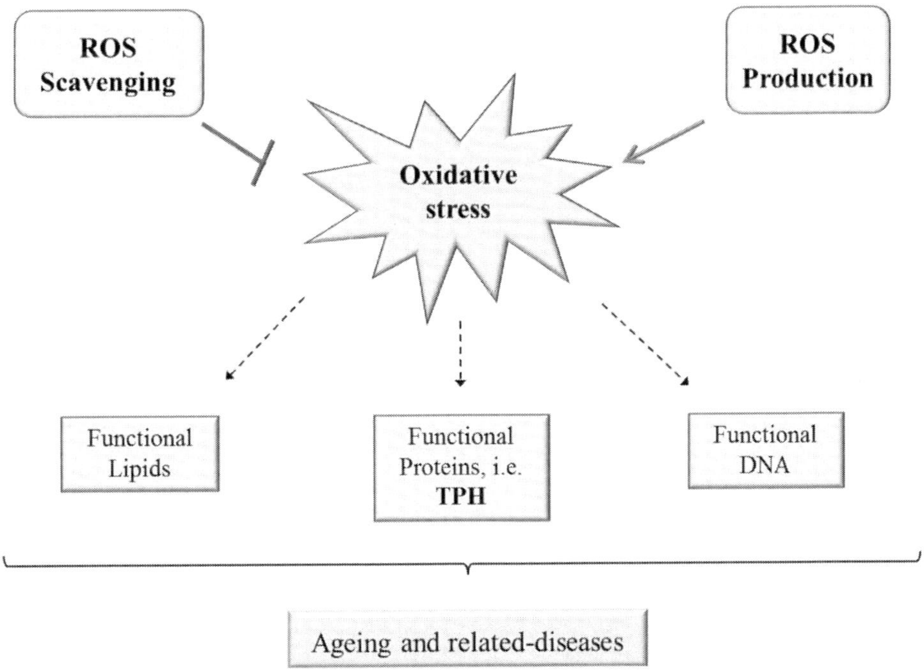

Figure 2. Accumulative oxidative damage to lipids, proteins, and nucleic acids. TPH: tryptophan hydroxylase.

Oxidative damage is considered a likely cause of age-associated brain dysfunction because the brain is particularly vulnerable to oxidative stress [58-60]. Neural tissue, although constituting only a small percentage of the body weight, exhibits a high utilization of O_2 (the source of many damaging species) with a consequent elevated rate of free radical generation.

Besides the greater oxygen consumption other factors contribute causing damage, such as a larger number of mitochondria, abundant nitric oxide, and polyunsaturated fatty acid rich cell membranes (readily peroxidated), without suitable levels of antioxidant defenses compared with other somatic tissues [59-63]. Several authors suggested that the increase in oxidative stress in aging brain result in a relative decrease in antioxidant enzyme activities [64-66]. Cognitive impairment in aged subjects has been correlated to the accumulation of oxidative damage to lipids, proteins, and nucleic acids (Figure 2) [59, 67] and also with a decrease in brain and plasma antioxidants [68-71]. Some studies suggest a correlation between deficits of endogenous antioxidants (such as vitamin E) and memory disorders [72]. Thus, the free radical theory of aging provides a tool to investigate the effects of antioxidant nutrients [57, 73]. In this manner, several studies showed that supplementing diet with fruit or vegetables has slowed and, in some cases, even reversed deficits in brain function, motor performance, learning and memory in old experimental animals [74-76].

4. Age-Related Decrease of Indoleamines Synthesis. Consequences on Brain Function

During aging, the synthesis of serotonin decreases [6, 19, 21, 31, 77] and the nocturnal production of melatonin is also reduced (Figure 3). Whereas the 24h rhythm of melatonin production is robust in young subjects, the cycle deteriorates during aging, reducing the amount of circulating melatonin, suggesting a potential association between the loss of melatonin and the signs of aging [78-85]. It can be partly explained by a decrease of serum/plasma tryptophan concentration related with aging and associated with an enhanced indoleamine (2,3)-dioxygenase (IDO) activity, which degrades tryptophan to kynurenine derivatives [49]. In addition, TPH activity is reduced in aging brain, which could be due to an inefficient phosphorylation and/or an oxidative damage of the serotonin synthesis limiting enzyme [86, 87]. In this manner, TPH has been shown to be susceptible to oxidative damage by reactive oxygen species [88], which could explain the reduce TPH activity with a consequent decline in the brain content of indoleamines serotonin and melatonin [89] as mentioned above. On the other hand, the age-related decline of pineal melatonin production also could be related to degenerative changes of the neural structures (serotonergic and noradrenergic neuron systems) innervating the pineal gland and the suprachiasmatic nuclei rather than to the degeneration of the pineal tissue itself [90].

Besides the well-known function of the pineal hormone melatonin on circadian rhythm transduction namely a transducer of the light signal, melatonin has other important roles in the aging process. Because of its several functions, melatonin achieves this widespread protection by means of its ubiquitous direct and indirect actions to reduce oxidative damage. Melatonin can act as a potent direct scavenging of toxic free radicals and reactive intermediates, a likely cause of age-associated brain dysfunction [61, 91-93]. In addition to the functions linked to melatonin, many of the melatonin metabolites (Figure 4), which may appear in elevated concentrations after melatonin synthesis or exogenous administration, possess biological properties [94-100]. Altogether, and related to the free radical theory of aging, this suggests that there may be physiological alterations when secretion of melatonin is reduced on old age. Moreover the activity of antioxidant enzymes undergoes a significant age-related decrease in brain that may lead to an increase in reactive oxygen species concentrations [61, 73], which in turn may cause the oxidative damage to TPH (Figure 2) and tyrosine hydroxylase (TH). TH has been shown highly homologous [101] and is a rate-limiting enzyme for dopamine (DA) and noradrenaline (NA) synthesis (monoaminergic neurotransmitters also important for several cognitive abilities).

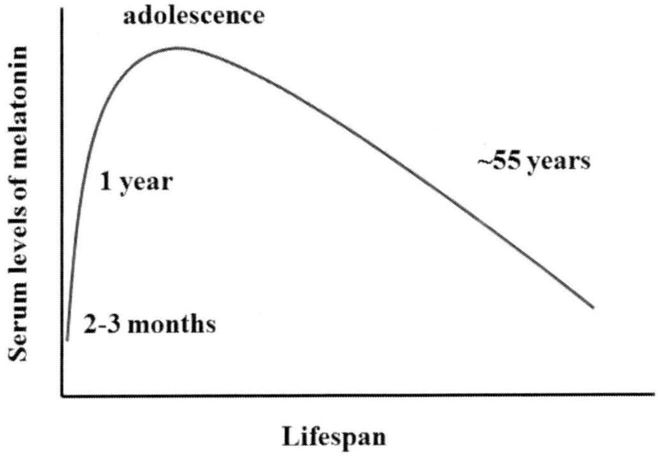

Figure 3. Reduction in the nocturnal production of melatonin across lifespan.

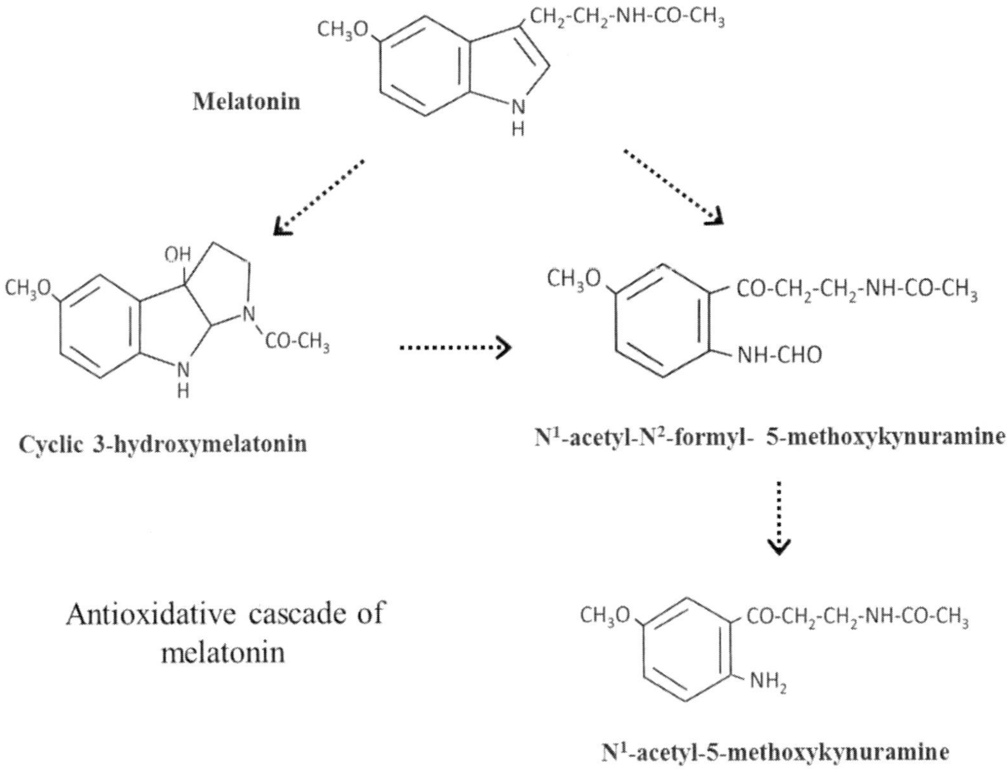

Figure 4. Melatonin metabolites produced during the scavenger of radical agents.

TPH is an oxygen-liable neuroenzyme, and maintenance of adequate oxygen levels is essential for maintaining its enzymatic activity [102]. Similarly, TH enzyme was inactivated by oxidation in the substantia nigra of old rats [103]. Consequently, the age-related decrease in serotonin and other monoamines neurotransmitters with aging could be explained by a reduction in their synthesis due to the low TPH and TH activities. This decline in neurotransmitter function is thought to be partially responsible for memory disorders [6, 7, 30, 87]. In this way, it was observed an improvement in cognitive functions associated to increased levels of serotonin as well as other monoamines in the hippocampus and striatum, brain regions involved in memory processing and motor coordination, after several treatments in aged rats [6, 7].

5. Preventive Treatments: Health Effects of Indoleamines

5.1. Serotonin and L-Tryptophan

Serotonin release increases during working memory task [104] and extensive evidence suggests that different serotonin receptors have a role in learning and memory [31]. Serotonin receptors are classified into seven distinct receptor classes (5-HT_1 to 5-HT_7) and growing evidence is emerging on the role of 5-HT_4 receptor in various forms of cognitive processes in the mammalian brain, particularly different learning tasks [105-108]. Serotonin synthesis depends on brain tryptophan levels, which in turn depends on blood tryptophan concentration. Elevated tryptophan in the brain produces physiologically important changes in serotonergic system [19], which has an important role in cognitive processes [26, 31]; for example, an increase of serotonin in the brain can improve cognitive performance [19, 20]. The administration of L-tryptophan to old animals increased serotonin synthesis, irrespective of the time of administration. This suggests that, TPH was always far from being saturated by its tryptophan substrate. Moreover, the metabolite 5-HIAA was strongly increased after L-tryptophan ingestion, but only during light-time suggesting an increased use of the serotonin for melatonin synthesis at night [109, 110]. Assuming that the synthesis of serotonin and melatonin can be modulated by tryptophan ingestion [19], the decrease in serotonin and melatonin which normally occurs during aging could be prevented if tryptophan intake increases, and perhaps some complaints of aging could also be delayed. In this regard, the administration of tryptophan increased serotonin metabolism in different regions of the brain. The increase in serotonin metabolism was, however largest in the hippocampus than in any other region of brain, showing also a greater conversion of tryptophan to serotonin in this region which is conceived as a key structure involved in cognitive function such as learning and memory [24, 27, 29, 111-114]. Moreover, the repeated treatment with L-tryptophan improved the descent in serotonin neurotransmission as well as the TPH activity that normally occurs as a consequence of aging. It was also accompanied to a restorative effect of the motor and cognitive functions that were impaired as a consequence of aging [7] probably mediated by an increase in the synthesis of serotonin and melatonin. Accordingly, L-tryptophan administration could be also used as a therapy for other diseases involving reductions of indoleamines. Given the low toxicity of ingested tryptophan, an adequate dietary supply of tryptophan (Figure 5) would be of paramount importance and might aid to improve some age-related degenerative conditions. There are several food sources of tryptophan [115-117]: eggs, meat (pork, turkey, chicken, beef, lamb, rabbit), fish (cod, salmon, perch, tuna) and shrimp; it is also present in milk and derived products (cheese, yogurt), and cocoa and chocolate among others. This essential amino acid also is detected in several crops (chickpeas, soybeans, sesame seeds, sunflower seeds, mustard seed, oats, wheat flour, rice, quinoa, potatoes, and beans) and fruits such as banana. On the other hand, serotonin also is found in fruits and vegetables (such as walnuts, pineapples, banana, kiwifruit, avocado, plums, cherries, grapefruit, date, eggplant, potato, spinach, broccoli, black olives and tomatoes), some meat and fish [118-121], and in chocolate with varying cocoa contents [117] (Figure 5).

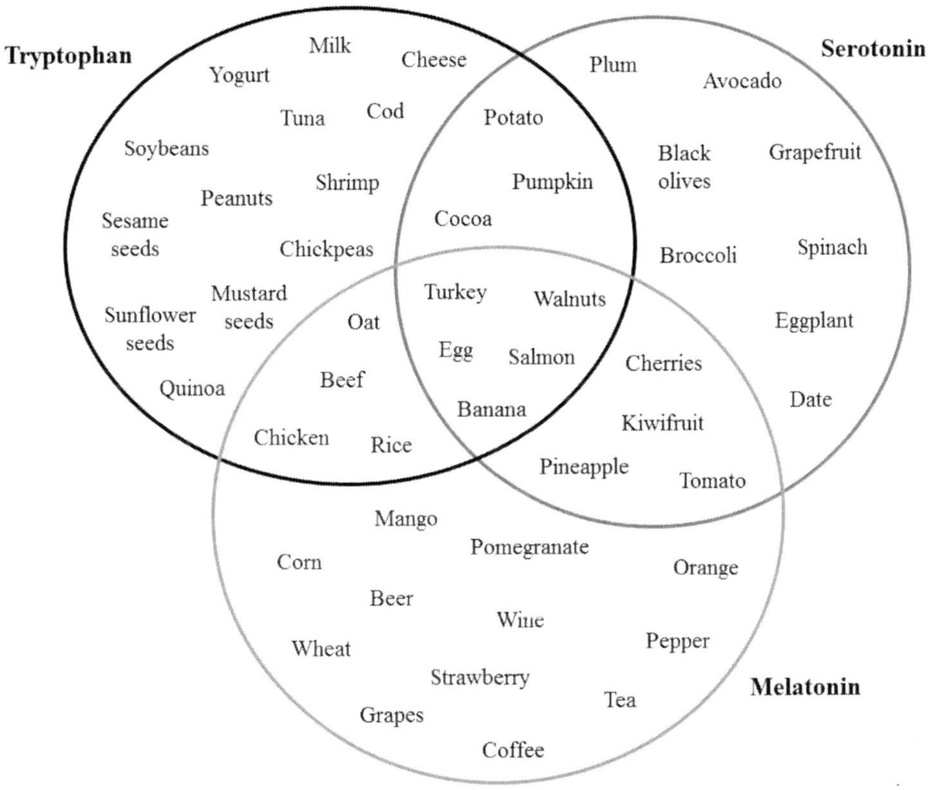

Figure 5. Tryptophan, serotonin and melatonin presence in food.

5.2. Melatonin

Melatonin is a highly conserved and widely distributed molecule. It is found throughout the plant and animal kingdoms, i.e., perhaps in every living organism [93, 122-125]. It has the capability of penetrating all morphophysiological barriers and entering all subcellular compartments due to its amphiphilic nature [92, 126]. As noted above, melatonin has an important role in the aging process due to its ability to reduce oxidative damage [61]. The role of melatonin in scavenging destructive free radicals was initially uncovered by Tan et al. [127] who documented that melatonin directly detoxifies the highly damaging hydroxyl radical (•OH). When melatonin interacts with •OH, it initiates a scavenging cascade reaction in which the metabolites produced, i.e., cyclic 3-hydroxymelatonin, N1-acetyl-N2-formyl-5-methoxykynuramine and N1-acetyl-5-methoxykynuramine (Figure 4), also function in neutralizing of free radicals [94-97, 99, 100]. Melatonin and its metabolites have strong antioxidant properties (Figure 6), which act as endogenous buffering agents against oxidative stress [98, 128, 129]. Since then, numerous publications have appeared which have verified the ability of melatonin to neutralize a variety of free radicals and related reactants (including reactive oxygen and nitrogen species), confirming its ability to protect macromolecules (i.e., proteins, lipids and both nuclear and mitochondrial DNA) from oxidative damage in all subcellular compartments [94, 95]. In addition, melatonin increases enzymes that modulate glutathione metabolism (another important antioxidant), i.e., glutathione peroxidase and glutathione reductase, as well as other antioxidant defense enzymes [44, 97].

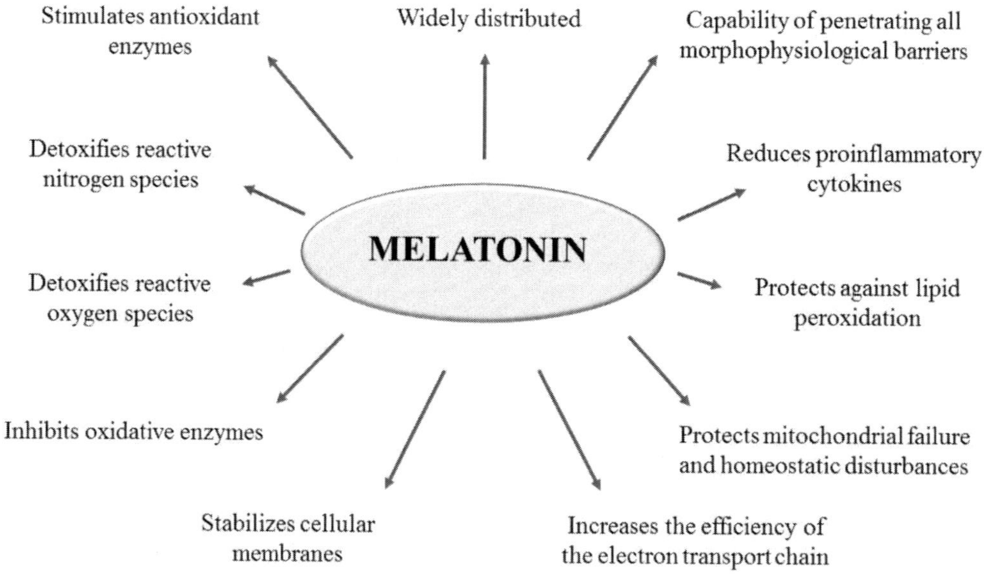

Figure 6. Main antioxidant properties of melatonin.

Aging and related-diseases are also accompanied by structural changes in mitochondria, with corresponding alterations in the biophysical properties of the membrane such as asymmetry and decreased fluidity; alteration in the electron transport chain (ETC) complexes activities also occur causing mitochondrial failure and an energy imbalance. These perturbations endanger cellular homeostasis and enhance vulnerability to additional oxidative stress [130, 131]. The progressive failure of homeostasis leads to physiological malfunction manifested as a general functional decline, diseases and ultimately death [132]. Although the concentrations of melatonin vary in subcellular compartments, the highest levels are found within membranes and mitochondria, where it interacts with lipids, stabilizes all cellular membranes [133, 134] and reduces lipid peroxidation [44]. These effects are reflected in the ability of melatonin to improve ETC activity in the inner mitochondrial membrane and to enhance ATP production [130]. Due to its antioxidant and free radical scavenging capacity, melatonin protects proteins of the ETC and mitochondrial DNA from oxidative stress [135]. Thus, melatonin has also been shown to increase the efficiency of the ETC and, as a consequence, to reduce electron leakage and the generation of free radicals (Figure 6). Finally, there is evidence that melatonin stabilizes cellular membranes, thereby probably helping them resist oxidative damage. This is especially important during brain aging [61, 136], as neuronal death during brain aging results, at least in part, from the disruption of synaptic connectivity caused by oxidative stress. When the nocturnal production of melatonin decreases with aging, all the strong antioxidative properties decrease, and this in turn suggests significant neurological changes.

These multiple neuroprotective actions of melatonin make this indoleamine a potentially useful agent in the treatment of neurological disorders that have oxidative damage as part of their etiological basis. In recent years, it have been published works suggesting that the exogenouous administration of melatonin could prevent the increased production of age-related lipid peroxidation products and might have a potential role for retardation of age-related oxidative events. Accordingly, melatonin exhibit abilities to protect brain from beta-amyloid toxicity, Parkinsonism, excitotoxicity, nitric oxide toxicity, aminolevulinic acid, lipopolysaccharide, hyperbaric hyperoxia, L-cysteine, cyanide and ischemia/reperfusion injury [137-140].

Given the absence of side effects in relation to the intake of melatonin, an adequate dietary supply of melatonin would be useful as therapy to improve some neurodegenerative conditions. Remarkably high melatonin concentrations have been measured in crops (corn, rice, wheat, barley, and oats), fruits and vegetables (oranges, apples, bananas, pineapples, grapes, strawberries, cherries, pomegranates, kiwifruits, mangos, walnuts, tomatoes, peppers) and popular beverages (coffee, tea, wine, and beer) [121, 141-150]. Recently, melatonin has also been detected in eggs, lamb, beef, pork, chicken, and salmon [151] (Figure 5).

5.3. Other Dietary Antioxidants

Accordingly with the free radical theory of aging, dietary manipulations are expected to further lower the rate of production of free radical and its associated damages [57, 73, 152].

An antioxidant is any substance that delays, prevents or removes oxidative damage to a target molecule [153]. Age-related memory impairment, at least in part, is correlated with a decrease in brain and plasma antioxidants [68-70] and might be associated with a progressive imbalance between antioxidant defense and intracellular concentrations of ROS [61]. As mentioned above, a common feature in the neurodegenerative diseases is the existence of mitochondrial damage due to oxidative stress [130], suggesting the use of antioxidants in the treatment of these disorders [57, 73, 152].

Antioxidants such as vitamins A, C, and E, are essential nutrients necessaries for optimal growth, development, and reproduction in animals [154, 155]. Among them, vitamin E is a potent lipid-soluble, chain-breaking antioxidant [155], which regulates neuronal functions and maintains cellular integrity [156].

Vitamin E is a well-known antioxidant substance very frequent on Mediterranean diet [157]. The α-tocopherol, the most active form of vitamin E, is a lipophilic antioxidant that resides within the lipid bilayer of membranes, and is primarily responsible for limiting/inhibiting endogenous lipid peroxidation [158-162]. It is a required nutrient for humans because it is necessary for the prevention of deficiency symptoms of vitamin E, including peripheral neuropathy and hemolytic anemia [155]. In this regard, vitamin E deficiency has been associated with cognitive impairments [72] while long-term dietary supplementation of vitamin E could improve cognitive function during aging [74, 163-165] and reduce the degeneration of hippocampal cells [166] through a decrease of lipid peroxidation [164]. Furthermore, long-term treatment with vitamin E could also tend to increase vitamin E concentrations within the brain and enhance the effectiveness of this compound as an antioxidant in neurodegenerative diseases [167]. In addition, positive effects of vitamin E on motor capabilities have been associated with their antioxidant properties [168]. In this regard, concentrations of vitamin E in cerebellum are the lowest of brain, and during vitamin E deficiency it is depleted faster than other parts of brain, being cerebellum especially susceptible to oxidative damage [167, 169, 170].

Several lines of evidence indicate that NO and $ONOO^-$ directly inactivate the enzymatic activity of TPH since exposure of this enzyme to oxidizing conditions rapidly destroys its catalytic function [171-174] with a consequent reduction on brain indoleamines. However, when large amounts and varieties of radicals are produced, administration of vitamin E may restore TPH activity as well as serotonergic functions in brain [33]. Cerebral hypoxia induces down-regulation of TPH-1 and TPH-2, as well as of serotonin functions; nevertheless, repeated injections with vitamin E substantially restored TPH expression and serotonin concentrations [33]. Thus, these results suggest that increasing the antioxidant status can prevent hypoxia-induced down-regulation of the serotonergic system.

Thus, vitamin E shows beneficial properties on the function of the brain and the vulnerability to age-dependent pathology. This antioxidant can delay the cognitive decline and can increase the TPH activity, which suggest an improvement the levels of serotonin and melatonin in aging brain. The recommended daily amount of vitamin E is 10 - 15 mg/day according to the Food and Nutrition Board of the Institute of Medicine [175], and is found in foods such as vegetable oils, margarines, fruits, seeds and vegetables [157, 175]. A web-page hosted by the US Department of Agriculture gives a detailed listing of the vitamin E content of various foods (http://www.ars.usda.gov/ba/bhnrc/ndl).

Besides vitamin E, other potent antioxidants present in the diet have been investigated. Among them, numerous studies indicate the great antioxidant role of polyphenols. These compounds are secondary metabolites of plants, essential to plant physiology, contributing to their pigmentation, growth, reproduction, and resistance to ultraviolet radiation, pathogens and predators [176, 177]. About 10,000 phenolic compounds of plant origin have been characterized [176], being resveratrol and flavonoids (such as quercetin, catechins, naringenin…) the most studied and the most abundant presents in foods [178]. Foods high in dietary polyphenols, found in colourful fruits, vegetables, spices, teas, and wines, are becoming popular with their powerful antioxidant and anti-inflammatory activities in the brain [178, 179]. In particular, polyphenols have garnered special attention for their broad spectrum of molecular and cellular actions against neurological degeneration, due to their ability to cross the blood-brain barrier [177, 180]. Polyphenols can modulate synaptic transmission, enhance cognitive function with their antioxidant properties and engage signalling pathways that link molecules that act at the interface between cellular metabolism and synaptic plasticity [178]. Besides, polyphenols can improve cognitive function in models of neurodegenerative diseases [181, 182]. In this way, chronic resveratrol treatment further increased serotonin content in the brain through a protective effect on limiting enzymes TPH-1 and TPH-2 [183] which was accompanied by delay memory disability associated with healthy aging. It suggests that treatment with trans-resveratrol might aid to prevent the descent in indoleamines that normally occurs as a consequence of aging.

CONCLUSION

As the synthesis of serotonin, and consequently the production of melatonin, can be modulated by treatments with tryptophan and different potent antioxidants (such as vitamin E and resveratrol), the decrease in serotonin and melatonin that normally occurs during aging could be prevented, and perhaps could help to promote healthy brain aging. The enhancement in the levels of serotonin in brain regions involved in memory processing, might have a restorative effect of the cognitive functions that are impaired as a consequence of aging. Thus, dietary intake of foods rich in tryptophan, serotonin, melatonin and other antioxidants could be used as a therapy to mitigate the cognitive decline associated with aging, and in diseases that involve reductions in these indoleamines.

ACKNOWLEDGMENTS

This study was supported by grant (Programa Pont La Caixa-UIB 7/2014). Ramis M.R. was supported by FPU fellowship (MEC) and Sarubbo F. was supported by UIB fellowship.

REFERENCES

[1] Jiang CH, Tsien JZ, Schultz PG, Hu Y. The effects of aging on gene expression in the hypothalamus and cortex of mice. *Proc Natl Acad Sci USA*. 2001; 98(4):1930-4.

[2] Bishop NA, Lu T, Yankner BA. Neural mechanisms of ageing and cognitive decline. *Nature* 2010; 464(7288):529-35.

[3] Mandavilli BS, Santos JH, Van Houten B. Mitochondrial DNA repair and aging. *Mutat Res*. 2002; 509(1-2):127-51.

[4] Viña J, Borrás C, Miquel J. Theories of ageing. *IUBMB Life*. 2007; 59(4-5):249-54.

[5] Park DC, Reuter-Lorenz P. The adaptive brain: aging and neurocognitive scaffolding. *Annu Rev Psychol*. 2009; 60:173-96.

[6] Esteban S, Garau C, Aparicio S, Moranta D, Barceló P, Ramis M, et al. Improving effects of long-term growth hormone treatment on monoaminergic neurotransmission and related behavioral tests in aged rats. *Rejuvenation Res*. 2010; 13(6):707-16.

[7] Esteban S, Garau C, Aparicio S, Moranta D, Barceló P, Fiol MA, et al. Chronic melatonin treatment and its precursor L-tryptophan improve the monoaminergic neurotransmission and related behavior in the aged rat brain. *J Pineal Res*. 2010; 48(2):170-7.

[8] Pieramico V, Esposito R, Cesinaro S, Frazzini V, Sensi SL. Effects of non-pharmacological or pharmacological interventions on cognition and brain plasticity of aging individuals. *Front Syst Neurosci*. 2014; 8:153.

[9] Bopp KL, Verhaeghen P. Aging and verbal memory span: a meta-analysis. *J Gerontol B Psychol Sci Soc Sci.* 2005; 60(5):P223-33.

[10] Payer D, Marshuetz C, Sutton B, Hebrank A, Welsh RC, Park DC. Decreased neural specialization in old adults on a working memory task. *Neuroreport.* 2006; 17(5):487-91.

[11] Borella E, Carretti B, De Beni R. Working memory and inhibition across the adult life-span. *Acta Psychol* (Amst). 2008; 128(1):33-44.

[12] Bizon JL, Foster TC, Alexander GE, Glisky EL. Characterizing cognitive aging of working memory and executive function in animal models. *Front Aging Neurosci.* 2012; 4:19.

[13] Park DC, Lautenschlager G, Hedden T, Davidson NS, Smith AD, Smith PK. Models of visuospatial and verbal memory across the adult life span. *Psychol Aging.* 2002; 17(2):299-320.

[14] Takeuchi H, Taki Y, Kawashima R. Effects of working memory training on cognitive functions and neural systems. *Rev Neurosci.* 2010; 21(6):427-49.

[15] Morrison AB, Chein JM. Does working memory training work? The promise and challenges of enhancing cognition by training working memory. *Psychon Bull Rev.* 2011; 18(1):46-60.

[16] Shipstead Z, Redick TS, Engle RW. Is working memory training effective? *Psychol Bull.* 2012; 138(4):628-54.

[17] Morrison RG, Doumas LA, Richland LE. A computational account of children's analogical reasoning: balancing inhibitory control in working memory and relational representation. *Dev Sci.* 2011; 14:516-529.

[18] Andrews-Hanna JR, Snyder AZ, Vincent JL, Lustig C, Head D, Raichle ME, et al. Disruption of large-scale brain systems in advanced aging. *Neuron.* 2007; 56(5):924-35.

[19] Haider S, Khaliq S, Ahmed SP, Haleem DJ. Long-term tryptophan administration enhances cognitive performance and increases 5HT metabolism in the hippocampus of female rats. *Amino Acids.* 2006; 31(4):421-5.

[20] Haider S, Khaliq S, Haleem DJ. Enhanced serotonergic neurotransmission in the hippocampus following tryptophan administration improves learning acquisition and memory consolidation in rats. *Pharmacol Rep.* 2007; 59(1):53-7.

[21] Tsunemi A, Utsuyama M, Seidler BK, Kobayashi S, Hirokawa K. Age-related decline of brain monoamines in mice is reversed to young level by Japanese herbal medicine. *Neurochem Res.* 2005; 30(1):75-81.

[22] Diamond DM, Park CR, Woodson JC. Stress generates emotional memories and retrograde amnesia by inducing an endogenous form of hippocampal LTP. *iippocampus.* 2004; 14(3):281-91.

[23] Collier TJ, Greene JG, Felten DL, Stevens SY, Collier KS. Reduced cortical noradrenergic neurotransmission is associated with increased neophobia and impaired spatial memory in aged rats. *Neurobiol Aging.* 2004; 25(2):209-21.

[24] Burgess N, Maguire EA, O'Keefe J. The human hippocampus and spatial and episodic memory. *Neuron.* 2002; 35(4):625-41.

[25] Peters R. Ageing and the brain. *Postgrad Med J.* 2006; 82(964):84-8.

[26] Schmitt JA, Wingen M, Ramaekers JG, Evers EA, Riedel WJ. Serotonin and human cognitive performance. *Curr Pharm Des.* 2006; 12(20):2473-86.

[27] González-Burgos I, Feria-Velasco A. Serotonin/dopamine interaction in memory formation. *Prog Brain Res.* 2008; 172:603-23.

[28] Del Arco A, Segovia G, de Blas M, Garrido P, Acuña-Castroviejo D, Pamplona R, et al. Prefrontal cortex, caloric restriction and stress during aging: studies on dopamine and acetylcholine release, BDNF and working memory. *Behav Brain Res.* 2011; 216(1):136-45.

[29] McNeill TH, Koek LL, Haycock JW. The nigrostriatal system and aging. *Peptides.* 1984; 5 Suppl 1:263-8.

[30] Luine V, Bowling D, Hearns M. Spatial memory deficits in aged rats: contributions of monoaminergic systems. *Brain Res.* 1990; 537(1-2):271-8.

[31] Meneses A. 5-HT system and cognition. *Neurosci Biobehav Rev.* 1999; 23(8):1111-25.

[32] Volkow ND, Logan J, Fowler JS, Wang GJ, Gur RC, Wong C, et al. Association between age-related decline in brain dopamine activity and impairment in frontal and cingulate metabolism. *Am J Psychiatry.* 2000; 157(1):75-80.

[33] Rahman MS, Thomas P. Restoration of tryptophan hydroxylase functions and serotonin content in the Atlantic croaker hypothalamus by antioxidant treatment during hypoxic stress. *Front Neurosci.* 2014; 8:130.

[34] Walther DJ, Peter JU, Bashammakh S, Hörtnagl H, Voits M, Fink H, et al. Synthesis of serotonin by a second tryptophan hydroxylase isoform. *Science.* 2003; 299(5603):76.

[35] Zhang X, Beaulieu JM, Sotnikova TD, Gainetdinov RR, Caron MG. Tryptophan hydroxylase-2 controls brain serotonin synthesis. *Science.* 2004; 305(5681):217.

[36] Reiter RJ. Pineal melatonin: cell biology of its synthesis and of its physiological interactions. *Endocr Rev.* 1991; 12(2):151-80.

[37] Huether G, Poeggeler B, Reimer A, George A. Effect of tryptophan administration on circulating melatonin levels in chicks and rats: evidence for stimulation of melatonin synthesis and release in the gastrointestinal tract. *Life Sci.* 1992; 51(12):945-53.

[38] Tan DX, Manchester LC, Reiter RJ, Qi WB, Zhang M, Weintraub ST, et al. Identification of highly elevated levels of melatonin in bone marrow: its origin and significance. *Biochim Biophys Acta.* 1999; 1472(1-2):206-14.

[39] Stefulj J, Hörtner M, Ghosh M, Schauenstein K, Rinner I, Wölfler A, et al. Gene expression of the key enzymes of melatonin synthesis in extrapineal tissues of the rat. *J Pineal Res.* 2001; 30(4):243-7.

[40] Slominski A, Pisarchik A, Semak I, Sweatman T, Wortsman J, Szczesniewski A, et al. Serotoninergic and melatoninergic systems are fully expressed in human skin. *FASEB J.* 2002; 16(8):896-8.

[41] Lanoix D, Guérin P, Vaillancourt C. Placental melatonin production and melatonin receptor expression are altered in preeclampsia: new insights into the role of this hormone in pregnancy. *J Pineal Res.* 2012; 53(4):417-25.

[42] Acuña-Castroviejo D, Escames G, Venegas C, Díaz-Casado ME, Lima-Cabello E, López LC, et al. Extrapineal melatonin: sources, regulation, and potential functions. *Cell Mol Life Sci.* 2014; 71(16):2997-3025.

[43] Stehle JH, Saade A, Rawashdeh O, Ackermann K, Jilg A, Sebestény T, et al. A survey of molecular details in the human pineal gland in the light of phylogeny, structure, function and chronobiological diseases. *J Pineal Res.* 2011; 51(1):17-43.

[44] García JJ, López-Pingarrón L, Almeida-Souza P, Tres A, Escudero P, García-Gil FA, et al. Protective effects of melatonin in reducing oxidative stress and in preserving the fluidity of biological membranes: a review. *J Pineal Res.* 2014; 56(3):225-37.

[45] Klein DC, Coon SL, Roseboom PH, Weller JL, Bernard M, Gastel JA, et al. The melatonin rhythm-generating enzyme: molecular regulation of serotonin N-acetyltransferase in the pineal gland. *Recent Prog Horm Res.* 1997; 52:307-58.

[46] Reiter RJ. Neuroendocrine effects of light. *Int J Biometeorol.* 1991; 35(3):169-75.

[47] Sugden D. Melatonin biosynthesis in the mammalian pineal gland. *Experientia.* 1989; 45(10):922-32.

[48] Reiter RJ. The pineal gland: an intermediary between the environment and the endocrine system. *Psychoneuroendocrinology.* 1983; 8(1):31-40.

[49] Frick B, Schroecksnadel K, Neurauter G, Leblhuber F, Fuchs D. Increasing production of homocysteine and neopterin and degradation of tryptophan with older age. *Clin Biochem.* 2004; 37(8):684-7.

[50] Patnaik SK, Patnaik R. Age-related differential induction of tryptophan pyrrolase by hydrocortisone in the liver of male rats. *Biochem Int.* 1989; 18(6):1221-8.

[51] Buydens-Branchey L, Branchey M, Worner TM, Zucker D, Aramsombatdee E, Lieber CS. Increase in tryptophan oxygenase activity in alcoholic patients. *Alcohol Clin Exp Res.* 1988; 12(1):163-7.

[52] Ball HJ, Yuasa HJ, Austin CJ, Weiser S, Hunt NH. Indoleamine 2,3-dioxygenase-2; a new enzyme in the kynurenine pathway. *Int J Biochem Cell Biol.* 2009; 41(3):467-71.

[53] Poormasjedi-Meibod MS, Jalili RB, Hosseini-Tabatabaei A, Hartwell R, Ghahary A. Immuno-regulatory function of indoleamine 2,3 dioxygenase through modulation of innate immune responses. *PLoS One.* 2013; 8(8):e71044.

[54] Harman D. Aging: a theory based on free radical and radiation chemistry. *J Gerontol.* 1956; 11(3):298-300.

[55] Brieger K, Schiavone S, Miller FJ, Krause KH. Reactive oxygen species: from health to disease. *Swiss Med Wkly.* 2012; 142:w13659.

[56] Eddy DE, Harman D. Free radical theory of aging: effect of age, sex and dietary precursors on rat-brain docosahexanoic acid. *J Am Geriatr Soc.* 1977; 25(5):220-9.

[57] Sastre J, Pallardó FV, Viña J. Mitochondrial oxidative stress plays a key role in aging and apoptosis. *IUBMB Life.* 2000; 49(5):427-35.

[58] Floyd RA, Hensley K. Oxidative stress in brain aging. Implications for therapeutics of neurodegenerative diseases. *Neurobiol Aging.* 2002; 23(5):795-807.

[59] Mariani E, Polidori MC, Cherubini A, Mecocci P. Oxidative stress in brain aging, neurodegenerative and vascular diseases: an overview. *J Chromatogr B Analyt Technol Biomed Life Sci.* 2005; 827(1):65-75.

[60] Freeman LR, Keller JN. Oxidative stress and cerebral endothelial cells: regulation of the blood-brain-barrier and antioxidant based interventions. *Biochim Biophys Acta.* 2012; 1822(5):822-9.

[61] Reiter RJ. Oxidative damage in the central nervous system: protection by melatonin. *Prog Neurobiol.* 1998; 56(3):359-84.

[62] Sohal RS, Arnold LA, Sohal BH. Age-related changes in antioxidant enzymes and prooxidant generation in tissues of the rat with special reference to parameters in two insect species. *Free Radic Biol Med.* 1990; 9(6):495-500.

[63] Coyle JT, Puttfarcken P. Oxidative stress, glutamate, and neurodegenerative disorders. *Science.* 1993; 262(5134):689-95.

[64] Tian L, Cai Q, Wei H. Alterations of antioxidant enzymes and oxidative damage to macromolecules in different organs of rats during aging. *Free Radic Biol Med.* 1998; 24(9):1477-84.

[65] Rodrigues Siqueira I, Fochesatto C, da Silva Torres IL, Dalmaz C, Alexandre Netto C. Aging affects oxidative state in hippocampus, hypothalamus and adrenal glands of Wistar rats. *Life Sci.* 2005; 78(3):271-8.

[66] Pietá Dias C, Martins de Lima MN, Presti-Torres J, Dornelles A, Garcia VA, Siciliani Scalco F, et al. Memantine reduces oxidative damage and enhances long-term recognition memory in aged rats. *Neuroscience.* 2007; 146(4):1719-25.

[67] Butterfield DA, Abdul HM, Newman S, Reed T. Redox proteomics in some age-related neurodegenerative disorders or models thereof. *NeuroRx.* 2006; 3(3):344-57.

[68] Perrig WJ, Perrig P, Stähelin HB. The relation between antioxidants and memory performance in the old and very old. *J Am Geriatr Soc.* 1997; 45(6):718-24.

[69] Berr C. Cognitive impairment and oxidative stress in the elderly: results of epidemiological studies. *Biofactors.* 2000; 13(1-4):205-9.

[70] Rinaldi P, Polidori MC, Metastasio A, Mariani E, Mattioli P, Cherubini A, et al. Plasma antioxidants are similarly depleted in mild cognitive impairment and in Alzheimer's disease. *Neurobiol Aging.* 2003; 24(7):915-9.

[71] Dröge W, Schipper HM. Oxidative stress and aberrant signaling in aging and cognitive decline. *Aging Cell.* 2007; 6(3):361-70.

[72] Fukui K, Omoi NO, Hayasaka T, Shinnkai T, Suzuki S, Abe K, et al. Cognitive impairment of rats caused by oxidative stress and aging, and its prevention by vitamin E. *Ann N Y Acad Sci.* 2002; 959:275-84.

[73] Maxwell SR. Prospects for the use of antioxidant therapies. *Drugs.* 1995; 49(3):345-61.

[74] Joseph JA, Shukitt-Hale B, Denisova NA, Prior RL, Cao G, Martin A, et al. Long-term dietary strawberry, spinach, or vitamin E supplementation retards the onset of age-related neuronal signal-transduction and cognitive behavioral deficits. *J Neurosci.* 1998; 18(19):8047-55.

[75] Galli RL, Shukitt-Hale B, Youdim KA, Joseph JA. Fruit polyphenolics and brain aging: nutritional interventions targeting age-related neuronal and behavioral deficits. *Ann N Y Acad Sci.* 2002; 959:128-32.

[76] Youdim KA, Qaiser MZ, Begley DJ, Rice-Evans CA, Abbott NJ. Flavonoid permeability across an in situ model of the blood-brain barrier. *Free Radic Biol Med.* 2004; 36(5):592-604.

[77] Míguez JM, Aldegunde M, Paz-Valiñas L, Recio J, Sánchez-Barceló E. Selective changes in the contents of noradrenaline, dopamine and serotonin in rat brain areas during aging. *J Neural Transm.* 1999; 106(11-12):1089-98.

[78] Reiter RJ, Richardson BA, Johnson LY, Ferguson BN, Dinh DT. Pineal melatonin rhythm: reduction in aging Syrian hamsters. *Science.* 1980; 210(4476):1372-3.

[79] Reiter RJ, Craft CM, Johnson JE, King TS, Richardson BA, Vaughan GM, et al. Age-associated reduction in nocturnal pineal melatonin levels in female rats. *Endocrinology.* 1981; 109(4):1295-7.

[80] Reiter RJ. The ageing pineal gland and its physiological consequences. *Bioessays.* 1992; 14(3):169-75.

[81] Pang SF, Tang F, Tang PL. Negative correlation of age and the levels of pineal melatonin, pineal N-acetylserotonin, and serum melatonin in male rats. *J Exp Zool.* 1984; 229(1):41-7.

[82] Nair NP, Hariharasubramanian N, Pilapil C, Isaac I, Thavundayil JX. Plasma melatonin--an index of brain aging in humans? *Biol Psychiatry.* 1986; 21(2):141-50.

[83] Poeggeler B, Reiter RJ, Tan DX, Chen LD, Manchester LC. Melatonin, hydroxyl radical-mediated oxidative damage, and aging: a hypothesis. *J Pineal Res.* 1993; 14(4):151-68.

[84] Pierpaoli W, Regelson W. Pineal control of aging: effect of melatonin and pineal grafting on aging mice. *Proc Natl Acad Sci U S A.* 1994; 91(2):787-91.

[85] Touitou Y. Human aging and melatonin. Clinical relevance. Exp Gerontol. 2001; 36(7):1083-100.

[86] Hussain AM, Mitra AK. Effect of aging on tryptophan hydroxylase in rat brain: implications on serotonin level. *Drug Metab Dispos.* 2000; 28(9):1038-42.

[87] Haider S, Saleem S, Perveen T, Tabassum S, Batool Z, Sadir S, et al. Age-related learning and memory deficits in rats: role of altered brain neurotransmitters, acetylcholinesterase activity and changes in antioxidant defense system. *Age (Dordr).* 2014; 36(3):9653.

[88] Cash CD. Why tryptophan hydroxylase is difficult to purify: a reactive oxygen-derived species-mediated phenomenon that may be implicated in human pathology. *Gen Pharmacol.* 1998; 30(4):569-74.

[89] Moranta D, Barceló P, Aparicio S, Garau C, Sarubbo F, Ramis M, et al. Intake of melatonin increases tryptophan hydroxylase type 1 activity in aged rats: Preliminary study. *Exp Gerontol.* 2014; 49:1-4.

[90] Rúzsás C, Mess B. Melatonin and aging. A brief survey. *Neuro Endocrinol Lett.* 2000; 21(1):17-23.

[91] Reiter RJ, Pablos MI, Agapito TT, Guerrero JM. Melatonin in the context of the free radical theory of aging. *Ann N Y Acad Sci.* 1996; 786:362-78.

[92] Reiter RJ. Melatonin: Lowering the High Price of Free Radicals. *News Physiol Sci.* 2000; 15:246-50.

[93] Tan DX, Manchester LC, Terron MP, Flores LJ, Reiter RJ. One molecule, many derivatives: a never-ending interaction of melatonin with reactive oxygen and nitrogen species? *J Pineal Res.* 2007 ; 42(1):28-42.

[94] Reiter RJ, Tan DX, Manchester LC, Qi W. Biochemical reactivity of melatonin with reactive oxygen and nitrogen species: a review of the evidence. *Cell Biochem Biophys.* 2001; 34(2):237-56.

[95] Tan DX, Reiter RJ, Manchester LC, Yan MT, El-Sawi M, Sainz RM, et al. Chemical and physical properties and potential mechanisms: melatonin as a broad spectrum antioxidant and free radical scavenger. *Curr Top Med Chem.* 2002; 2(2):181-97.

[96] Allegra M, Reiter RJ, Tan DX, Gentile C, Tesoriere L, Livrea MA. The chemistry of melatonin's interaction with reactive species. *J Pineal Res.* 2003; 34(1):1-10.

[97] Rodriguez C, Mayo JC, Sainz RM, Antolín I, Herrera F, Martín V, et al. Regulation of antioxidant enzymes: a significant role for melatonin. *J Pineal Res.* 2004; 36(1):1-9.

[98] Manda K, Ueno M, Anzai K. Space radiation-induced inhibition of neurogenesis in the hippocampal dentate gyrus and memory impairment in mice: ameliorative potential of the melatonin metabolite, AFMK. *J Pineal Res.* 2008; 45(4):430-8.

[99] Galano A, Tan DX, Reiter RJ. On the free radical scavenging activities of melatonin's metabolites, AFMK and AMK. *J Pineal Res.* 2013; 54(3):245-57.

[100] Tan DX, Hardeland R, Manchester LC, Galano A, Reiter RJ. Cyclic-3-hydroxymelatonin (C3HOM), a potent antioxidant, scavenges free radicals and suppresses oxidative reactions. *Curr Med Chem.* 2014; 21(13):1557-65.

[101] Grenett HE, Ledley FD, Reed LL, Woo SL. Full-length cDNA for rabbit tryptophan hydroxylase: functional domains and evolution of aromatic amino acid hydroxylases. *Proc Natl Acad Sci U S A.* 1987; 84(16):5530-4.

[102] Kuhn DM, Ruskin B, Lovenberg W. Tryptophan hydroxylase. The role of oxygen, iron, and sulfhydryl groups as determinants of stability and catalytic activity. *J Biol Chem.* 1980; 255(9):4137-43.

[103] De La Cruz CP, Revilla E, Venero JL, Ayala A, Cano J, Machado A. Oxidative inactivation of tyrosine hydroxylase in substantia nigra of aged rat. *Free Radic Biol Med.* 1996; 20(1):53-61.

[104] Karakuyu D, Herold C, Güntürkün O, Diekamp B. Differential increase of extracellular dopamine and serotonin in the 'prefrontal cortex' and striatum of pigeons during working memory. *Eur J Neurosci.* 2007; 26(8):2293-302.

[105] Fontana DJ, Daniels SE, Wong EH, Clark RD, Eglen RM. The effects of novel, selective 5-hydroxytryptamine (5-HT)4 receptor ligands in rat spatial navigation. *Neuropharmacology.* 1997; 36(4-5):689-96.

[106] Marchetti-Gauthier E, Roman FS, Dumuis A, Bockaert J, Soumireu-Mourat B. BIMU1 increases associative memory in rats by activating 5-HT4 receptors. *Neuropharmacology.* 1997; 36(4-5):697-706.

[107] Kulla A, Manahan-Vaughan D. Modulation by serotonin 5-HT(4) receptors of long-term potentiation and depotentiation in the dentate gyrus of freely moving rats. *Cereb Cortex.* 2002; 12(2):150-62.

[108] Kemp A, Manahan-Vaughan D. The 5-hydroxytryptamine4 receptor exhibits frequency-dependent properties in synaptic plasticity and behavioural metaplasticity in the hippocampal CA1 region in vivo. *Cereb Cortex.* 2005; 15(7):1037-43.

[109] Garau C, Aparicio S, Rial RV, Nicolau MC, Esteban S. Age-related changes in circadian rhythm of serotonin synthesis in ring doves: effects of increased tryptophan ingestion. *Exp Gerontol.* 2006; 41(1):40-8.

[110] Esteban S, Nicolaus C, Garmundi A, Rial RV, Rodríguez AB, Ortega E, et al. Effect of orally administered L-tryptophan on serotonin, melatonin, and the innate immune response in the rat. *Mol Cell Biochem.* 2004; 267(1-2):39-46.

[111] Eichenbaum H. Is the rodent hippocampus just for 'place'? *Curr Opin Neurobiol.* 1996; 6(2):187-95.

[112] Nadel L, Moscovitch M. The hippocampal complex and long-term memory revisited. *Trends Cogn Sci.* 2001; 5(6):228-30.

[113] Bannerman DM, Rawlins JN, McHugh SB, Deacon RM, Yee BK, Bast T, et al. Regional dissociations within the hippocampus--memory and anxiety. *Neurosci Biobehav Rev.* 2004; 28(3):273-83.

[114] Adams W, Kusljic S, van den Buuse M. Serotonin depletion in the dorsal and ventral hippocampus: effects on locomotor hyperactivity, prepulse inhibition and learning and memory. *Neuropharmacology.* 2008; 55(6):1048-55.

[115] Chang SF, Ayres JW, Sandine WE. Analysis of cheese for histamine, tyramine, tryptamine, histidine, tyrosine, and tryptophane. *J Dairy Sci.* 1985; 68(11):2840-6.
[116] USDA National Nutrient Database for Standard Reference, Release 26 on May 19, 2014. *Chart displays levels of tryptophan in various foods.*
[117] Guillén-Casla V, Rosales-Conrado N, León-González ME, Pérez-Arribas LV, Polo-Díez LM. Determination of serotonin and its precursors in chocolate samples by capillary liquid chromatography with mass spectrometry detection. *J Chromatogr A.* 2012; 1232:158-65.
[118] Feldman JM, Lee EM. Serotonin content of foods: effect on urinary excretion of 5-hydroxyindoleacetic acid. *Am J Clin Nutr.* 1985; 42(4):639-43.
[119] Waalkes TP, Sjoerdsma A, Creveling CR, Weissbach H, Udenfriend S. Serotonin, norepinephrine, and related compounds in bananas. . Science 1958; 127(3299):648-50.
[120] Bruce DW. Serotonin in pineapple. *Nature* 1960; 188:147.
[121] González-Gómez D, Lozano M, Fernández-León MF, Ayuso MC, Bernalte MJ, Rodríguez AB. Detection and quantification of melatonin and serotonin in eight Sweet Cherry cultivars (Prunus avium L.). *European Food Research and Technology* 2009; 229(2): 223-229.
[122] Dubbels R, Reiter RJ, Klenke E, Goebel A, Schnakenberg E, Ehlers C, et al. Melatonin in edible plants identified by radioimmunoassay and by high performance liquid chromatography-mass spectrometry. *J Pineal Res.* 1995; 18(1):28-31.
[123] Manchester LC, Poeggeler B, Alvares FL, Ogden GB, Reiter RJ. Melatonin immunoreactivity in the photosynthetic prokaryote Rhodospirillum rubrum: implications for an ancient antioxidant system. *Cell Mol Biol Res.* 1995; 41(5):391-5.
[124] Hardeland R, Fuhrberg B. Ubiquitous melatonin presence and effects in unicells, plants and animals. *Trends Comp Biochem;* 1996. 2:25-44.
[125] Reiter RJ, Tan DX, Rosales-Corral S, Manchester LC. The universal nature, unequal distribution and antioxidant functions of melatonin and its derivatives. *Mini Rev Med Chem.* 2013; 13(3):373-84.
[126] Menendez-Pelaez A, Reiter RJ. Distribution of melatonin in mammalian tissues: the relative importance of nuclear versus cytosolic localization. *J Pineal Res.* 1993; 15(2):59-69.
[127] Tan DX, Chen LD, Poeggeler B, Manchester LC, Reiter RJ. Melatonin: a potent, endogenous hydroxyl radical scavenger. *Endocr J* 1993; 1:57-60.
[128] Peyrot F, Ducrocq C. Potential role of tryptophan derivatives in stress responses characterized by the generation of reactive oxygen and nitrogen species. *J Pineal Res.* 2008; 45(3):235-46.
[129] Hardeland R, Tan DX, Reiter RJ. Kynuramines, metabolites of melatonin and other indoles: the resurrection of an almost forgotten class of biogenic amines. *J Pineal Res.* 2009; 47(2):109-26.
[130] Acuña-Castroviejo D, Martín M, Macías M, Escames G, León J, Khaldy H, et al. Melatonin, mitochondria, and cellular bioenergetics. *J Pineal Res.* 2001; 30(2):65-74.
[131] Eckmann J, Eckert SH, Leuner K, Muller WE, Eckert GP. Mitochondria: mitochondrial membranes in brain ageing and neurodegeneration. *Int J Biochem Cell Biol.* 2013; 45(1):76-80.
[132] Rattan SI. Theories of biological aging: genes, proteins, and free radicals. *Free Radic Res.* 2006; 40(12):1230-8.

[133] Martín M, Macías M, Escames G, León J, Acuña-Castroviejo D. Melatonin but not vitamins C and E maintains glutathione homeostasis in t-butyl hydroperoxide-induced mitochondrial oxidative stress. *FASEB J.* 2000; 14(12):1677-9.

[134] Paradies G, Petrosillo G, Paradies V, Reiter RJ, Ruggiero FM. Melatonin, cardiolipin and mitochondrial bioenergetics in health and disease. *J Pineal Res.* 2010; 48(4):297-310.

[135] Karbownik M, Tan D, Manchester LC, Reiter RJ. Renal toxicity of the carcinogen delta-aminolevulinic acid: antioxidant effects of melatonin. *Cancer Lett.* 2000; 161(1):1-7.

[136] Reiter RJ. Oxidative processes and antioxidative defense mechanisms in the aging brain. *FASEB J.* 1995; 9(7):526-33.

[137] Manev H, Uz T, Kharlamov A, Joo JY. Increased brain damage after stroke or excitotoxic seizures in melatonin-deficient rats. *FASEB J.* 1996; 10(13):1546-51.

[138] Reiter RJ, Guerrero JM, Garcia JJ, Acuña-Castroviejo D. Reactive oxygen intermediates, molecular damage, and aging. Relation to melatonin. *Ann N Y Acad Sci.* 1998; 854:410-24.

[139] Antolín I, Mayo JC, Sainz RM, del Brío MeL, Herrera F, Martín V, et al. Protective effect of melatonin in a chronic experimental model of Parkinson's disease. *Brain Res.* 2002; 943(2):163-73.

[140] Matsubara E, Bryant-Thomas T, Pacheco Quinto J, Henry TL, Poeggeler B, Herbert D, et al. Melatonin increases survival and inhibits oxidative and amyloid pathology in a transgenic model of Alzheimer's disease. *J Neurochem.* 2003; 85(5):1101-8.

[141] Reiter RJ, Manchester LC, Tan DX. Melatonin in walnuts: influence on levels of melatonin and total antioxidant capacity of blood. *Nutrition.* 2005; 21(9):920-4.

[142] Okazaki M, Ezura H. Profiling of melatonin in the model tomato (Solanum lycopersicum L.) cultivar Micro-Tom. *J Pineal Res.* 2009; 46(3):338-43.

[143] Posmyk MM, Janas KM. Melatonin in plants. *Acta Physiol Plant;* 2009; 31(1): 1-11.

[144] Iriti M, Varoni EM, Vitalini S. Melatonin in traditional Mediterranean diets. *J Pineal Res.* 2010; 49(2):101-5.

[145] Howatson G, Bell PG, Tallent J, Middleton B, McHugh MP, Ellis J. Effect of tart cherry juice (Prunus cerasus) on melatonin levels and enhanced sleep quality. *Eur J Nutr.* 2012; 51(8):909-16.

[146] Tan DX, Hardeland R, Manchester LC, Korkmaz A, Ma S, Rosales-Corral S, et al. Functional roles of melatonin in plants, and perspectives in nutritional and agricultural science. *J Exp Bot.* 2012; 63(2):577-97.

[147] Vitalini S, Gardana C, Simonetti P, Fico G, Iriti M. Melatonin, melatonin isomers and stilbenes in Italian traditional grape products and their antiradical capacity. *J Pineal Res.* 2013; 54(3):322-33.

[148] Fernández-Pachón MS, Medina S, Herrero-Martín G, Cerrillo I, Berná G, Escudero-López B, et al. Alcoholic fermentation induces melatonin synthesis in orange juice. *J Pineal Res.* 2014; 56(1):31-8.

[149] Riga P, Medina S, García-Flores LA, Gil-Izquierdo Á. Melatonin content of pepper and tomato fruits: effects of cultivar and solar radiation. *Food Chem.* 2014; 156:347-52.

[150] Feng X, Wang M, Zhao Y, Han P, Dai Y. Melatonin from different fruit sources, functional roles, and analytical methods.: *Trends in Food Science & Technology* 2014; 37(1): 21-31.

[151] Tan DX, Zanghi BM, Manchester LC, Reiter RJ. Melatonin identified in meats and other food stuffs: potentially nutritional impact. *J Pineal Res.* 2014; 57(2):213-8.

[152] Schapira AH. Mitochondrial involvement in Parkinson's disease, Huntington's disease, hereditary spastic paraplegia and Friedreich's ataxia. Biochim *Biophys Acta.* 1999; 1410(2):159-70.

[153] Gutteridge JM, Halliwell B. Antioxidants: Molecules, medicines, and myths. *Biochem Biophys Res Commun.* 2010; 393(4):561-4.

[154] Evans HM, Bishop KS. On the existence of a hitherto unrecognized dietary factor essential for reproduction. *Science.* 1922; 56(1458):650-1.

[155] Traber MG, Stevens JF. Vitamins C and E: beneficial effects from a mechanistic perspective. *Free Radic Biol Med.* 2011; 51(5):1000-13.

[156] Muller DP. Vitamin E and neurological function. *Mol Nutr Food Res.* 2010; 54(5):710-8.

[157] Schneider C. Chemistry and biology of vitamin E. *Mol Nutr Food Res.* 2005; 49(1):7-30.

[158] Tayyaba K, Hasan M. Vitamin E protects against metasystox-induced adverse effect on lipid metabolism in the rat brain and spinal cord. *Acta Pharmacol Toxicol (Copenh).* 1985; 57(3):190-6.

[159] Yoshida S, Busto R, Watson BD, Santiso M, Ginsberg MD. Postischemic cerebral lipid peroxidation in vitro: modification by dietary vitamin E. *J Neurochem.* 1985; 44(5):1593-601.

[160] Haider SS, el-Fakhri M. Action of alpha-tocopherol on vanadium-stimulated lipid peroxidation in rat brain. *Neurotoxicology.* 1991; 12(1):79-85.

[161] Kappus H, Diplock AT. Tolerance and safety of vitamin E: a toxicological position report. *Free Radic Biol Med.* 1992; 13(1):55-74.

[162] Niki E, Noguchi N, Gotoh N. Dynamics of lipid peroxidation and its inhibition by antioxidants. *Biochem Soc Trans.* 1993; 21(2):313-7.

[163] Socci DJ, Crandall BM, Arendash GW. Chronic antioxidant treatment improves the cognitive performance of aged rats. *Brain Res.* 1995; 693(1-2):88-94.

[164] Murray CA, Lynch MA. Dietary supplementation with vitamin E reverses the age-related deficit in long term potentiation in dentate gyrus. *J Biol Chem.* 1998; 273(20):12161-8.

[165] Navarro A, Gómez C, Sánchez-Pino MJ, González H, Bández MJ, Boveris AD, et al. Vitamin E at high doses improves survival, neurological performance, and brain mitochondrial function in aging male mice. *Am J Physiol Regul Integr Comp Physiol.* 2005; 289(5):R1392-9.

[166] Hara H, Kato H, Kogure K. Protective effect of alpha-tocopherol on ischemic neuronal damage in the gerbil hippocampus. *Brain Res.* 1990; 510(2):335-8.

[167] Vatassery GT, Bauer T, Dysken M. High doses of vitamin E in the treatment of disorders of the central nervous system in the aged. *Am J Clin Nutr.* 1999; 70(5):793-801.

[168] Anderson DK, Waters TR, Means ED. Pretreatment with alpha tocopherol enhances neurologic recovery after experimental spinal cord compression injury. *J Neurotrauma.* 1988; 5(1):61-7.

[169] Vatassery GT. Vitamin E. Neurochemistry and implications for neurodegeneration in Parkinson's disease. *Ann N Y Acad Sci.* 1992; 669:97-109.

[170] Vatassery GT, Angerhofer CK, Peterson FJ. Vitamin E concentrations in the brains and some selected peripheral tissues of selenium-deficient and vitamin E-deficient mice. *J Neurochem.* 1984; 42(2):554-8.
[171] Kuhn DM, Arthur RE. Inactivation of brain tryptophan hydroxylase by nitric oxide. *J Neurochem.* 1996; 67(3):1072-7.
[172] Kuhn DM, Arthur R. Molecular mechanism of the inactivation of tryptophan hydroxylase by nitric oxide: attack on critical sulfhydryls that spare the enzyme iron center. *J Neurosci.* 1997; 17(19):7245-51.
[173] Kuhn DM, Arthur RE. Inactivation of tryptophan hydroxylase by nitric oxide: enhancement by tetrahydrobiopterin. *J Neurochem.* 1997; 68(4):1495-502.
[174] Kuhn DM, Geddes TJ. Peroxynitrite inactivates tryptophan hydroxylase via sulfhydryl oxidation. Coincident nitration of enzyme tyrosyl residues has minimal impact on catalytic activity. *J Biol Chem.* 1999; 274(42):29726-32.
[175] Horwitt MK. Critique of the requirement for vitamin E. *Am J Clin Nutr.* 2001; 73(6):1003-5.
[176] Rossi L, Mazzitelli S, Arciello M, Capo CR, Rotilio G. Benefits from dietary polyphenols for brain aging and Alzheimer's disease. *Neurochem Res.* 2008; 33(12):2390-400.
[177] Pandey KB, Rizvi SI. Plant polyphenols as dietary antioxidants in human health and disease. *Oxid Med Cell Longev.* 2009; 2(5):270-8.
[178] Gomez-Pinilla F, Nguyen TT. Natural mood foods: the actions of polyphenols against psychiatric and cognitive disorders. *Nutr Neurosci.* 2012; 15(3):127-33.
[179] Arts IC, Hollman PC. Polyphenols and disease risk in epidemiologic studies. *Am J Clin Nutr.* 2005; 81(1 Suppl):317S-25S.
[180] Harikumar KB, Aggarwal BB. Resveratrol: a multitargeted agent for age-associated chronic diseases. *Cell Cycle.* 2008; 7(8):1020-35.
[181] Ma J, Yang WQ, Zha H, Yu HR. [Effect of naringenin on learning and memory ability on model rats with Alzheimer disease]. *Zhong Yao Cai.* 2013; 36(2):271-6.
[182] Khan MB, Khan MM, Khan A, Ahmed ME, Ishrat T, Tabassum R, et al. Naringenin ameliorates Alzheimer's disease (AD)-type neurodegeneration with cognitive impairment (AD-TNDCI) caused by the intracerebroventricular-streptozotocin in rat model. *Neurochem Int.* 2012; 61(7):1081-93.
[183] Sarubbo F, Ramis MR, Aparicio S, Ruiz L, Esteban S, Miralles A, et al. Improving effect of long-term trans-resveratrol treatment on central monoamines synthesis and cognition in aged rats. *Submitted to Neurobiol Aging;* 2014.

In: Indoleamines
Editor: Ángel Catalá

ISBN: 978-1-63482-097-4

Chapter 6

MELATONIN AS AN OCULAR ANTI-INFLAMMATORY

Ruth E. Rosenstein[1*], Diego C. Fernandez[1], Mónica S. Chianelli[1], María José Del Sole[2], Daniel A. Sáenz[1] and Pablo H. Sande[1]

[1] Laboratory of Retinal Neurochemistry and Experimental Ophthalmology, Department of Human Biochemistry, School of Medicine/CEFyBO, University of Buenos Aires/CONICET, Buenos Aires, Argentina

[2] Laboratory of Nervous System Physiology and Endocrinology, Department of Physiopathology, School of Veterinary Science, National University of the Center of Buenos Aires Province, Tandil, Argentina

ABSTRACT

Uveitis is a prevalent intraocular inflammatory disease, and one of the most damaging ocular conditions. Currently available therapies are relatively effective in the treatment of uveitis, but are often associated with severe side effects. In that context, uveitis remains a challenging field to ophthalmologists and represents a significant public health concern. Oxidative stress-induced inflammation is a major contributor to uveitis. In addition, altered nitric oxide (NO) metabolism, and increased prostaglandins (PGs) and tumor necrosis factorα (TNFα) levels also play a significant role in the pathogenesis of uveitis. Therefore, an antioxidant and anti-nitridergic therapy, as well as reducing PG and TNFα levels may be useful strategies to diminish tissue damage and ocular dysfunction induced by uveitis. Several lines of evidence strongly support that melatonin is a potent antioxidant molecule which is effective in scavenging free radicals that are generated in ocular tissues, and that it is able to reduce the retinal nitridergic pathway activity along with PG and TNFα levels. In view of the fact that melatonin lacks significant adverse collateral effects even at high doses, the application of melatonin could potentially protect ocular tissues by effectively scavenging free radicals, and reducing excessive amounts of NO, PGs, and TNFα generated in the uveitic eye. This chapter summarizes recent investigations showing that melatonin not only prevents clinical, biochemical, histological, ultrastructural, and functional consequences of

* Corresponding Author: Dr. Ruth E. Rosenstein; Department of Human Biochemistry, School of Medicine, CEFyBO; University of Buenos Aires, CONICET; Paraguay 2155, 5th floor, (1121); Buenos Aires, ARGENTINA; Tel: 54-11-45083672 (ext 37); Fax: 54-11-45083672 (ext 317); e-mail: ruthr@fmed.uba.ar.

experimental uveitis, but it is also capable of suppressing the actively ongoing ocular inflammatory response. These results strongly support the role of melatonin as a therapeutic strategy for uveitis treatment.

Keywords: Melatonin, uveitis, ocular inflammation, nitric oxide, oxidative damage

ABBREVIATIONS

AA	arachidonic acid
AA-NAT	arylalkylamine N-acetyltransferase
BOB	blood-ocular barrier
COX	cyclooxygenase
EAU	experimental autoimmune uveitis
EIU	endotoxin induced uveitis
ERG	electroretinogram
HIOMT	hydroxyl-indole-O-methyltransferase
IL	interleukin
iNOS	inducible isoform of NO synthase
IOP	intraocular pressure
LPS	lipopolysaccharide
NOS	NO synthase
NSAIDs	non-steroidal anti-inflammatory drugs
NFκB	nuclear factor-kappa B
PGs	prostaglandins
ROS	reactive oxygen species
TNFα	tumor necrosis factor α

MELATONIN IN THE EYE

Melatonin (5-methoxy-N-acetyltryptamine) is a ubiquitous natural substance which was first isolated and discovered by Lerner and co-workers in the late 1950s. [1-3] Melatonin is widely distributed in nature, being found both in plants and animals, [4-6] and it is probably one of the first biologically significant compounds that appeared in living organisms. Although it is well-established that melatonin is a major regulator of circadian rhythm, abundant evidence suggests that this hormone may be also involved in the regulation of a variety of physiological processes, such as sleep, reproduction, immune and vascular response, etc. (see [7-9] for reviews).

In vertebrates, the synthesis of melatonin takes place primarily in the pineal gland, but its synthesis also occurs in various parts of the eye, i.e., the retina, [10-13] the ciliary body [14] and the lachrymal gland.[15] Melatonin has been detected in retinas of various orders of vertebrates from fish to mammals by high performance liquid chromatography and immunohistochemistry [16-22]. The concentration of melatonin in the retina maintains at a relatively high level even following pinealectomy, [12, 23-25] suggesting that melatonin may be synthesized within the retina, independently from the pineal gland. Intraocularly, melatonin is synthesized through the same pathway which occurs in the pineal gland. [26] Tryptophan taken up from the blood is converted into serotonin, which is subsequently metabolized to N-acetyl serotonin by the enzyme arylalkylamine N-acetyltransferase (AA-NAT). N-acetyl serotonin is then converted into melatonin by the enzyme hydroxyindole-O-methyltransferase (HIOMT).

The earliest finding supporting melatonin's biosynthetic pathway in the mammalian retina was the description of a retinal HIOMT activity [10] and the demonstration that labeled serotonin is converted into melatonin in the rat retina. [11] The presence of HIOMT in the chicken retina at both protein and mRNA level has been confirmed. [27,28] Retinal AA-NAT levels show a circadian rhythm, peaking at night in the chicken and rat. [29,30] Moreover, the presence of AA-NAT in the human eye has been well documented. [31] In the rhesus monkey, retinal AA-NAT activity shows more than a four-fold increase at night. [32] In the hamster retina, the retinal melatonin biosynthesis is regulated by the light/dark cycle. [33] The finding that isolated *Xenopus* photoreceptor cells rhythmically secrete melatonin suggests that photoreceptors contain an endogenous circadian clock which regulates melatonin biosynthesis. [34] This hypothesis has been confirmed in the golden hamster and mouse retinas. [35,36] At present, it is not known whether the circadian rhythms in mammalian photoreceptors are driven by inner retinal clocks or by paracrine outputs. Experiments with the rodless mouse have shown that melatonin synthesis is not abolished by the complete loss of photoreceptors, but that its circadian expression disappears. [36,37] This finding suggests that rods are necessary for the rhythmic synthesis of melatonin. However, it has been demonstrated that chick retinal ganglion cells are also able to rhythmically synthesize melatonin, even in isolated conditions. [38] Regulated by the interaction between a circadian clock and the photic information, retinal melatonin levels rise rapidly after the onset of darkness and decrease after light exposure in the golden hamster and rat. [33,39]

The degradation of melatonin involves several enzymatic reactions, and ~90% of melatonin is cleaned out through a single passage through the liver. [40] Although there is no evidence that the degradation of melatonin occurs within the pineal gland, studies on cultured eye cups of several vertebrates have shown that the retina has the capacity for rapid metabolism of melatonin. [16,41,42] A two-step biological reaction may be responsible for melatonin degradation in the retina: [16] the initial step is deacetylation to 5-methoxytryptamine by arylacylamidase (aryl-acylamideamidohydrolase); in the subsequent step, 5-methoxytryptamine is deaminated by monoamine oxidase, producing 5-methoxyindole acetaldehyde, which is either oxidated to 5-methoxyindoleacetic acid or reduced to 5-methoxytryptophol. It should be noted that the metabolism of exogenous melatonin is undetectable in retinas of rat and pig, suggesting that melatonin deacetylation activity is very low or absent in mammalian retina, [41] even though 5-methoxytryptamine level rhythms have been detected in rat and hamster. [43, 44]

Whereas pineal melatonin acts as an endocrine hormone, retinal melatonin acts locally as a paracrine neuromodulator ([45], and reviewed in [46]). Ocular melatonin which is locally synthesized or entering from the circulation may contribute to the regulation of retinomotor movements, rod outer segment disc shedding, dopamine synthesis and release, and intraocular pressure (IOP). Retinal melatonin acts as a neuromodulator that mediates dark adaptive regulation of retinomotor movements, [47] and regulates dopamine release from amacrine cells. [48] Both dopamine and melatonin are key signaling agents in the regulation of retinal rhythmicity. These two substances are mutually inhibitory, acting as signals for day and night, respectively. It was demonstrated that picomolar concentrations of melatonin selectively inhibit the calcium-dependent release of dopamine from rabbit retina.[48] The action of melatonin on the rod pathway at the level of horizontal and amacrine cells has been proposed as a unique mechanism by which the retina adapts to low light intensities. [49] A correlation between melatonin levels and the electroretinographic response has been shown in humans which suggest that daily melatonin and electroretinogram (ERG) cycling are associated. [50] On the other hand, melatonin decreases retinal cAMP accumulation [33], regulates the activity of the retinal glutamate/glutamine cycle activity in the golden hamster, [51] and the GABAergic system in the rat retina. [52]

MELATONIN AS A RETINAL ANTIOXIDANT

The possibility that melatonin could detoxify reactive oxygen species (ROS) was originally suggested by Ianas et al. (1991) [53]. Shortly thereafter, Poeggeler et al. [54] and Reiter and coworkers [55], using spin trapping and electron resonance spectroscopy, demonstrated melatonin's capacity to directly scavenge the highly reactive hydroxyl radical. Since then, several reports have shown that melatonin acts as a free radical scavenger and an efficient antioxidant. [5, 56-59] Not only melatonin, but also several of its metabolites generated during its free radical scavenging action act as antioxidants. [60] The kynurenic pathway of melatonin metabolism includes a series of radical scavengers with the possible sequence: Melatonin → cyclic 3-hydroxymelatonin → N1-acetyl-N2-formyl-5-methoxykynuramine (AFMK) → N1-acetyl-5-methoxykynuramine. In the metabolic step from melatonin to AFMK, up to four free radicals can be consumed [60-63]. Because of this pathway, melatonin's efficacy as an antioxidant is greatly increased.

Melatonin has been shown to protect against oxidative stress in various, highly divergent experimental systems. Melatonin's functions as an antioxidant include: a) direct free radical scavenging, b) stimulation of antioxidative enzymes, c) increasing the efficiency of mitochondrial oxidative phosphorylation and reducing electron leakage (thereby lowering free radical generation), and d) augmenting the efficiency of other antioxidants. [64] Melatonin has been shown to scavenge free radicals generated in mitochondria, reduce electron leakage from the respiratory complexes and improve ATP synthesis. [65,66] Moreover, melatonin preserves mitochondrial glutathione levels, thereby enhancing the antioxidant potential. [67]

The retina is particularly susceptible to oxidative stress because of its high consumption of oxygen, its high proportion of polyunsaturated fatty acids, and its direct exposure to light. It was demonstrated that melatonin decreases lipid peroxidation of polyunsaturated fatty acids located in rod outer segment membranes,[68] it may protect retinal pigment epithelial cells from oxidative stress, [69] and it prevents retinal oxidative damage from ischemia/reperfusion injury. [70] In cultures of human retinal neuronal cells subjected to experimental hypoxanthine/xanthine oxidase-induced injury, melatonin is able to rescue retinal neurons. [71] Furthermore, in dark-adapted single photoreceptors isolated from the frog retina after illumination with saturating light, melatonin at picomolar and low nanomolar concentrations was shown to be 100 times more potent in inhibiting light-induced oxidative processes than vitamin E. [72] We have shown that melatonin increases superoxide dismutase activity and reduced glutathione levels, whereas it decreases lipid peroxidation in the rat retina. [52] Taken together, these results strongly support the role of melatonin as an effective retinal antioxidant.

MELATONIN AND THE RETINAL NITRIDERGIC PATHWAY

In isolated conditions, melatonin may directly react with nitric oxide (NO) yielding at least one stable product, N-nitrosomelatonin, which was characterized by nuclear magnetic resonance (NMR) and X-ray spectroscopy. [73,74] In addition to a direct scavenging of NO, low concentrations of melatonin significantly decrease hamster retinal NOS activity. Melatonin behaves as a non-competitive inhibitor of this enzyme, since it decreases the Vmax, with no changes on the Km. [75] This result agrees with the previously described effect of melatonin on NOS from other neural structures such as hypothalamus, [76] cerebellum, [77] striatum, and cerebral frontal cortex. [78] However, while the inhibitory effect of melatonin in those tissues was evident up to 1 nM, a much higher sensitivity to the methoxyindole is evident in the retina, as it is effective even at 1 pM. In addition, while in other systems melatonin is active when added to crude homogenates, in the hamster retina, it is only effective when the whole tissues are preincubated in the presence of the methoxyindole. The high sensitivity of this inhibitory effect together with the fact that the tissue integrity seems to be necessary, suggest that in the retina, melatonin modulates NOS activity through a receptor-mediated mechanism, but not through its binding to calmodulin. The reasons of this discrepancy are not clear. It should be noted that the retina, unlike the cerebellum, the hypothalamus, or the striatum, is able to synthesize melatonin. Being a highly lipophilic molecule, melatonin has the capability of crossing all physiological barriers. However, these results suggest that the effect of melatonin on NOS activity could be restricted to those cell types that co-express melatonin receptors and NOS.

Besides NOS, another limiting step in NO synthesis is the availability of its precursor, L-arginine, which depends on the presence and activity of a specific uptake system. Physiological concentrations of melatonin significantly decrease retinal L-arginine influx. [75] The lower Vmax of this mechanism observed in the presence of the methoxyindole suggests that melatonin decreases the number of active transporter molecules, without changing their affinity. Since melatonin decreases both L-arginine influx and its conversion to NO, a decrease in L-arginine-induced response is expectable in the presence of the methoxyindole. Indeed, the increase in cGMP accumulation induced by L-arginine is significantly reduced by melatonin. [75] Despite the fact that NO donors, such as sodium nitroprusside releases NO in a NOS-independent way, we showed that melatonin also decreases its effect on cGMP accumulation, probably through its capacity to directly scavenge NO. [75] In agreement with these results, it was shown that uncontrolled NO elevation causes morphological and nuclear changes in the retina and that melatonin significantly suppresses the NO induced increase in mean inner retinal layer thickness, nuclei in the inner nuclear layer, as well as the number of retinas expressing hyperchromatic expression, and apoptotic retinal ganglion cells. [79] In addition, it was demonstrated that melatonin reduces NO-induced lipid peroxidation in rat retinal homogenates. [80]

Peroxynitrite is a toxic oxidant formed from the reaction of superoxide and NO under conditions of inflammation and oxidative stress, and it was shown that melatonin is a scavenger of peroxynitrite. [81] Furthermore, it was demonstrated that melatonin is protective against oxidative damage in situations where NO is known to account for molecular destruction. [82]

Uveitis

Uveitis is the term used to describe many forms of intraocular inflammation involving the uveal tract (iris, ciliary body, and choroid) and adjacent ocular structures (retina, vitreous, and optic nerve). Uveitis is derived from the Latin *uva*, or "grape"; since a peeled blue grape has a bluish vein structure that resembles the middle, vascular layer of the eye, the uvea.

Uveitis is a major cause of severe visual impairment, being the fourth most frequent reason of blindness in the working-age population in the developed world. [83, 84] Uveitis is highly heterogeneous, varying in etiology, pattern, tissue involved, and extent. Many cases of uveitis do indeed involve inflammation of the uvea (which comprises the iris, ciliary body, and choroid), but may also involve adjacent structures such as the retina or vitreous. Inflammation of the anterior uvea is termed anterior uveitis or iridocyclitis. If only the iris is involved, the condition is called iritis. Inflammation of the ciliary body is cyclitis, and choroidal inflammation is termed choroiditis. In clinical usage, uveitis has come to describe most forms of intraocular inflammation. If a layer contiguous with the uvea is predominantly involved in an inflammatory process, the term used for the intraocular inflammation is modified accordingly, that is, retinochoroiditis, retinitis, or vitritis. [85] Hogan and co-workers [86] have used the classification anterior, posterior and diffuse uveitis (or panuveitis). Anterior uveitis consists of either iritis or iridocyclitis, whereas uveitis involving the middle portion of the eye is termed either cyclitis or intermediate uveitis. The classification of posterior uveitis depends on which intraocular layer is predominantly involved. It is termed choroiditis or retinitis when just the choroid or the retina is involved, whereas retinochoroiditis or chorioretinitis is used when both layers are involved; in the latter situation the first portion of the term denotes the most intensely involved layer of the eye. Panuveitis is a generalized inflammation of not only the whole uveal tract, but also involves the retina and vitreous humor, and is associated with the worst prognosis. Uveitis can be acute sudden onset, chronic long lasting, or recurrent relapsing. The most important uveitic complications are cataract formation, cystoid macular edema, band keratopathy, secondary glaucoma, vitreous opacities, retinal detachment, retinoschisis, "retinitis pigmentosa-like" changes, and dragged disk vessels (reviewed by [87]). Typical symptoms and signs of uveitis include eye redness, eye pain, light sensitivity, blurred vision, floaters, and decreased vision. The etiology of uveitis includes many different causes, both systemic disorders with ocular involvement, and disorders that are primarily located in the eye, such as ocular infections. However, in many cases, there is not a specific cause recognized, and then uveitis is characterized as idiopathic. [87] The assessment of uveitis is based on a complete ophthalmologic examination, which includes visual acuity examination, slit-lamp examination, indirect ophthalmoscopy and evaluation of the posterior pole after pupil dilatation, and evaluation of the intraocular pressure. Furthermore, a complete medical history should be acquired, and laboratory tests may be performed in order that possible coexisting systemic diseases are recognized. Vision often decreases as a result of posterior segment inflammation, due to opacities in the vitreous, or from inflammation or vascular occlusions of the macular area (fovea) or the optic nerve. Inflammatory single cells or aggregates in the vitreous cavity are perceived by a patient as "floaters" and may be detected either by a slit-lamp examination or with an ophthalmoscope. The retina, which is normally transparent, appears cloudy or white when inflamed. Inflammation involving only the choroid appears as yellow, white or grey patches with well-defined borders. Histologically, there are lymphocytes cuffing the inflamed retinal vessels. Increased vascular permeability of retinal vessels in either the macula or the optic nerve, leads to edema and diminution of vision. Persistent macular edema results in permanent visual disability due to the formation of macular cysts. Optic atrophy may occur as a consequence of chronic inflammation of the optic nerve (reviewed by [88])

Current Therapy for Uveitis

Corticosteroids are well known for their potent anti-inflammatory effects and have been used to treat uveitis since 1950. [89,90] Their anti-inflammatory mechanism is complex, only partly understood [91], and include direct binding of the glucocorticoid/receptor complex to genes involved in the inflammatory cascade; indirect effects on other transcription factors such as activating protein-1 or nuclear factor-kappa B (NFκB); inhibitory effects on many inflammation-associated molecules such as cytokines, chemokines, arachidonic acid (AA) metabolites, and adhesion molecules, and upregulation of many anti-inflammatory mediators. Whilst this diverse mechanism of action drives an excellent anti-inflammatory response, it also causes significant unwanted side effects. [92] Ocular side effects (more common with topical or local administration) include accelerated cataract formation and increase in IOP. [93,94] Systemic side effects (more common with systemic administration) include hypertension, diabetes, Cushing's syndrome, osteoporosis, and disorders of sleep, mood, and appetite. [95] Other immunomodulatory drugs such as nonsteroidal anti-inflammatory drugs (NSAIDs) are used in uveitis therapy. [96] NSAIDs block the synthesis of PGs, [97] suppress polymorphonuclear cell locomotion and chemotaxis, [98] decrease expression of inflammatory cytokines, [99] and may act as free radical scavengers. [100] In addition, several other drugs have been used for uveitis treatment, such as cyclosporine, antimetabolites (e.g. azathioprine and methotrexate), alkylating agents (e.g. cyclophosphamide), interferons, and anti-TNFα drugs (e.g. infliximab and adalimumab), among many others (reviewed by [101]). While often effective, these drugs have potential serious side effects and some of them may compromise protective immunity to pathogens. Therefore, there is urgent need to develop effective strategies that should be more specific and non-toxic.

Mechanisms Involved in Uveitis

Uveitis is characterized by a disruption of the blood-ocular barrier (BOB), accompanied by protein leakage and leukocyte infiltration into the aqueous humor. Resident cells are activated and inflammatory cells are recruited. Despite that uveitis is one of the main causes of eye morbidity and loss of visual functions, the complexity of the biochemical and immune mechanisms involved in its generation and development remain unknown. Unraveling which are the most critical mechanisms is unlikely to be achieved in studies limited to the clinically observable ocular changes in humans with uveitis. Far more detailed and invasive studies are required, preferably in a readily available animal model. In this sense, several animal models such as interleukin (IL)-1 induced uveitis, experimental autoimmune uveitis (EAU), and endotoxin induced uveitis (EIU), among others, were developed for understanding the disease pathogenesis and testing new therapies for ocular inflammatory diseases. [102] EIU is induced by the administration of bacterial lipopolysaccharide (LPS), a component of Gram-negative bacterial outer membranes. Several lines of evidence support that uveitis is due to damage generated by infiltrated leukocytes which release cytokines [103,104] and other inflammatory chemical mediators, like AA metabolites, [105] ROS, [106] NO, [107] and superoxide anion (O_2^-), [108] among many others. Oxygen-derived free radicals have been implicated in the inflammation associated with EIU in experimental animals. [109] Free radicals initiate retinal lipid peroxidation, and the resultant hydroperoxides serve as amplification factors, such as chemotaxis of inflammatory cells. [110,111] Oxidative stress in the photoreceptor mitochondria appears to be an early event in the development of uveitis, which initiates the subsequent irreversible retinal damage in experimental uveitis by attracting phagocytic cells to this initial site of injury. [112] In fact, there is evidence that free radical scavengers and antioxidants can act as anti-inflammatory agents, thus protecting the eye from inflammation-mediated tissue damage. [113]

Prostaglandins (PGs) can be synthesized through the activities of two cyclooxygenase (COX) isoforms. COX-1 is constitutively expressed in most tissues and its activity provides for the relative small amounts of PGs required for the mediation and modulation of normal physiological functions. In inflammatory conditions, COX-2 is rapidly induced by cytokines, growth factors, and LPS, and its enzymatic activity accounts for the large amounts of PGs produced during inflammation. [114] PGs in small doses administered topically or intraocularly produce some of the responses of injury and inflammation, such as hyperaemia, miosis, breakdown of the blood-aqueous barrier and ocular hypertension. Also, E-type prostaglandins administered topically together with histamine (but not the individual components) cause cellular infiltration and produce edema in conjunctival tissues. [115] Moreover, it was shown that AA metabolites regulate vascular permeability, chemotaxis, and contribute to uveitis amplification. [115, 116]

Studies from animal models of uveitis and human patients with uveitis implicate reactive oxygen/nitrogen radicals in the inflammatory process. Parks et al. [117] demonstrated that multiple intraperitoneal injections of NG-nitro-l-arginine methyl ester, a well-known NOS inhibitor, reduced clinical signs of uveitis in EIU, suggesting that NO could participate to the pathogenesis of experimental uveitis as a proinflammatory mediator. Furthermore, it has been shown that the hemodynamic and vascular permeability changes associated with EIU in the rat are in part due to increased production of NO. High levels of nitrite in the aqueous humor and vitreous of rats at the time of maximal ocular inflammation following injection with LPS, as well as and the presence of a calcium independent NOS activity in the anterior segment during EIU [118] have been demonstrated. Furthermore, reverse transcriptase-PCR assays show an increased expression of inducible isoform of NO synthase (iNOS) mRNA in the iris-ciliary body and in the retina of rats with EIU. [119] In situ hybridization revealed that epithelial cells of the iris-ciliary body and cells infiltrating (macrophages, polymorphonuclear leukocytes) the anterior segment and in the retina are the major source of NO. [108] Moreover, a significant correlation was obtained between nitrite levels in aqueous humor and clinical signs of EIU. The efficacy of treatment with NOS inhibitors in EIU in rat or rabbit suggests that treatments that reduce NO production or NO action would be beneficial and would provide new prospects for treatment of human uveitis. [120] Activated phagocytes synthesize large amounts of nitric oxide (NO) through a reaction catalyzed by the inducible isoform of NO synthase (iNOS). [119,121]

MELATONIN AS A THERAPEUTIC RESOURCE FOR UVEITIS TREATMENT

Kükner et al. [122] and Sande et al. [123] have investigated the effectiveness of melatonin as a therapeutic strategy in uveitis. In the Kukner study, intravitreal injections of bovine serum albumin were used to induce uveitis in the guinea pig. Melatonin, vitamin E, or aprotinin were intraperitoneally administered 3 days after albumin injection. A significant increase in leptin expression occurred in the retina, choroid, sclera, and episclera; retinal ganglion cells were edematous and inner plexiform layer thickness increased in uveitic eyes. In animals treated with melatonin, vitamin E, or aprotinin, leptin expression was similar to the control group and edema and histopathological changes were reduced. [122] Sande et al. [123] studied EIU, another established experimental model for uveitis, giving a single intravitreal injection of LPS in the golden hamster. In this study, a pellet of melatonin was implanted subcutaneously 2 h before the intravitreal injection of LPS. Twenty-four hours and 8 days after injection, the inflammatory response was evaluated in terms of: (i) integrity of blood-ocular barrier (BOB), (ii) clinical signs, (iii) histopathological studies, and (iv) retinal function. Melatonin reduces the leakage of proteins and cells in the anterior segment of LPS-injected eyes, decreases clinical signs, and protects the ultrastructure of BOB. In animals injected with LPS, a remarkable disorganization of rod outer segment membranous disks is observed, whereas no morphological changes in photoreceptor outer segments are observed in animals treated with LPS plus melatonin. Furthermore, melatonin prevents the decrease in the ERG activity induced by LPS. Melatonin significantly abrogates the LPS-induced increase in retinal NOS activity, TNFα, and NFκB p50 and p65 subunits nuclear levels [123]. These results indicate that melatonin prevents clinical, biochemical, histological, ultrastructural, and functional consequences of experimental uveitis in hamsters, probably through an NFκB-dependent mechanism.

We have also examined the effect of melatonin on experimentally-induced feline uveitis. For this purpose, melatonin was orally administered every 24 h, from 24 h before until 45 days after intravitreal injections of vehicle or LPS. [124] Eyes were evaluated by means of clinical evaluation, IOP, BOB integrity (*via* measurement of protein concentration and cell content in samples of aqueous humor), ERG, and histological examination of the retinas. The treatment with melatonin significantly decreases clinical signs and prevents the reduction in IOP induced by LPS. In LPS-injected eyes, melatonin significantly preserves the BOB integrity, as shown by a decrease in the number of infiltrating cells and protein concentration in the aqueous humor (AH). Mean amplitudes of scotopic electroretinographic a- and b-waves are significantly reduced in eyes injected with LPS, whereas melatonin significantly prevents the effect of LPS. At 45 days after injection, LPS induces alterations in photoreceptors and at the middle portion of the retina, whereas melatonin preserves the retinal structure. These results indicate that melatonin prevents clinical, biochemical, functional, and histological alterations induced by LPS injection in cats. [124] Thus, melatonin might constitute a useful tool for the treatment of feline uveitis.

More recently, we studied the anti-inflammatory effect of melatonin administered after the onset of ocular inflammation in the golden hamster. [125] For this purpose, hamster eyes were intravitreally injected with vehicle or LPS, and melatonin (10 mg/kg) was intraperitoneally supplied every 24 h, starting 12 h or 24 h post-LPS injection. The results indicate that both post-treatments with melatonin significantly decreased clinical symptoms, reduced the leakage of cells and proteins, and decreased PG levels in aqueous humor from eyes injected with LPS. In addition, melatonin avoided the decrease in scotopic electroretinogram (ERG) a- and b-wave amplitude, protected the retinal structure, and reduced the increase in NOS activity, lipid peroxidation, and TNFα levels induced by LPS. [125] These observations support that melatonin could be potentially effective in patients with fully established uveitis, since it was capable of suppressing actively ongoing inflammatory responses.

Conclusion

Uveitis, a disease with potentially blinding sequels remains a challenging field to ophthalmologists as the disease causes significant morbidity and the use of traditional forms of treatment is restricted by limited effectiveness and considerable side effects. Several lines of evidence support the involvement of oxidative damage, increased NO, and PGs production as pathogenic mechanisms of uveitis, and melatonin by itself exhibits both antioxidant and antinitridergic properties, and has the ability to reduce PG levels. Besides the already mentioned effects, several other mechanisms could be involved in the anti-inflammatory effect of melatonin, like inhibition of the rolling and adherence of leukocytes to endothelial layers, limiting cell migration, [126] reduction of pro-inflammatory [127] and increase in anti-inflammatory cytokines (such as IL-1, IL-6 and IL-12), [128] decrease in vascular endothelial growth factor [129] and chemokines levels, [130] and reduction of Fas ligand-mediated apoptosis, among many others, but up to now most of them were not confirmed at ocular level.

As mentioned before, ocular inflammation is mainly treated with topical and/or systemic application of corticosteroids. During long-term treatment with corticosteroids, however, care must be taken to guard against both ocular and systemic complications such as cataract, cortisonic glaucoma, diabetes, and immunosuppressive effect. Therefore, the establishment of additive anti-inflammatory approaches is desirable to decrease the rate and degree of these complications. The results obtained in hamsters and cats strongly support that melatonin, which lacks of adverse collateral effects even at high doses, could be a promissory resource in the management of uveitis. Alone or combined with corticosteroid therapy, the anti-inflammatory effects melatonin may benefit patients with chronic uveitis and decrease the rate and degree of corticosteroid-induced complications. Although in the mentioned studies in cats [124] only the preventive effect of melatonin in feline experimental uveitis was demonstrated, this "relative weakness" does not preclude its clinical relevance. Indeed, uveitis is a common consequence of ocular surgery (cataract surgery and vitrectomy, among others). Thus, even as preventive strategy, the present results suggest that melatonin, a very safe compound for human use, should be included in the ophthalmic therapeutic resources. Moreover, as already mentioned, a post-treatment with melatonin significantly attenuates ocular inflammation induced by LPS in golden hamsters. In summary, melatonin appears as a potentially useful anti-inflammatory drug, particularly in ophthalmology, and as an alternative or eventually as complement to glucocorticoids, since it resulted to be highly effective and it does not present the side effects of these steroids. Further investigations of the mechanisms by which melatonin affects its ocular targets may unveil new concepts in ocular therapeutics.

REFERENCES

[1] Lerner AB, Case JD, Takahashi Y, et al. Isolation of melatonin, the pineal gland factor that lightens melanocytes. *J Am Chem Soc.* 1958; 80: 2587.

[2] Lerner AB, Case JD, Heinzelman RV. Structure of melatonin. *J Am Chem Soc.* 1959; 81: 6084-6085.

[3] Lerner AB, Case JD, Takahashi Y. Isolation of melatonin and 5- methoxyindole-3-acetic acid from bovine pineal glands. *J Biol Chem.* 1960; 235: 1992-1997.

[4] Hardeland R, Poeggeler B. Non-vertebrate melatonin. *J Pineal Res.* 2003; 34: 233-241.

[5] Pandi-Perumal SR, Srinivasan V, Maestroni GJ, et al. Melatonin: Nature's most versatile biological signal? *FEBS J.* 2006; 273: 2813-2838.

[6] Paredes SD, Korkmaz A, Manchester LC, et al. Phytomelatonin: a review. *J Exp Bot.* 2009; 60: 57-69.

[7] Pandi-Perumal SR, Trakht I, Srinivasan V, et al. Physiological effects of melatonin: role of melatonin receptors and signal transduction pathways. *Prog Neurobiol.* 2008; 85: 335-353.

[8] Dubocovich ML, Delagrange P, Krause DN, et al. International Union of Basic and Clinical Pharmacology. LXXV. Nomenclature, classification, and pharmacology of G protein-coupled melatonin receptors. *Pharmacol Rev.* 2010;62: 343-380.

[9] Hardeland R, Cardinali DP, Srinivasan V, et al. Melatoninea pleiotropic, orchestrating regulator molecule. *Prog. Neurobiol.* 2011; 93: 350-384.

[10] Cardinali DP, Rosner JM. Metabolism of serotonin by the rat retina in vitro. *J Neurochem.* 1971;18: 1769-1770.

[11] Cardinali DP, Rosner JM. Retinal localization of the hydroxyindole-0-methyl transferase (HIOMT) in the rat. *Endocrinology* 1971; 89: 301-303.

[12] Faillace MP, Cutrera R, Sarmiento MI, Rosenstein RE. Evidence for local synthesis of melatonin in golden hamster retina. *Neuroreport* 1995; 6:2093-2095.

[13] Tosini G, Menaker M. The clock in the mouse retina: melatonin synthesis and photoreceptor degeneration. *Brain Res.* 1998; 789: 221-228.

[14] Martin XD, Malina HZ, Brennan MC, et al. The ciliary body--the third organ found to synthesize indoleamines in humans. *Eur J Ophthalmol.* 1992; 2: 67-72.

[15] Mhatre MC, Van Jaarsveld AS, Reiter RJ. Melatonin in the lacrimal gland: first demonstration and experimental manipulation. *Biochem Biophys Res Commun.* 1988; 153: 1186-1192.

[16] Cahill GM, Besharse JC. Retinal melatonin is metabolized within the eye of Xenopus laevis. *Proc. Natl. Acad. Sci. U S A* 1989; 86: 1098-1102.

[17] Chanut E, Nguyen-Legros J, Versaux-Botteri C, et al. Determination of melatonin in rat pineal, plasma and retina by high performance liquid chromatography with electrochemical detection. *J Chromatogr B Biomed Sci Appl.* 1998; 709: 11-18.

[18] Hall F, Tengerdy C, Morita M, Pautler E. Determination of bovine retinal melatonin with HPLC-EC. *Curr Eye Res.* 1985;4: 47-850.

[19] Mennenga K, Ueck M, Reiter RJ. Immunohistological localization of melatonin in the pineal gland and retina of the rat. *J Pineal Res.* 1991;10:159-164.

[20] Redburn DA, Mitchell CK. Darkness stimulates rapid synthesis and release of melatonin in rat retina. *Vis Neurosci.* 1989;3: 91-403.

[21] Reppert SM, Sagar SM. Characterization of the dayenight variation of retinal melatonin content in the chick. *Invest Ophthalmol Vis Sci.* 1983;24:294-300.

[22] Vivien-Roels B, Pevet P, Dubois MP, et al. Immunohistochemical evidence for the presence of melatonin in the pineal gland, the retina and the Harderian gland. *Cell Tissue Res.* 1981;17:105-115.

[23] Hamm HE, Menaker M. Retinal rhythms in chicks: circadian variation in melantonin and serotonin N-acetyltransferase activity. *Proc Natl Acad Sci U S A.* 1980;77:4998-5002.

[24] Reiter RJ, Richardson BA, Matthews SA, et al. Rhythms in immunoreactive melatonin in the retina and Harderian gland of rats: persistence after pinealectomy. *Life Sci.* 1983;32:1229-1236.

[25] Yu HS, Pang SF, Tang PL. Increase in the level of retinal melatonin and persistence of its diurnal rhythm in rats after pinealectomy. *J Endocrinol.* 1981;91:477-481.

[26] Axelrod J. The pineal gland: a neurochemical transducer. *Science* 1974; 184: 1341-1348.

[27] Bernard M, Guerlotte J, Greve P, et al. Melatonin synthesis pathway: circadian regulation of the genes encoding the key enzymes in the chicken pineal gland and retina. *Reprod Nutr Dev.* 1999; 39: 325-334.

[28] Liu C, Fukuhara C, Wessel JH III, et al. Localization of Aa-nat mRNA in the rat retina by fluorescence in situ hybridization and laser capture microdissection. *Cell Tissue Res.* 2004; 315: 197-201.

[29] Niki T, Hamada T, Ohtomi M, et al. The localization of the site of arylalkylamine N-acetyltransferase circadian expression in the photoreceptor cells of mammalian retina. *Biochem Biophys Res Commun.* 1998; 248:115-120.

[30] Iuvone PM, Brown AD, Haque R, et al. Retinal melatonin production: role of proteasomal proteolysis in circadian and photic control of arylalkylamine N-acetyltransferase. *Invest Ophthalmol Vis Sci.* 2002; 43: 564-572.

[31] Coon SL, Mazuruk K, Bernard M, et al. The human serotonin N-acetyltransferase (EC 2.3.1.87) gene (AANAT): structure, chromosomal localization, and tissue expression. *Genomics* 1996; 34: 76-84.

[32] Coon SL, Del OE, Young WS. III, Klein DC. Melatonin synthesis enzymes in Macaca mulatta: focus on arylalkylamine N-acetyltransferase (EC 2.3.1.87). *J Clin Endocrinol Metab.* 2002; 87: 4699-4706.

[33] Faillace MP, Sarmiento MI, Siri LN, Rosenstein RE. Diurnal variations in cyclic AMP and melatonin content of golden hamster retina. *J Neurochem.* 1994;62:1995-2000.

[34] Cahill GM, Besharse JC. Circadian clock functions localized in xenopus retinal photoreceptors. *Neuron* 1993; 10: 573-577.

[35] Tosini G, Menaker M. Circadian rhythms in cultured mammalian retina. *Science* 1996; 272: 419-421.

[36] Tosini G, Menaker M. The clock in the mouse retina: melatonin synthesis and photoreceptor degeneration. *Brain Res.* 1998; 789: 221-228.

[37] Tosini G. Melatonin circadian rhythm in the retina of mammals. *Chronobiol Int.* 2000; 17: 599-612.

[38] Garbarino-Pico E, Carpentieri AR, Contin MA, et al. Retinal ganglion cells are autonomous circadian oscillators synthesizing N-acetylserotonin during the day. *J Biol Chem.* 2004; 279: 51172-51181.

[39] Fukuhara C, Dirden JC, Tosini G. Photic regulation of melatonin in rat retina and the role of proteasomal proteolysis. *Neuroreport* 2001; 12: 3833-3837.

[40] Pardridge WM, Mietus LJ. Transport of albumin-bound melatonin through the blood-brain barrier. *J Neurochem.* 1980;34:1761-1763.

[41] Grace MS, Cahill GM, Besharse JC. Melatonin deacetylation: retinal vertebrate class distribution and Xenopus laevis tissue distribution. *Brain Res.* 1991;559:56-63.

[42] Zawilska JB, Nowak JZ. Regulatory mechanisms in melatonin biosynthesis in retina. *Neurochem Int.* 1992;20:23-36.

[43] Beck O, Pevet P. Analysis of melatonin, 5-methoxytryptophol and 5-methoxyindoleacetic acid in the pineal gland and retina of hamster by capillary column gas chromatography-mass spectrometry. *J Chromatogr.* 1984; 311:1-8.

[44] Li P, Pang SF, Tsang CW. Retinal 5-methoxytryptamine and 5- methoxyindole-3-acetic acid in the rat and quail: diurnal rhythms and interspecies differences. *Biochem Biophys Res Commun.* 1997;39:353-356.

[45] Iuvone PM, Tosini G, Pozdeyev N, et al. Circadian clocks, clock networks, arylalkylamine N-acetyltransferase, and melatonin in the retina. *Prog Retin Eye Res.* 2005; 24:433-456

[46] Huang H, Wang Z, Weng SJ, et al. Neuromodulatory role of melatonin in retinal information processing. *Prog Retin Eye Res.* 2013; 32:64-87.

[47] Pierce ME, Besharse JC. Circadian regulation of retinomotor movements. I. Interaction of melatonin and dopamine in the control of cone length. *J Gen Physiol.* 1985; 86: 671-689.

[48] Dubocovich ML. Melatonin is a potent modulator of dopamine release in the retina. *Nature* (London) 1983; 306: 782-784.

[49] Scher J, Wankiewicz E, Brown GM, Fujieda H. MT(1) melatonin receptor in the human retina: expression and localization. *Invest Ophthalmol Vis Sci.* 2002; 43: 889-897.

[50] Rufiange M, Dumont M, Lachapelle P. Correlating retinal function with melatonin secretion in subjects with an early or late circadian phase. *Invest Ophthalmol Vis Sci.* 2002; 43: 2491-2499.

[51] Sáenz DA, Goldin AP, Minces L, et al. Effect of melatoninon the retinal glutamate/glutamine cycle in the golden hamster retina. *FASEB J.* 2004;18:1912-1913.

[52] Belforte NA, Moreno MC, de Zavalía N, et al. Melatonin: a novel neuroprotectant for the treatment of glaucoma. *J Pineal Res.* 2010; 48:353-364.

[53] Ianas O, Olinescu R, Badescu I. Melatonin involvement in oxidative processes. *Endocrinologie* 1991; 29: 147-153.

[54] Poeggeler B, Reiter RJ, Tan DX, et al. Melatonin, hydroxyl radical-mediated oxidative damage, and aging: a hypothesis. *J Pineal Res.* 1993;14:151-168.

[55] Reiter RJ, Tan DX, Poeggeler B, et al. Melatonin as a free radical scavenger: implications for aging and age-related diseases. *Ann NY Acad Sci.* 1994; 719: 1-12.

[56] Hardeland R, Balzer I, Poeggeler B, et al. On the primary functions of melatonin in evolution: mediation of photoperiodic signals in a unicell, photooxidation, and scavenging of free radicals. *J Pineal Res.* 1995; 18: 104-111.

[57] Reiter RJ, Guerrero JM, Escames G, et al. Prophylactic actions of melatonin in oxidative neurotoxicity. *Ann NY Acad Sci.* 1997; 825: 70-78.

[58] Reiter RJ, Garcia JJ, Pie J. Oxidative toxicity in models of neurodegeneration: responses to melatonin. *Restor Neurol Neurosci.* 1998; 12: 135-142.

[59] Reiter RJ, Tan DX, Osuna C, Gitto E. Actions of melatonin in the reduction of oxidative stress. A review. *J Biomed Sci.* 2000; 7: 444-458.

[60] Tan DX, Manchester LC, Terron MP, et al. One molecule, many derivatives: a never-ending interaction of melatonin with reactive oxygen and nitrogen species? *J Pineal Res.* 2007; 42: 28-42.

[61] Adler LJ, Gyulai FE, Diehl DJ, et al. Regional brain activity changes associated with fentanyl analgesia elucidated by positron emission tomography. *Anesth Analg.* 1997; 84: 120-126.

[62] Guenther AL, Schmidt SI, Laatsch H, et al. Reactions of the melatonin metabolite AMK (N1-acetyl-5-methoxykynuramine) with reactive nitrogen species: formation of novel compounds, 3-acetamidomethyl-6-methoxycinnolinone and 3-nitro-AMK. *J Pineal Res.* 2005; 39: 251-260.

[63] Hardeland R. Antioxidative protection by melatonin: multiplicity of mechanisms from radical detoxification to radical avoidance. *Endocrine* 2005; 27: 119-130.

[64] Reiter RJ, Tan DX, Mayo JC, et al. Melatonin as an antioxidant: biochemical mechanisms and pathophysiological implications in humans. *Acta Biochim Pol.* 2003; 50:1129-1146.

[65] Acuña-Castroviejo D, Escames G, Leon J, et al. Mitochondrial regulation by melatonin and its metabolites. *Adv Exp Med Biol.* 2003; 527:549-557.

[66] León J, Acuña-Castroviejo D, Escames G, et al. Melatonin mitigates mitochondrial malfunction. *J Pineal Res.* 2005; 38: 1-9.

[67] León J, Acuña-Castroviejo D, Sainz RM, et al. Melatonin and mitochondrial function. *Life Sci* 2004; 75: 765-790.

[68] Guajardo MH, Terrasa AM, Catalá, A. Protective effect of indoleamines on in vitro ascorbate-Fe2+ dependent lipid peroxidation of rod outer segment membranes of bovine retina. *J Pineal Res.* 2003; 35: 276-282.

[69] Liang FQ, Green L, Wang C, et al. Melatonin protects human retinal pigment epithelial (RPE) cells against oxidative stress. *Exp Eye Res.* 2004; 78: 1069-1075.

[70] Celebi S, Dilsiz N, Yilmaz T, Kukner AS. Effects of melatonin, vitamin E and octreotide on lipid peroxidation during ischemia-reperfusion in the guinea pig retina. *Eur J Ophthalmol.* 2002; 12: 77-83.
[71] Lee MC, Chung YT, Lee JH, et al. Antioxidant effect of melatonin in human retinal neuron cultures. *Exp Neurol.* 2001; 172:407-415.
[72] Marchiafava PL, Longoni B. Melatonin as an antioxidant in retinal photoreceptors. *J Pineal Res.* 1999; 26: 184-189.
[73] Turjanski AG, Chaia Z, Doctorovich F, et al. N-Nitrosomelatonin. *Acta Crystallogr C.* 2000; 56: 682-683.
[74] Turjanski AG, Leonik F, Estrin DA, et al. Scavenging of NO by Melatonin. *J Am Chem Soc.* 2000; 122: 10468-10469.
[75] Sáenz DA, Turjanski AG, Sacca GB, et al. Physiological concentrations of melatonin inhibit the nitridergic pathway in the Syrian hamster retina. *J Pineal Res.* 2002; 33: 31-36.
[76] Bettahi I, Pozo D, Osuna C, et al. Melatonin reduces nitric oxide synthase activity in rat hypothalamus. *J Pineal Res.* 1996; 20: 205-210.
[77] Pozo D, Reiter RJ, Calvo JC, et al. Inhibition of cerebellar nitric oxide synhtase and cyclic GMP production by melatonin via complex formation with calmodulin. *J Cell Biochem.* 1997; 65:430-442.
[78] León J, Vives F, Crespo E, et al. Modification of nitric oxide synthase activity and neuronal response in rat striatum by melatonin and kynurenine derivatives. *J Neuroendocrinol.* 1998; 10: 297-302.
[79] Siu AW, Ortiz GG, Benitez-King G, et al. Effects of melatonin on the nitric oxide treated retina. *Br J Ophthalmol.* 2004; 88: 1078-1081.
[80] Siu AW, Reiter RJ, To CH. Pineal indoleamines and vitamin E reduce nitric oxide-induced lipid peroxidation in rat retinal homogenates *J Pineal Res.* 1999; 2:122-128.
[81] Gilad E, Cuzzocrea S, Zingarelli B, et al. Melatonin is a scavenger of peroxynitrite. *Life Sci.* 1997;60:PL169-174.
[82] Cuzzocrea S, Costantino G, Gitto E, et al. Protective effects of melatonin in ischemic brain injury. *J Pineal Res.* 2000; 29:217-227.
[83] Barisani-Asenbauer T, Maca SM, Mejdoubi L, et al. Uveitis- a rare disease often associated with systemic diseases and infections- a systematic review of 2619 patients. *Orphanet J Rare Dis.* 2012;7: 57.
[84] de Smet MD, Taylor SR, Bodaghi B, et al. Understanding uveitis: the impact of research on visual outcomes. *Prog Retin Eye Res.* 2011;30:452-470.
[85] Char DH, Schlaegel TF, Jr. General factors in uveitis, Chap 39, In: Duane TD (ed): Clinical Ophthalmology, Vol 4. Philadelphia, *Harper & Row,* 1982, pp 1-7.
[86] Hogan MJ, Kimura SJ, Thygeson P. Signs and symptoms of uveitis –I . Anterior uveitis. *Am J Ophthalmol.* 1959; 47: 155-176.
[87] Moschos MM, Gouliopoulos NS, Kalogeropoulos C. Electrophysiological examination in uveitis: a review of the literature. *Clin Ophthalmol.* 2014; 8:199-214.
[88] Lowder CY, Char DH Uveitis. A review. *West J Med.* 1984;140: 421-432.
[89] Carryer HM, Koelsche GA, Prickman LE, et al. Effects of cortisone on bronchial asthma and hay fever occurring in subjects sensitive to ragweed pollen. *Proc Staff Meet Mayo Clin.* 1950;25:482-486.

[90] Gordon DM. Effects of pituitary adrenocorticotrophic hormone (ACTH) therapy in ophthalmologic conditions. *J Am Med Assoc.* 1950;142:1271.

[91] van der Velden VH. Glucocorticoids: mechanisms of action and anti-inflammatory potential in asthma. *Mediators Inflamm.* 1998;7: 229-237.

[92] Gallego-Pinazo R, Dolz-Marco R, Martínez-Castillo S, et al. Update on the principles and novel local and systemic therapies for the treatment of non-infectious uveitis. *Inflamm Allergy Drug Targets* 2013;12:38-45.

[93] Pavesio CE, Decory HH. Treatment of ocular inflammatory conditions with loteprednoletabonate. *Br J Ophthalmol.* 2008; 92:455-459.

[94] Becker B, Mills DW. Elevated intraocular pressure following corticosteroid eye drops. *J Am Med Assoc.* 1963;185: 884-886.

[95] Jabs DA, Busingye J. Approach to the diagnosis of the uveitides. *Am J Ophthalmol.* 2013;156:228-236.

[96] Maggs DJ. Feline uveitis. An 'intraocular lymphadenopathy'. *J Feline Med Surg.* 2009;11:167-182.

[97] Giuliano EA. Nonsteroidal anti-inflammatory drugs in veterinary ophthalmology. *Vet Clin North Am Small Anim Pract.* 2004;34:707-723.

[98] Perianin A, Gougerot-Pocidalo MA, Giroud JP, Hakim J. Diclofenac sodium, a negative chemokinetic factor for neutrophil locomotion. *Biochem Pharmacol.* 1985;34:3433-3438.

[99] Leonardi A, Busato F, Fregona I, et al. Anti-inflammatory and antiallergic effects of ketorolac tromethamine in the conjunctival provocation model. *Br J Ophthalmol.* 2000;84:1228-1232.

[100] Flach AJ. Cyclo-oxygenase inhibitors in ophthalmology. *Surv Ophthalmol.* 1992;36:259-284.

[101] Barry RJ, Nguyen QD, Lee RW, Murray PI, Denniston AK. Pharmacotherapy for uveitis: current management and emerging therapy. *Clin Ophthalmol.* 2014 Sep 22;8:1891-911.

[102] Srivastava A, Rajappa M, Kaur J. Uveitis: Mechanisms and recent advances in therapy. *Clin Chim Acta.* 2010; 411:1165-1171.

[103] de Vos AF, Klaren VN, Kijlstra A. Expression of multiple cytokines and IL-1RA in the uvea and retina during endotoxin-induced uveitis in the rat. *Invest Ophthalmol Vis Sci.* 1994; 35: 3873-3883.

[104] Planck SR, Huang XN, Robertson JE, Rosenbaum JT. Cytokine mRNA levels in rat ocular tissues after systemic endotoxin treatment. *Invest Ophthalmol Vis Sci.* 1994;35:924-930.

[105] Bellot JL, Palmero M, Garcia-Cabanes C, et al. Additive effect of nitric oxide and prostaglandin-E2 synthesis inhibitors in endotoxin-induced uveitis in the rabbit. *Inflamm Res.* 1996; 45: 203-208.

[106] Rao N. Role of oxygen free radicals in retinal damage associated with experimental uveitis. Trans Am Ophthalmol Soc 1990; 88:797-850.

[107] Hashida M, Fukushima A, Zhang J, et al. Involvement of superoxide generated by polymorphonuclear leukocytes in endotoxin-induced uveitis. *Graefes Arch Clin Exp Ophthalmol.* 2000; 238: 359-365.

[108] Jacquemin E, De Kozak Y, Thillaye B, et al. Expression of inducible nitric oxide synthase in the eye from endotoxin-induced uveitis rats. *Invest Ophthalmol Vis Sci.* 1996; 37: 1187-1196.

[109] Satici A, Guzey M, Gurler B, et al. Malondialdehyde and antioxidant enzyme levels in the aqueous humor of rabbits in endotoxin-induced uveitis. *Eur J Ophthalmol.* 2003;13: 779-783.

[110] Goto H, Wu GS, Gritz DC, et al. Chemotactic activity of the peroxidized retinal membrane lipids in experimental autoimmune uveitis. Curr Eye Res. 1991;10:1009-1014.

[111] Wu GS, Sevanian A, Rao NA. Detection of retinal lipid hydroperoxides in experimental uveitis. *Free Radic Biol Med.* 1992;12:1927.

[112] Saraswathy S, Rao NA Photoreceptor mitochondrial oxidative stress in experimental autoimmune uveitis. *Ophthalmic Res.* 2008;40:160-164.

[113] Bosch-Morell F, Roma J, Marin N, et al. Role of oxygen and nitrogen species in experimental uveitis: anti-inflammatory activity of the synthetic antioxidant ebselen. *Free Radic Biol Med.* 2002;33: 669-675.

[114] Masferrer JL, Kulkarni PS. Cyclooxygenase-2 inhibitors: a new approach to the therapy of ocular inflammation. *Surv Ophthalmol.* 1997; 41:S35-40.

[115] Bhattacherjee P. Prostaglandins and inflammatory reactions in the eye. *Methods Find Exp Clin Pharmacol.* 1980; 2:17-31.

[116] Howes EL, Jr, McKay DG. The effects of aspirin and indomethacin on the ocular response to circulating bacterial endotoxin in the rabbit. *Invest Ophthalmol.* 1976;15: 648-651.

[117] Parks DJ, Cheung MK, Chan CC, Roberge FG. The role of nitric oxide in uveitis. *Arch Ophthalmol.* 1994;112: 544-546.

[118] Mandai M, Yoshimura N, Yoshida M, et al. The role of nitric oxide synthase in endotoxin-induced uveitis: effect of NG-nitro L-arginine. *Invest Ophthalmol Vis Sci.* 1994; 35: 3673-3681.

[119] Goureau O, Bellot J, Thillaye B, et al. Increased nitric oxide production in endotoxin-induced uveitis: reduction of uveitis by an inhibitor of nitric oxide synthase. *J Immunol.* 1995;154: 6518-6523.

[120] Becquet F, Courtois Y, Goureau O. Nitric oxide in the eye: multifaceted roles and diverse outcomes. *Surv Ophthalmol.* 1997;42: 71-82.

[121] Sonoki T, Nagasaki A, Gotoh T, et al. Coinduction of nitric-oxide synthase and arginase I in cultured rat peritoneal macrophages and rat tissues in vivo by lipopolysaccharide. *J Biol Chem.* 1997;272: 3689-3693.

[122] Kükner A, Colakoğlu N, Serin D, et al. Effects of intraperitoneal vitamin E, melatonin and aprotinin on leptin expression in the guinea pig eye during experimental uveitis. *Acta Ophthalmol Scand.* 2006;84:54-61.

[123] Sande PH, Fernandez DC, Aldana Marcos HJ, et al. Therapeutic effect of melatonin in experimentaluveitis. *Am J Pathol* 2008; 173:1702-1713.

[124] Del Sole MJ, Sande PH, Fernandez DC et al. Therapeutic benefit of melatonin in experimental feline uveitis. *J Pineal Res.* 2012;52:29-37.

[125] Sande PH, Dorfman D, Fernandez DC, Chianelli M, Domínguez Rubio AP, Franchi AM, Silberman DM, Rosenstein RE, Sáenz DA.The post-treatment with melatonin

attenuates experimental ocular inflammation. *Br J Pharmacol.* 2014. doi: 10.1111/bph.12873.

[126] Markus RP, Ferreira ZS, Fernandes PA, Cecon E. The immune-pineal axis: a shuttle between endocrine and paracrine melatonin sources. *Neuroimmunomodulation* 2007;14:126-33.

[127] Escames G, Acuña-Castroviejo D, López LC, et al. Pharmacological utility of melatonin in the treatment of septic shock: experimental and clinical evidence. *J Pharm Pharmacol.* 2006;58:1153-1165.

[128] Maestroni GJ. The immunotherapeutic potential of melatonin *Expert Opin Investig Drugs* 2001;10:467-476.

[129] Kaur C, Sivakumar V, Yong Z, et al. Blood-retinal barrier disruption and ultrastructural changes in the hypoxic retina in adult rats: the beneficial effect of melatonin administration. *J Pathol.* 2007; 212:429-439.

[130] Park HJ, Kim HJ, Ra J, et al. Melatonin inhibits lipopolysaccharide-induced chemokine subfamily gene expression in human peripheral blood mononuclear cells in a microarray analysis. *J Pineal Res.* 2007;43:121-129.

In: Indoleamines
Editor: Ángel Catalá

ISBN: 978-1-63482-097-4

Chapter 7

THE RELATIONS BETWEEN WATER-SOLUBLE B VITAMINS AND INDOLEAMINES

Robert Lalonde *

University of Rouen, Dept Psychology, ICONES Laboratory EA, Mont-Saint-Aignan France

ABSTRACT

B vitamin concentrations affect 5-hydroxytryptamine (5HT). 5HT neurons degenerate in the Wernicke-Korsakoff syndrome marked by thiamine deficiency. Likewise, pyrithiamine-induced thiamine-deficiency in mice reduced 5HT-immunoreactive cell numbers and tryptophan hydroxylase activity in medial and dorsal raphe nuclei along with their target structures. 5HT is also affected by folate. Indeed, a folate-deficient diet decreased brain 5HT levels in rats, probably mediated by S-adenosyl-methionine (SAM). Folate deficiency is prevalent in patients with major depression, an effect likely involving 5HT and SAM. As with folate, vitamin B_{12} deficiency was associated with low CSF 5HIAA levels, reflecting brain 5HT, an effect of possible relevance in major depression as well. Vitamin B_6-deficient rats had lower brain 5HT concentrations, probably because pyridoxal 5-phosphate acts as a cofactor for the 5HT-synthesizing enzyme, amino acid decarboxylase. The vitamin may be used in treating amino acid decarboxylase deficiency, an inborn error of metabolism leading to 5HT deficiency.

1. INTRODUCTION

Indoleamines are affected by changes in nutrition, notably water-soluble vitamins B_1 (thiamine), B_6 (pyridoxine), B_9 (folate), and B_{12} (cobalamin). The main purpose of the present review is to summarize findings on the effects of deficiencies or excessive intake of these vitamins on brain levels and function of indoleamines, mainly 5-hydroxytryptamine (5HT,

* TEL +33 (0)2 35 14 61 08; Fax +33 (0)2 35 14 63 49. Email: robert.lalonde@univ-rouen.fr

serotonin) but also melatonin. Such findings are relevant in discovering the effects of nutrition on neurobehavioral processes associated with 5HT such as mood and anxiety.

2. Thiamine (Vitamin B_1)

Brain thiamine is mainly found in the form of thiamine diphosphate (TDP), but also in monophosphate (TMP) or triphosphate (TTP) forms [1]. TDP acts as a coenzyme for decarboxylation reactions in energy and amino acid metabolism. In particular, TDP facilitates acetyl-CoA synthesis from pyruvate to form succinyl-CoA in the citric acid cycle [2]. Acute thiamine deficiency (TD) causes beriberi, leading to weight loss, emotional disturbances, and impaired sensory perception. In the Western world, TD mainly results from chronic alcoholism (Wernicke-Korsakoff syndrome) but also from malnutrition [3]. Ethanol reduces thiamine uptake, increases its excretion, and impairs its activation to TDP as well as the ability of the liver to store it [4].

2.1. Thiamine Deficiency in Animals

Experimental TD in animals may be initiated by a diet deficient in the vitamin combined or not with the injection of pyrithiamine, an inhibitor of thiamine pyrophosphorylation [5]. The behavioral effects of 2 week-long pyrithiamine-induced thiamine-deficiency (PTD) include loss of the righting reflex, hyperreactivity, piloerection, flat-body posturing, gait anomalies, spasms, and generalized seizures [1]. During PTD recovery, rats had lower than normal TDP, TMP, and TTP concentrations in thalamus and neocortex [1] and diencephalic lesions [6].

PTD reduced 5HT-immunoreactive cell numbers in medial and dorsal raphe along with their target structures in mice [7]. Moreover, TD reduced tryptophan hydroxylase activity in medial and dorsal raphe of mice [8]. Others have found that PTD increased 5HT concentrations in several brain regions of rats [9,10], or else the main metabolite of 5HT, 5-hydroxy-indole acetic acid (5HIAA) [11]. On one hand, it is possible that the mouse is more subject to the neurotoxic action of thiamine deficiency on 5HT neurons. On the other, an increase in 5HT levels in the rat may eventually lead to a decrease and neurodegeneration. After 4-6 weeks on a thiamine-free diet, basal Purkinje cell firing frequency was unaffected in rats [12]. But whereas controls had varied Purkinje cell responses, the responses of the deficient group to iontophoretically-applied 5-HT were solely inhibitory. These results indicate an up-regulation of 5-HT receptors on Purkinje cells caused by thiamine deficiency-induced impairment of indoleamine input to the cerebellum from raphe and related nuclei.

2.3. Thiamine Deficiency in Humans

2.3.1. Wernicke-Korsakoff Syndrome

The Wernicke-Korsakoff syndrome is characterized by neuronal losses in the dicencephalon, including thalamus and mamillary bodies, as well as the cerebellum, whereas the neocortex is relatively spared [13]. This regionally selective vulnerability is probably due to higher thiamine turnover rates in the most vulnerable brain regions [14]. As reported in mouse brain [7], neuronal losses have also been reported in brainstem 5HT neurons [15],

In addition to 5HT, there is evidence of an effect of Korsakoff's psychosis on metatonin. Indeed, the urinary excretion rate of the metatonin metabolite, 6-hydroxymelatonin, was reduced in this condition, an indication of reduced metatonin synthesis [16]. The effect was reversed following treatment with the α2-noradrenergic agonist, clonidine.

2.3.2. Other Neural Diseases

Low cerebrospinal fluid (CSF) levels of thiamine characterize patients with spinocerebellar atrophy and Friedreich's ataxia [17, 18]. In comparison with controls, patients with Friedreich's ataxia had lower CSF levels of 5HIAA [17]. In patients with low blood and CSF thiamine from various aetiologies, CSF 5HIAA levels were also lower than normal [19]. In epileptic patients, the duration of treatment with phenytoin, an antiepileptic medication with neurotoxic potential, was related with low CSF but not blood thiamine and so not explainable by bad nutrition [20]. CSF thiamine levels were directly correlated CSF 5HIAA levels. Both phenytoin treatment duration and CSF thiamine levels were correlated with brain atrophy. Thus, as with the Wernicke-Korsakoff syndrome, low CSF thiamine probably leads to low brain 5HT in Friedreich's ataxia and epilepsy, opening up the possibility of treating these disorders with the vitamin.

3. VITAMIN B_9 (FOLATE)

Folates are actively transported through the gastro-intestinal tract, the choroid plexus, the placenta, the CSF, and neurons. Methylenetetrahydrofolate reductase (MTHFR) catalyzes the reduction of 5,10-methylenetetrahydrofolate to 5-methyltetrahydrofolate (5-methyl-THF), the carbon donor for homocysteine metabolism and other methylation reactions. In the cell methionine cycle, 5-methyl-THF is demethylated in a reaction catalyzed by vitamin B_{12}-requiring methione synthase to tetrahydrofolate (THF), its active enzymatic form. To prevent accumulation, homocysteine must either be converted to methionine with 5-methyl-THF, requiring methionine synthase, or to cystathionine via the transsulfuration pathway, requiring cystathionine β-synthase (CBS) with vitamin B_6 used as a co-factor [21-23]. Because of these interrelations, plasma homocysteine levels are influenced by all three B vitamins [24,25]. The folate concentration in the central nervous system is tightly regulated, CSF values being typically 4-fold higher than in plasma. Severe folate deficiency affects DNA synthesis and alters gene expression, its capacity for repair, and amino acid metabolism, as well as the synthesis of phospholipids and S-adenosyl-methionine (SAM), a universal methyl donor [26].

3.1. Folate Deficiency in Animals

Selective effects of folate manipulations have been obtained in regard to monoamines. A folate-deficient diet decreased whole brain 5HT levels in rats [27] and dopamine levels in the medial dorsal striatum of mice [28] as well as potentiating losses in dopamine neurons of the substantia nigra following the administration of the neurotoxic agent, methyl-phenyl-tetrahydropyridine (MPTP) [29]. On the contrary, a folate-deficient diet had no effect on cholinergic or GABA systems in rats [30]. SAM may mediate the effects of folate on 5HT, since a folate-deficient diet in mice [29] or rats [31] reduced brain SAM levels, while SAM administration increased rat brain 5HT levels [32].

A folate-deficient diet also decreased levels of melatonin in the pineal gland of rats [33], an effect which may explain sleep disorders seen in humans with this type of deficiency [34]. The reduction in metatonin is probably mediated by SAM, because the final step in melatonin synthesis requires methylation of N-acetyl-serotonin and the remethylation of homocysteine to methionine requires folic acid.

3.2. Folate Administration in Animals

Folate administration affects behavior. Some of these effects may be mediated by 5HT. In particular, folate administration decreased immobility time in forced swimming and tail suspension tests of depressant-like behavior in mice [35]. The anti-immobility action of folate during forced swimming was prevented by a 5HT synthesis inhibitor as well as 5HT and noradrenergic receptor blockers, consistent with a role of these neurotransmitters on vitamin-related mood disorders in humans [27, 36].

3.3. Folate Deficiency in Humans

When folate-deficient patients were given folate replacement therapy, those with responsive neurologic symptoms had higher CSF folate levels than those with non-responsive symptoms [37,38]. During folate therapy, no linear correlation was established between the rise in serum and the rise in CSF folate, which indicates that CSF folate is independently and actively regulated, so that more folate enters the brain when serum levels are low.

3.3.1. Neuropsychiatric Disorders in Adults

In an analogous manner to the effects of a folate-deficient diet on 5HT levels in rat brains [27], folate-responsive patients with various neurologic disorders had lower CSF 5HIAA levels than controls, a finding of relevance in the depressive symptoms observed during folate deficiency [27,38]. Indeed, folate deficiency is prevalent in patients with major depression [39]. In depressed patients without macrocytic anaemia, RBC and CSF folate levels were lower than those of controls and plasma homocysteine levels higher than controls [40]. In a high homocysteine subgroup with greater depression, CSF concentrations of SAM and 5HIAA were lower than normal, as were homovanillic acid (HVA) and methoxy-hydroxy-propylene-glycol (MHPG), the main metabolites of dopamine and noradrenaline, respectively. Serum folate levels were also lower in Parkinson patients with depression relative to those without depression [41]. In epileptic patients, the length of phenytoin treatment was related to low serum and CSF folate levels [20]. Lengthy phenytoin treatment and low serum or CSF folate were related with brain atrophy. Low CSF folate was also related with low CSF 5HIAA and MHPG.

Thus, the relation between folate and 5HT likely has an impact on several neuropsychiatric disorders, including major depression or brain atrophy resulting from antiepileptic therapy. Under certain conditions, folate may be used as an antidepressive agent via its effect on 5HT. It may also counter the neurotoxic effects of phenytoin. Its role in the symptomatology of Parkinson's disease remains to be established.

3.4. Folate Administration in Humans

Folates in the form of 5-methyl-THF or folinic acid (5-formyltetrahydrofolate) have been successfully used as adjunct therapy of depressive symptoms along with 5HT reuptake blockers [36, 42, 43]. However, a single dose of folate (30 mg PO) had no effect on CSF 5HIAA levels on diverse psychiatric patients [44]. Moreover, folate administration (100-200 μg PO) for 4 months caused no change in mood or 5HT blood levels in normal subjects [45]. Thus, more research is needed to determine the long-term impact of folate administration on the 5HT system.

4. Vitamin B_6 (Pyridoxine)

4.1. Background

Vitamin B_6 (pyridoxine, pyridoxal or pyridoxamine) is a cofactor for enzymes involved in amino acid metabolism and its dietary requirement of 1 mg/day is linked with protein intake [2, 46, 47]. In particular, pyridoxal phosphate is a cofactor of cystathionine β-synthase (CBS), which, as mentioned above, catalyzes the conversion of homocysteine to cysteine. It is also a cofactor of amino acid decarboxylase, required for monoamine synthesis, as well as glutamic acid decarboxylase-1 and GABA transaminase, required respectively for GABA synthesis and degradation. Like folate and vitamin B_{12} and because of its role as a cofactor for CBS, low vitamin B_6 intake may elevate homocysteine levels, thereby leading to suboptimal methyl-group metabolism. But because the affinity of CBS to its substrate is ten-fold higher than that of methionine synthase, vitamin B_6 is less effective than the other two B vitamins in counteracting hyperhomocysteinemia [46].

Symptoms associated with vitamin B_6 deficiency include retarded growth, pain in the extremities, hair loss, dermatitis, microcytic anemia, convulsions, depression, and confusion [2]. Specific vitamin B_6 deficiency is extremely rare in adults, although convulsions or at least changes in EEG patterns have been reported in children exposed to prolonged intake of canned milk.

4.2. Vitamin B_6 Deficiency in Animals

Vitamin B_6-deficient adult rats had lower brain 5HT concentrations than controls [48-52] along with changes in 5HT receptor binding [48, 51, 53]. These results are probably due to the fact that pyridoxal 5-phosphate acts as a cofactor for amino acid decarboxylase, an enzyme necessary for 5HT synthesis. On the contrary, vitamin B_6 deficiency caused no change in brain 5HT concentrations in weanling rats [54]. Thus, developmental patterns should be more closely examined. In contrast to 5HT, vitamin B_6 deficiency increased concentrations of pro-convulsant tryptophan metabolites in adult rats, which may underlie the convulsions marking such a diet [55, 56]. In addition to 5HT, vitamin B_6-deficient adult rats had lower pineal melatonin concentrations than controls [50, 57]. These results are also likely due to inhibition of amino acid decarboxylase activity because 5HT is the metabolic precursor of melatonin.

4.3. Vitamin B_6 Administration in Animals

Pyridoxine (100 mg/kg IP) administration for 7 days increased brain 5HT concentrations in rats [58]. The same effect was found in rats treated with 10 mg/kg IP pyridoxine for 20 days [59]. Likewise, 10 mg/kg IV pyridoxine increased 5HT synthesis in the dorsal striatum of monkeys [60]. In addition to amino acid decarboxylase activity, these effects may be mediated by a pyridoxine-induced increase in tryptophan uptake in brain [61]. However, long-term diets containing excessive pyridoxine had no effect on 5HT concentrations, perhaps because of the effects of the vitamin on branched-chain amino acids competing with tryptophan [62, 63].

4.4. Vitamin B_6 Deficiency in Humans

Low vitamin B_6 intake was associated with a higher risk of Parkinson's disease, perhaps mediated by lower activity of vitamin B_6-dependent amino acid decarboxylase [64]. Another link between the vitamin and dopamine metabolism is its effectiveness in treating neuroleptic-induced tardive dyskinesia [65].These effects may also be mediated by GABA neurons interacting with dopamine at the level of the basal ganglia [66].

4.4. Vitamin B_6 Administration in Human Subjects

Since vitamin B_6 is a cofactor of amino acid decarboxylase, the vitamin may be used in treating amino acid decarboxylase deficiency, an inborn error of metabolism leading to 5HT deficiency [67, 68].

Despite a lack of vitamin B_6 deficiency in the premenstrual syndrome [61], high dose (40-200 mg/day PO) vitamin B_6 has been used to treat it [69] along with 5HT reuptake blockers [70]. Moreover, 100 mg/day PO vitamin B_6 is effective in treating the side effects of oestrogen-progestagen contraceptives [69]. The factors underlying these relationships remain to be determined.

As opposed to the effects of vitamin B_6 deficiency in rats, 3 mg/kg pyridoxine IV increased plasma melatonin levels in children [71]. This result might lead to night-time benefits in sleep quality. On the contrary, 100 mg/kg pyridoxine PO had no effect on serum melatonin levels or sleep quality in men [72].

5. Vitamin B_{12} (Cobalamin)

5.1. Background

Adequate levels of vitamin B_{12} for erythropoiesis and neuronal function are mostly supplied by meat [2, 73]. Deficiency in the healthy population is rare because relative to daily requirements (2.4 μg in adults), body stores are large, mostly in the liver (0.6-3.9 mg). But after prolonged deficiencies in subjects eating vegan diets or newborns from vitamin B_{12}-depleted mothers, symptoms such as celiac disease, ileal resection, and pernicious anaemia appear [3]. Like pyridoxine, cobalamin is a cofactor of methionine synthase, converting homocysteine to methionine; it is also a cofactor of methylmalonyl-CoA synthase, converting methylmalonyl-CoA to succinyl CoA [74]. Like the lack in vitamins B_6 and B_9, vitamin B_{12} deficiency raises serum homocysteine levels [75] along with causing defective synthesis of DNA and choline-containing phospholipids, while the increase in methylmalonyl-CoA damages myelin sheaths due to the incorporation of non-physiological fatty acids.

5.2. Vitamin B_{12} Deficiency in Animals

Vitamin B_{12} deficiency decreased brain noradrenaline levels in rats [76]. However, this effect was not found with 5HT. The relation between vitamin B_{12} and 5HT has been insufficiently explored.

5.3. Vitamin B_{12} Administration in Animals

The effects of subcutaneous vitamin B_{12} administration for 2 months were evaluated in rats with respect to suppressed serum melatonin during light onset [77]. Vitamin B_{12} inhibited the suppressive action of melatonin under light pulses. The acute administration of intraperitoneal administration of methylcobalamin before night-time activity onset in rats induced phase advances of free-running activity rhythms in a dose-dependent fashion [78]. The injection of the vitamin prior to melatonin injection induced larger phase advances than saline pre-injected controls, whereas the injection of the vitamin in combination with saline did not induce a phase advance. Moreover, the vitamin increased melatonin content in the pineal gland collected 2 hr after activity onset. Pinealectomy abolished the phase-shifting effect, indicating the pineal gland as the site of action. These results indicate that the vitamin amplifies melatonin-induced phase advances.

5.4. Vitamin B_{12} Deficiency In Humans

Like folate deficiency [27], vitamin B_{12} deficiency was associated with low CSF 5HIAA levels, reflecting brain 5HT [38]. Once again, these results may have an impact on depressive symptoms. Indeed, low vitamin B_{12} status was associated with the onset of depression in geriatric subjects [79].

5.5. Vitamin B_{12} Administration in Humans

Vitamin B_{12} has been examined in regard to melatonin levels and biorhythms. An oral dose of 3 mg vitamin B_{12} for 2 weeks reduced urinary melatonin levels and sleep time [80]. An intravenous dose of 0.5 mg vitamin B_{12} for 11 days enhanced the phase shift of plasma melatonin to a bright light [81]. The modulatory role of the vitamin was next examined in a blind adolescent girl with a free-running sleep-wake rhythm [82]. An oral dose of 1.5 mg vitamin B_{12} for a short period led to entrainment of the sleep-wake rhythm. A beneficial action also occurred in a 55 year-old man suffering from delayed sleep phase syndrome marked by slowed sleep onset during the dark phase of the diurnal cycle [82], results in agreement with those found in rats in which the vitamin amplified melatonin-induced phase advances [78]. However, oral 3 mg/day methylcobalamin for one month was ineffective versus placebo on delayed sleep phase syndrome in a multi-center double-blind study [83], probably because vitamin B_{12} levels did not cross a threshold of deficiency.

Vitamin B_{12} has been examined in regard to melatonin levels and tinnitus (ringing of the ears) in aged subjects [84]. Subjects with tinnitus had lower plasma levels of vitamin B_{12} and melatonin than those without tinnitus. It is unknown whether there is a causative relation between tinnitus and either of these measures.

CONCLUSION

Relations have been drawn between 5HT and all four water-soluble B vitamins. Degeneration of 5HT neurons has been described in the Wernicke-Korsakoff syndrome marked by thiamine deficiency. Pyrithiamine-induced thiamine-deficiency in mice reduced 5HT-immunoreactive cell numbers and tryptophan hydroxylase activity in medial and dorsal raphe nuclei.

Moreover, low CSF thiamine probably leads to low brain 5HT in Friedreich's ataxia and likely has an impact on brain atrophy caused by antiepiletic medication. Thus, thiamine administration may prevent brain atrophy under these conditions.

Folate deficiency is prevalent in patients with major depression, an effect likely mediated by 5HT. Moreover, the relation between folate and 5HT likely has an impact on brain atrophy during chronic antiepileptic therapy. Under certain conditions, folate may be used as an antidepressive agent via its effect on 5HT. Like folate deficiency, vitamin B_{12} deficiency was associated with low CSF 5HIAA levels, reflecting brain 5HT, an effect which might also be of relevance in major depression.

Vitamin B_6-deficient rats have lower brain 5HT concentrations, probably because pyridoxal 5-phosphate acts as a cofactor for the 5HT-requiring enzyme, amino acid decarboxylase. Since vitamin B_6 is a cofactor of amino acid decarboxylase, the vitamin may be used in treating amino acid decarboxylase deficiency, an inborn error of metabolism leading to 5HT deficiency.

REFERENCES

[1] Héroux M, Butterworth RF. Regional alterations of thiamine phosphate esters and of thiamine diphosphate-dependent enzymes in relation to function in experimental Wernicke's encephalopathy. *Neurochem. Res.* 1995;20:87-93.

[2] Institute of Medicine. Thiamine: dietary reference intakes for thiamin, riboflavin, niacin, vitamin B6, folate, vitamin B12, panthotenic acid, biotin and choline. Washington: National Academy Press, 1998.

[3] Jeffery DR. Nutrition and Diseases of the nervous system. In: Shils ME, Olson JA, Shike M, Ross CA, eds. Modern nutrition in health and disease. Baltimore: Williams & Wilkins, 1999.

[4] McCormick DB. Thiamine. In: Shils ME, Young VR, editors. Modern nutrition in health and disease. Philadelphia: Lea and Febiger, 1988.

[5] Savage LM, Hall JM, Resende LS. Translational rodent models of Korsakoff syndrome reveal the critical neuroanatomical substrates of memory dysfunction and recovery. *Neuropsychol. Rev.* 2012;22:195-209.

[6] Rao VL, Mousseau DD, Butterworth RF. Nitric oxide synthase activities are selectively decreased in vulnerable brain regions in thiamine deficiency. *Neurosci. Lett.* 1996;208:17-20.

[7] Matsushita H, Takeuchi Y, Kosaka K, Fushiki S, Kawata M, Sawada T. Changes in serotonergic neurons in the brain of pyrithiamine-induced acute thiamine-deficient mice. *Acta Neuropathol.* 1999;98:614-21.

[8] Nakagawasai O, Murata A, Arai Y, Ohba A, Wakui K, Mitazaki S, et al. Enhanced head-twitch response to 5-HT-related agonists in thiamine-deficient mice. *J. Neural Transm.* 2007;114:1003-10.

[9] Mousseau DD, Rao VL, Butterworth RF. Alterations in serotonin parameters in brain of thiamine-deficient rats are evident prior to the appearance of neurological symptoms. *J. Neurochem.* 1996;67:1113-23.

[10] Langlais PJ, Mair RG, Anderson CD, McEntee WJ. Monoamines and metabolites in cortex and subcortical structures: normal regional distribution and the effects of thiamine deficiency in the rat. *Brain Res.* 1987;421:140-9.

[11] Vigil FA, Oliveira-Silva Ide F, Ferreira LF, Pereira SR, Ribeiro AM. Spatial memory deficits and thalamic serotonergic metabolite change in thiamine deficient rats. *Behav. Brain Res.* 2010;210:140-2.

[12] Lee RS, Strahlendorf HK, Strahlendorf JC. Enhanced sensitivity of cerebellar Purkinje cells to iontophoretically-applied serotonin in thiamine deficiency. *Brain Res.* 1985;327:249-58.

[13] Victor M, Adams RD, Collins GH. The Wernicke-Korsakoff syndrome and related neurological disorders due to alcoholism and malnutrition. 2nd ed. Philadelphia: FA Davis, 1989.

[14] Rindi G, Patrini C, Comincioli V, Reggiani C. Thiamine content and turnover rates of some rat nervous regions, using labeled thiamine as a tracer. *Brain Res.* 1980;181:369-80.

[15] Halliday G, Ellis J, Heard R, Caine D, Harper C. Brainstem serotonergic neurons in chronic alcoholics with and without the memory impairment of Korsakoff's psychosis. *J Neuropathol. Exp. Neurol.* 1993;52:567-79.
[16] Martin PR, Higa S, Burns RS, Tamarkin L, Ebert MH, Markey SP. Decreased 6-hydroxymelatonin excretion in Korsakoff's psychosis. *Neurology* 1984;34:966-8.
[17] Botez MI, Young SN. Biogenic amine metabolites and thiamine in cerebrospinal fluid in heredo-degenerative ataxias. *Can. J. Neurol. Sci.* 2001;28:134-40.
[18] Pedraza OL, Botez MI. Thiamine status in inherited degenerative ataxias. *J. Neurol. Neurosurg. Psychiatry* 1992;55:136-7.
[19] Botez MI, Young SN, Bachevalier J, Gauthier S. Thiamine deficiency and cerebrospinal fluid 5-hydroxyindoleacetic acid: a preliminary study. *J. Neurol. Neurosurg. Psychiatry* 1982;45:731-3.
[20] Botez MI, Young SN. Effects of anticonvulsant treatment and low levels of folate and thiamine on amine metabolites in cerebrospinal fluid. *Brain* 1991;114:333-48.
[21] Mattson MP, Kruman, II, Duan W. Folic acid and homocysteine in age-related disease. *Ageing Res. Rev.* 2002;1:95-111.
[22] Morris MS. Homocysteine and Alzheimer's disease. *Lancet Neurol.* 2003;2:425-8.
[23] Reynolds E. Vitamin B12, folic acid, and the nervous system. *Lancet Neurol.* 2006;5:949-60.
[24] Nilsson K, Gustafson L, Hultberg B. Folate and cobalamin levels as determinants of plasma homocysteine in different age groups of healthy controls and psychogeriatric patients. *Clin. Chem. Lab. Med.* 2003;41:681-5.
[25] Selhub J, Jacques PF, Wilson PW, Rush D, Rosenberg IH. Vitamin status and intake as primary determinants of homocysteinemia in an elderly population. *JAMA* 1993;270:2693-8.
[26] Djukic A. Folate-responsive neurologic diseases. *Pediatr. Neurol.* 2007;37:387-97.
[27] Botez MI, Young SN, Bachevalier J, Gauthier S. Folate deficiency and decreased brain 5-hydroxytryptamine synthesis in man and rat. *Nature* 1979;278:182-3.
[28] Gospe SM Jr, Gietzen DW, Summers PJ, Lunetta JM, Miller JW, Selhub J, et al. Behavioral and neurochemical changes in folate-deficient mice. *Physiol. Behav.* 1995;58:935-41.
[29] Duan W, Ladenheim B, Cutler RG, Kruman, II, Cadet JL, Mattson MP. Dietary folate deficiency and elevated homocysteine levels endanger dopaminergic neurons in models of Parkinson's disease. *J. Neurochem.* 2002;80:101-10.
[30] Botez MI, Bachevalier J, Tunnicliff G. Dietary folic acid and the activity of brain cholinergic and gamma-aminobutyric acid (GABA) enzymes. *Can. J. Neurol. Sci.* 1980;7:133-4.
[31] Ordonez LA, Wurtman RJ. Folic acid deficiency and methyl group metabolism in rat brain: effects of L-dopa. *Arch. Biochem. Biophys.* 1974;160:372-6.
[32] Curcio M, Catto E, Stramentinoli G, Algeri S. Effect of S-adenosyl-L-methionine on serotonin metabolism in rat brain. *Prog. Neuropsychopharmacol.* 1978;2:65-71.
[33] Fournier I, Ploye F, Cottet-Emard JM, Brun J, Claustrat B. Folate deficiency alters melatonin secretion in rats. *J. Nutr.* 2002;132:2781-4.
[34] Kelly GS. Folates: supplemental forms and therapeutic applications. *Altern. Med. Rev.* 1998;3:208-20.

[35] Brocardo PS, Budni J, Kaster MP, Santos ARS, Rodrigues ALS. Folic acid administration produces an antidepressant-like effect in mice: evidence for the involvement of the serotonergic and noradrenergic systems. N*europharmacology* 2008;54:464-73.
[36] Reynolds EH. Methylfolate as adjunctive treatment in depression. *Am. J. Psychiatry* 2013, 170, 560.
[37] Botez MI, Bachevalier J. The blood-brain-barrier and folate-deficiency. *Am. J. Clin.* Nutr 1981;34:1725-30.
[38] Botez MI, Young SN, Bachevalier J, Gauthier S. Effect of folic acid and vitamin B12 deficiencies on 5-hydroxyindoleacetic acid in human cerebrospinal fluid. *Ann. Neurol.* 1982;12:479-84.
[39] Young SN, Ghadirian AM. Folic acid and psychopathology. *Prog Neuropsychopharmacol Biol. Psychiatry* 1989;13:841-63.
[40] Bottiglieri T, Laundy M, Crellin R, Toone BK, Carney MWP, Reynolds EH. Homocysteine, folate, methylation, and monoamine metabolism in depression 1. *J. Neurol. Neurosurg. Psychiatry* 2000;69:228-32.
[41] Triantafyllou NI, Nikolaou C, Boufidou F, Angelopoulos E, Rentzos M, Kararizou E, et al. Folate and vitamin B12 levels in levodopa-treated Parkinson's disease patients: their relationship to clinical manifestations, mood and cognition. *Parkinsonism Relat. Disord.* 2008;14:321-5.
[42] Alpert JE, Mischoulon D, Rubenstein GE, Bottonari K, Nierenberg AA, Fava M. Folinic acid (Leucovorin) as an adjunctive treatment for SSRI-refractory depression. *Int. J. Geriatr. Psychiatry* 2013;28:925-32.
[43] Papakostas GI, Shelton RC, Zajecka JM, Etemad B, Rickels K, Clain A, et al. L-methylfolate as adjunctive therapy for SSRI-resistant major depression: results of two randomized, double-blind, parallel-sequential trials. *Am. J. Psychiatry* 2012;169:1267-74.
[44] Hunter R, Barnes j, Curzon G, Kantamaneni BD, Duncan C. Effect of folic acid by mouth on cerebrospinal fluid homovanillic acid and 5-hydroxyindoleacetic acid concentration. *J. Neurol. Neurosurg. Psychiatry* 1971;34: 571–575.
[45] Williams E, Stewart-Knox B, Bradbury I, Rowland I, Pentieva K, Helander A, McNulty H. Effect of folic acid supplementation on mood and serotonin response in healthy males. *Br. J. Nutr.* 2005;94:602-8.
[46] Bender DA. Non-nutritional uses of vitamin B-6. *Br. J. Nutr.* 1999;81:7-20.
[47] Cooper JR, Bloom FE, Roth RH. The biochemical basis of neuropharmacology. 7 ed: Oxford University Press, 1996.
[48] Dakshinamurti K, Paulose CS. Consequences of decreased brain serotonin in the pyridoxine-deficient young rat. *Prog. Neuropsychopharmacol. Biol. Psychiatry* 1983;7:743-6.
[49] Dakshinamurti K, LeBlancq WD, Herchl R, Havlicek V. Nonparallel changes in brain monoamines of pyridoxine-deficient growing rats. *Exp. Brain. Res.* 1976;26:355-66.
[50] Dakshinamurti K, Paulose CS, Viswanathan M, Siow YL. Neuroendocrinology of pyridoxine deficiency. *Neurosci. Biobehav. Rev.* 1988;12:189-93.
[51] Sharma SK, Dakshinamurti K. Effects of serotonergic agents on plasma prolactin levels in pyridoxine-deficient adult male rats. *Neurochem. Res.* 1994;19:687-92.

[52] Siow YL, Dakshinamurti K. Effect of pyridoxine deficiency on aromatic L-amino acid decarboxylase in adult rat brain. *Exp. Brain Res.* 1985;59:575-81.

[53] Paulose CS, Dakshinamurti K. Effect of pyridoxine deficiency in young rats on high-affinity serotonin and dopamine receptors. *J. Neurosci. Res.* 1985;14:263-70.

[54] Bhagavan HN, Coursin DB. Effects of pyridoxine deficiency and DL-p-chlorophenylalanine administration to rats on 5-hydroxytryptamine and noradrenaline concentrations in brain and 5-hydroxytryptamine concentration in blood. *Biochem. J.* 1973;134:763-7.

[55] Guilarte TR. Vitamin B6 and cognitive development: recent research findings from human and animal studies. *Nutr. Rev.* 1993;51:193-8.

[56] Guilarte TR, Wagner HN, Jr. Increased concentrations of 3-hydroxykynurenine in vitamin B6 deficient neonatal rat brain. *J. Neurochem.* 1987;49:1918-26.

[57] Viswanathan M, Siow YL, Paulose CS, Dakshinamurti K. Pineal indoleamine metabolism in pyridoxine-deficient rats. *Brain Res.* 1988;473:37-42.

[58] Dakshinamurti K, Sharma SK, Bonke D. Influence of B vitamins on binding properties of serotonin receptors in the CNS of rats. *Klin. Wochenschr.* 1990;68:142-5.

[59] Calderón-Guzmán D, Hernández-Islas JL, Espitia-Vázquez I, Barragán-Mejía G, Hernández-García E, Santamaría-del Angel D, et al. Pyridoxine, regardless of serotonin levels, increases production of 5-hydroxytryptophan in rat brain. *Arch. Med. Res.* 2004;35:271-4.

[60] Hartvig P, Lindner KJ, Bjurling P, Laengstrom B, Tedroff J. Pyridoxine effect on synthesis rate of serotonin in the monkey brain measured with positron emission tomography. *J. Neural Transm. Gen. Sect.* 1995;102:91-7.

[61] Bender DA, Totoe L. High doses of vitamin B6 in the rat are associated with inhibition of hepatic tryptophan metabolism and increased uptake of tryptophan into the brain. *J. Neurochem.* 1984;43:733-6.

[62] Lee NS, Muhs G, Wagner GC, Reynolds RD, Fisher H. Dietary pyridoxine interaction with tryptophan or histidine on brain serotonin and histamine metabolism. *Pharmacol. Biochem. Behav.* 1988;29:559-64.

[63] Schaeffer MC, Gretz D, Gietzen DW, Rogers QR. Dietary excess of vitamin B-6 affects the concentrations of amino acids in the caudate nucleus and serum and the binding properties of serotonin receptors in the brain cortex of rats. *J. Nutr.* 1998;128:1829-35.

[64] de Lau LM, Koudstaal PJ, Witteman JC, Hofman A, Breteler MM. Dietary folate, vitamin B12, and vitamin B6 and the risk of Parkinson disease. *Neurology* 2006;67:315-8.

[65] Lerner V, Miodownik C, Kaptsan A, Bersudsky Y, Libov I, Sela BA, et al. Vitamin B6 treatment for tardive dyskinesia: a randomized, double-blind, placebo-controlled, crossover study. *J. Clin. Psychiatry* 2007;68:1648-54.

[66] Dakshinamurti K. Neurobiology of pirodoxine. In: Reynolds RD, Leklem JE, eds. Vitamin B6: its role in health nad disease. New York: Alan R Liss, 1985; 99-121.

[67] Maller A, Hyland K, Milstien S, Biaggioni I, Butler IJ. Aromatic L-amino acid decarboxylase deficiency: clinical features, diagnosis, and treatment of a second family. *J. Child Neurol.* 1997;12:349-54.

[68] Montioli R, Oppici E, Cellini B, Roncador A, Dindo M, Voltattorni CB. S250F variant associated with aromatic amino acid decarboxylase deficiency: molecular defects and intracellular rescue by pyridoxine. *Hum. Mol. Genet.* 2013;22:1615-24.

[69] Wyatt KM, Dimmock PW, Jones PW, O'Brien PMS. Efficacy of vitamin B-6 in the treatment of premenstrual syndrome: systematic review 1. *Br. Med. J.* 1999;318:1375-81.

[70] Wyatt KM, Dimmock PW, O'Brien PM. Selective serotonin reuptake inhibitors for premenstrual syndrome. *Cochrane Database Syst. Rev.* 2002;CD001396.

[71] Muñoz-Hoyos A, Amorós-Rodríguez I, Molina-Carballo A, Uberos-Fernández J, Acuña-Castroviejo D. Pineal response after pyridoxine test in children. *J. Neural Transm.* 1996;103:833-42.

[72] Luboshitzky R, Ophir U, Nave R, Epstein R, Shen-Orr Z, Herer P. The effect of pyridoxine administration on melatonin secretion in normal men. *Neuro Endocrinol Lett.* 2002;23:213-7.

[73] NIH office of dietary supplements. https:/ods.od.nih.gov/factsheets/VitaminB12/Quickfacts/

[74] Miller A, Korem M, Almog R, Galboiz Y. Vitamin B12, demyelination, remyelination and repair in multiple sclerosis. *J. Neurol. Sci.* 2005;233:93-97.

[75] Lindenbaum J, Savage DG, Stabler SP, Allen RH. Diagnosis of cobalamin deficiency: II. Relative sensitivities of serum cobalamin, methylmalonic acid, and total homocysteine concentrations. *Am. J. Hematol.* 1990;34:99-107.

[76] Deana R, Vincenti E, Deana AD. Levels of neurotransmitters in brain of vitamin B12 deficient rats. *Int. J. Vitam. Nutr. Res.* 1977;47:119-22.

[77] Nakamura T, Uchida K, Moriguchi Y, Okamoto N, Morita Y. Transient fluctuation of serum melatonin rhythm is suppressed centrally by vitamin B12. *Chronobiol. Int.* 1997;14:549-60.

[78] Ikeda M, Asai M, Moriya T, Sagara M, Inoué S, Shibata S. Methylcobalamin amplifies melatonin-induced circadian phase shifts by facilitation of melatonin synthesis in the rat pineal gland. *Brain Res.* 1998;795:98-104.

[79] Heok KE, Ho R. The many faces of geriatric depression. *Curr. Opin. Psychiatry* 2008;21:540-5.

[80] Mayer G, Kröger M, Meier-Ewert K. Effects of vitamin B_{12} on performance and circadian rhythm in normal subjects. *Neuropsychopharmacology* 1996;15:456-64.

[81] Hashimoto S, Kohsaka M, Morita N, Fukuda N, Honma S, Honma K. Vitamin B12 enhances the phase-response of circadian melatonin rhythm to a single bright light exposure in humans. *Neurosci. Lett.* 1996;220:129-32.

[82] Okawa M, Mishima K, Nanami T, Shimizu T, Iijima S, Hishikawa Y, Takahashi K. Vitamin B12 treatment for sleep-wake rhythm disorders. *Sleep* 1990;13:15-23.

[83] Okawa M, Takahashi K, Egashira K, Furuta H, Higashitani Y, Higuchi T, et al. Vitamin B12 treatment for delayed sleep phase syndrome: a multi-center double-blind study. *Psychiatry Clin. Neurosci.* 1997;51:275-9.

[84] Lasisi AO1, Fehintola FA, Lasisi TJ. The role of plasma melatonin and vitamins C and B12 in the development of idiopathic tinnitus in the elderly. *Ghana Med. J.* 2012;46:152-7.

In: Indoleamines
Editor: Ángel Catalá

ISBN: 978-1-63482-097-4

Chapter 8

MELATONIN IN NEURODEGENERATIVE, PSYCHIATRIC AND SYSTEMIC INFLAMMATORY DISORDERS

***George Anderson*[1,*] *and Michael Maes, PhD, MD*[2]**
[1]CRC Scotland and London, Davies Court, London, UK
[2]Deakin University, Department of Psychiatry, Geelong, Australia

ABSTRACT

The release of melatonin from the pineal gland has a long-standing association with the regulation of the circadian rhythm and, via this, with a plethora of medical conditions. Alterations in the melatonergic pathways are associated with a) neurodegenerative conditions, including: Alzheimer's disease, multiple sclerosis and Parkinson's disease; b) psychiatric conditions, including: bipolar disorder, depression and schizophrenia; and c) a wide array of other medical conditions, including obesity, cancers and migraines. Recent work has suggested that melatonin effects are not simply mediated via circadian regulation, but are also driven by variations in local mitochondrial melatonin synthesis by a wide array of cells, including central glia and systemic immune cells. In glia and immune cells, melatonin decreases reactivity and increases phagocytosis, suggesting that melatonin has significant impacts on such a wide array of medical conditions via the modulation of systemic and central immune system responses. As such, the melatonergic pathways are important in the etiology, course and treatment of a host of medical conditions. In this chapter, we look at the role of the melatonergic pathways in neurodegenerative, psychiatric and wider medical conditions, emphasizing its interactions with oxidative and nitrosative stress, immune inflammation and the tryptophan catabolite pathway.

Keywords: Melatonin; neurodegeneration; psychiatric; immune inflammation; cancer; tryptophan catabolites; mitochondria

* E-mail: anderson.george@rocketmail.com.

INTRODUCTION

Relevant changes in many neurodegenerative and psychiatric conditions have classically emphasized alterations in neurons and neuronal synapses as being the biological underpinnings of these disorders. Such a classically conceptualized neuroncentric perspective has recently been challenged, with more current conceptualizations of CNS associated medical conditions emphasizing the contribution of the immune system, including astrocytes and microglia, in the etiology and course of such disorders.[1]

Although conditions such as Alzheimer's disease, Parkinson's disease, multiple sclerosis (MS), Huntington's disease, schizophrenia, bipolar disorder, depression, autism, stress, obesity, migraine, fibromyalgia and cancer have a diverse array of seemingly distinct presentations, there are a number of significant commonalities. These diverse medical conditions show decreased melatonin levels and/or genetic susceptibility genes in melatonergic pathways, involving either melatonin synthesis or melatonin receptors.[2-13] As to how variations in the melatonergic pathways modulate such conditions has generally been linked to alterations in pineal melatonin synthesis and circadian rhythm alterations.

Melatonin is a powerful antioxidant and anti-inflammatory that significantly optimizes mitochondrial functioning. Melatonin is also a significant inducer of endogenous antioxidants, via its activation of nuclear erythroid 2 p45-related factor-2 (Nrf2).[14] The above array of medical conditions are also associated with decreased endogenous antioxidants, increased levels of oxidative and nitrosative stress (O&NS) and increased activation of immune inflammatory pathways, often in association with increased levels of clinical depression.[15-18] Increased O&NS and immune inflammatory pathways drive further commonalities, including the activation of the tryptophan catabolites (TRYCATs) pathway, predominantly by the induction of indoleamine 2,3-dioxygenase (IDO) in glia and immune cells, which takes tryptophan away from serotonin, N-acetylserotonin (NAS) and melatonin synthesis. TRYCATs synthesized, such as kynurenic acid and quinolinic acid, are neuroregulatory, with quinolinic acid also being neuro-excitotoxic. The synthesis of such TRYCATs by glia drive central changes in neuronal activity across these diverse conditions, in turn driving the processes underlying the increased levels of depression in this wide array of medical conditions.[18, 19] As such, melatonin levels are intimately connected to levels of O&NS, immune inflammation and TRYCATs as well as with levels of depression and often coupled with changes in mitochondrial functioning.

Increased O&NS and immune inflammatory cytokines also increase levels of autoimmunity, a common emergent condition that is driven by lipid peroxidation damaging the plasma membrane, in turn exposing neoepitopes.[20] O&NS also damages DNA, thereby increasing poly(ADP-ribose) polymerase (PARP)1, which, by depleting nicotinamide, drives down the synthesis of sirtuin-1, sirtuin-3 and peroxisome proliferator-activated receptor gamma coactivator-1alpha (PGC-1alpha), thereby contributing to mitochondrial dysfunction and the association of suboptimal mitochondrial activity with the diverse array of medical conditions above, including autoimmunity [21]. The beneficial effects of a wide range of medications, anti-oxidants and dietary factors across these different disorders also hint at important commonalities.[22-24]

In this chapter, we review the data on the above factors and their association with such a wide range of medical conditions, proposing that the biological commonalities in etiology, course and treatment are powerfully influenced by the impacts of O&NS and immune-inflammatory pathways on levels of melatonin and NAS synthesis by local immune cells, including in central astrocytes and microglia, in the course of the inflammatory response. This is likely to have significant impacts on the conceptualization, prevention, etiology, course and management of the diverse range of currently poorly treated disorders detailed above.

MELATONINERGIC PATHWAYS

Melatonin is synthesized when tryptophan is utilized to produce serotonin, which is converted to NAS by arylalkylamine N-acetyltransferase (AA-NAT), with NAS then converted to melatonin by acetylserotonin O-methyltransferase (ASMT), also referred to as hydroxyindole O-methyltransferase (HIOMT). Norepinephrine (NE) is a significant driver of pineal melatonin synthesis. NAS is also released with melatonin and, like melatonin, is a powerful antioxidant, with both pineal NAS and melatonin effluxes readily crossing into cells, due to their amphiphilic nature. Relatively little is known about the relevance of pineal NAS efflux or the NAS/melatonin ratio and is therefore little referred to in this chapter. However, NAS is known to activate the brain derived neurotrophic factor (BDNF) receptor tropomyosin receptor kinase B (TrkB), suggesting that it may have important and different tropic effects in comparison to melatonin.[25]

Melatonin and Immunity

Melatonin and NAS are released by the pineal gland in the absence of light at night, with melatonin being an important entrainer of the circadian rhythms, both by its systemic effects and by its modulation of the activity of the suprachiasmatic nucleus, the master circadian regulator. Alterations in night-time functioning of the immune system and in the regulation of circadian genes are regulated by pineal melatonin efflux in a variety of animals, including humans.[26-27] Pineal melatonin generally has an immuno-suppressive effect, increasing levels of immune cell phagocytic activity and decreasing levels of inflammatory responses [28]. However, melatonin can increase levels of T helper (Th)-1 cytokines as well as the activity of natural killer cells, [29] suggesting that melatonin may help to optimize the generally more beneficial Th1 response, minimizing the activation of the more damaging and prolonged Th17 response, which is highly associated with autoimmunity.[30] These effects of pineal melatonin are partly mediated by its induction of endogenous antioxidants and the optimization of mitochondrial functioning. [31]

Pineal Melatonin and Medical Conditions

Pineal melatonin release at night is commonly decreased across a range of disorders, including Alzheimer's disease, [32] MS, [33] schizophrenia, [3] bipolar disorder, [34] depression, [35] fibrobmyalgia, [36] migraine, [4] cancer, [37] and obesity [38]. Some of these effects will be mediated by the high levels of comorbid depression in all of these conditions, [15, 39] driving down serotonin levels and therefore levels of NAS and melatonin. All of these conditions are also associated with comorbid sleep dysregulation, which will interact with pineal melatonin synthesis.

Single nucleotide polymorphisms (SNPs) in melatoninergic pathway molecules and receptors increase the risk of a number of medical conditions, including diabetes and obesity,[40, 41] autism, [42] MS, [43] bipolar disorder, [44] schizophrenia, [45] breast cancer, [46] depression, [47, 48] cardiovascular disorders, [49] and seasonal affect disorder.[50] These susceptibility effects of melatonergic alleles are modelled as being mediated via pineal melatonin dysregulation and thereby on the circadian rhythm. However, it is not unlikely that non-circadian effects of such susceptibility alleles occur, especially via the regulation of immune and glia cell reactivity responses, as detailed below.

Melatonin Treatment

As a consequence of such decreases in melatonin and SNP susceptibility alleles across a host of medical conditions, melatonin has been used in the treatment of many of these conditions, usually with beneficial effects. Melatonin has shown efficacy in the treatment of schizophrenia and bipolar disorder, decreasing antipsychotic metabolic side-effects in both conditions, [51, 52] thereby decreasing the changes that occur over time in psychosis that contribute to treatment resistance, which is termed neuroprogression and is evident in schizophrenia, [53] and in bipolar disorder.[54] Melatonin has also shown treatment efficacy in Alzheimer's disease,[55, 56] mild cognitive impairment, [57] Parkinson's disease, [58] depression, [59] migraine, [60] cancer, [61, 62] and fibromyalgia, [63] among many other medical conditions.[12]

Models of melatonin treatment efficacy include decreasing the toxicity of amyloid B in Alzheimer's disease, whilst improving sleep and decreasing depression in mild cognitive impairment as well as in Alzheimer's disease.[55, 57] In Parkinson's disease, melatonin decreases levels and protofibril formation, oligomerization, and secondary structure transitions of alpha-synuclein.[64] Melatonin increases antioxidants as well as improving sleep and patient quality of life in MS patients.[65, 66] This is coupled to wider general improvements in circadian and sleep regulation. Melatonin's anti-oxidant, anti-inflammatory and anti-nociceptive effects, coupled to improvements in mitochondrial oxidative phosphorylation are thought to mediate these widely beneficial effects across diverse disorders, including in rodent models of MS.[67, 68] As indicated above, melatonin effects on the sirtuins and PGC-1alpha are likely to mediate many of its beneficial effects on the functioning of mitochondria, which are a major source for ROS generation and the regulation of immune cell reactivity.[69]

Neurogenesis and Melatonin

Neurogenesis is thought to be decreased in many human medical conditions, which is strongly supported by physiological measurements in the animal models of such disorders, including Alzheimer's disease, [70] Parkinson's disease, [71] multiple sclerosis, [72] and depression.[73] Decreased neurogenesis may be an early event in neurodegenerative and psychiatric conditions, being another mechanism whereby comorbid depression may modulate the etiology and course of neurodegenerative processes.[15, 18] Melatonin significantly increases neurogenesis and this may be another means by which depression, in decreasing serotonin, contributes to decreased neurogenesis.[74] It should be noted that in the early stages of the inflammatory process, the pro-inflammatory milieu may actually contribute to increasing neurogenesis, although this seems transient and certainly inadequate to compensate the loss of neurons and glia occurring in neurodegenerative disorders.[75] This parallels the often opposing effects of acute and chronic stress.[76, 77]

Stress commonly associates with the above medical conditions. Chronic stress decreases neurogenesis, with the effects thought to be mediated by increases in the stress hormone cortisol, which melatonin prevents in animal models.[78] The effects of chronic stress are at least partly mediated by increased monoamine oxidase, [77] which, unlike acute stress, [76] leads to an increased breakdown of monoamines, including serotonin. As a consequence there is a decrease in the availability of serotonin for the synthesis of NAS and melatonin. In the context of local melatonin synthesis in the CNS, increases in chronic stress are likely to raise levels of monoamine oxidase in astrocytes, decreasing the proposed influence of astrocyte melatonin and NAS on neurogenesis.[79] This is detailed below in regard to the putative role of locally synthesized melatonin in the CNS, but is relevant to the biological processes underpinning decreased neurogenesis and increased depression across a host of medical conditions. Sirtuin-1 is also a significant regulator of neurogenesis, [80] suggesting that melatonin induction of sirtuin-1 is likely to contribute to its modulation of neurogenesis.

Immune-Pineal Axis

It is important to note that the level of circadian NAS and melatonin synthesis may be intimately associated with the degree of immuno-inflammation and levels of circulating pro-inflammatory cytokines. Recent work suggests that pineal melatonin production may be actively switched off during the course of inflammation, specifically by circulating tumor necrosis factor (TNF)-alpha effects in pinealocytes and by the activation of toll-like receptor (TLR)4 in pineal gland associated microglia, which leads to local TNF-alpha production.[81] In pregnant women that have undergone caesarean section, the resultant inflammation increase induces TNF-alpha, which then acts to switch off pineal melatonin production for around two weeks. Circadian pineal melatonin is reinstated once the peripheral inflammation is resolved and TNF-alpha levels return to normal.[82] As to how relevant this is to the decreased pineal melatonin release in the plethora of medical conditions listed above requires investigation. Certainly, increased levels of circulating TNF-alpha are evident in Alzheimer's disease, [83] especially when comorbid with depression [84]. Increased levels of serum TNF-alpha are also evident in MS patients, where it correlates with disability level, [85] as well as being increased in patients with Parkinson's disease, [86] bipolar disease, [87] fibromyalgia and depression.[88, 89] Given that pro-inflammatory cytokines tend to induce one another, it requires investigation as to how specific TNF-alpha is among the cytokines to such pineal melatonin suppressing effects and as to whether TNF-alpha levels correlate with levels of pineal melatonin suppression in the medical conditions listed above. Patients with schizophrenia tend to show decreased serum TNF-alpha, [90] although it requires investigation as to whether other factors, including central TNF-alpha , are more important to pineal melatonin suppression in this and other medical conditions. Cerebrospinal fluid levels of TNF-alpha are increased in migraine patients, without significant increases evident in the serum, [91] highlighting that central and peripheral TNF-alpha levels are not always synchronized.

Likewise the suppressive effects of microglia TLR4 activation on pineal melatonin production would suggest that ligands for TLR4 are potential regulators of circadian pineal melatonin synthesis.

As well as its classical activation by lysopolysaccharide (LPS), there are many TLR4 agonists, including many of the heat shock proteins, free fatty acids, high-mobility group box-1 (HMGB1) and the *mu*-opioid receptor.[92-94] As to whether such a variety of TLR4 ligands differentially contribute to the suppression of circadian pineal melatonin synthesis and release across different medical conditions requires investigation. For example, increases in gut permeability, which are discussed below, are known to increase levels of circulating LPS and immuno-inflammatory factors, suggesting that gut permeability may contribute to the modulation of pineal melatonin and NAS efflux.

Activation of the *mu*-opioid receptor increases LPS induced microglia activation, including the release of TNF-alpha, [95] suggesting that variations in exogenous and endogenous *mu*-opioid receptor ligands may interact with the effects of other TLR4 ligands in the regulation of microglia reactivity, including in the modulation of pineal melatonin release. However, activation of the *mu*-opioid receptor on pinealocytes can increase melatonin release in some animal studies, [96] via the cAMP pathways.[97] Stress-induced cortisol, both directly and indirectly, increases levels of the mu-opioid receptor and astrocyte preproenkephalin, an endogenous *mu*-opioid receptor ligand.[98, 99] We proposed previously that the mu-opioid receptor plays a role centrally in mediating alterations in attachment and the effects of stress, including social stressors such as racial discrimination.[100] With alterations in the opioid receptor being evident in Alzheimer's disease and regulating susceptibility to depression and social rejection, [101, 102] it requires clarification as to how opioids and the *mu*-opioid receptor modulate microglia, TLR4 and pinealocytes in human studies, including in the modulation of melatonin and NAS circadian efflux, especially in the context of systemic inflammation.

Overall, the presence of an immune-pineal axis suggests that evolutionary processes have acted to suppress pineal melatonin and NAS synthesis under conditions of systemic inflammation, involving a number of interacting pathways, including TLR4, endogenous opioids and the specific nature of the immuno-inflammatory response. This suggests that the commonly observed decrease in melatonin across a range of central and systemic disorders is an evolved response that decreases the immuno-suppressive effects of released pineal melatonin, [103] allowing local factors to determine the appropriate immune inflammatory response.

However, a consequence of this is the loss of pineal melatonin's influence on the circadian rhythm, as well as its wider antioxidant, anti-inflammatory and mitochondria regulating effects. Such impacts of systemic inflammation on melatonin and the circadian rhythm gives a simple model as to how low level immune-inflammatory activity interacts with the etiology and course of a wide range of medical conditions. Interestingly, the level of local melatonin and NAS synthesis in activated immune cells at sites of inflammation dampen the immune response, [104] in effect switching on the pineal gland's circadian synthesis of melatonin and NAS.

LOCAL MELATONIN AND IMMUNO-INFLAMMATION

Melatonin is produced by many immune cells, including macrophages, [104] T cells and mast cells, [105-106] as well as by astrocytes, [107] fibroblasts and many other cells and tissues, especially in the gut, where melatonin levels are many times higher than that produced by the pineal gland.[108, 109] It requires investigation as to whether microglia and oligodendrocytes synthesize melatonin and NAS. This does not seem unlikely as it has recently been proposed that all mitochondria containing cells synthesize melatonin to some degree [110]. Mitochondria are derived from melatonin synthesizing bacteria, [110] with melatonergic pathway mRNA long-known to be present in mitochondria.[111]

Generally, melatonin increases an anti-inflammatory (M2 macrophage) and phagocytic phenotype, including by its autocrine effects in macrophages.[104] The activation of specific NF-kB subunits in macrophages leads to melatonin, and presumably NAS, synthesis, which then acts in an autocrine manner to induce a phagocytic, anti-inflammatory macrophage phenotype.[104] As such, local autocrine or paracrine melatonin secretion contributes to the dampening of immuno-inflammation, instead inducing an anti-inflammatory and debris clearing macrophage phenotype. It is this dampening of the immuno-inflammatory response that triggers the re-emergence of pineal circadian melatonin and NAS synthesis, probably involving a decrease in the levels of TNF-alpha. As such many of the medical conditions listed above that show melatonin SNP genetic susceptibilities that decrease melatonin and NAS synthesis, will show changes not only in pineal melatonin and/or NAS synthesis, but also alteration in the nature, and in the closure, of immuno-inflammatory responses driven by immune and/or glia cells.

The above suggests that factors modulating local melatonin and NAS synthesis in glia and immune cells will contribute to varying their reactivity threshold and duration, in turn determining the levels and persistence of pro-inflammatory cytokines. Given that pro-inflammatory cytokines interact with levels of O&NS and neuroregulatory TRYCAT production, local melatonin production may be intimately involved with a host of neurodegenerative, psychiatric and general medical conditions, as detailed above. As to how other immune regulating factors, including epigenetic factors such as microRNAs, interact with local melatonin synthesis, and therefore the nature of the immune response, will be important to determine. This is exemplified in psychiatric conditions where neuroprogression is thought to be driven by a prolonged immuno-inflammatory response in association with increased O&NS.[54] It is interesting to look at the effects of melatonin across a range of conditions in this context.

Local Melatonin and Alzheimer's Disease

Yin yang (YY)1 significantly induces the melatoninergic pathway enzymes in some cell types.[112] In astrocytes, YY1 is only expressed in the reactive, but not quiescent, state, [113] suggesting that should YY1 induce the melatoninergic pathways in astrocytes, it will only be under conditions of inflammation and astrocyte reactivity. Given that YY1 is also expressed in reactive microglia, [114] it is not unlikely that microglia YY1 will increase melatonin in these cells also, in turn regulating microglia reactivity. The role of YY1 requires investigation in both astrocytes and microglia and may be important, given that YY1 is downstream of NF-kB and is a significant regulator of an important peptide that is classically regarded as defining in Alzheimer's disease, namely amyloid B.[115]

YY1 is also a significant inducer of Beta-site amyloid precursor protein-cleaving enzyme (BACE1) in astrocytes, thereby raising levels of the peptide most strongly associated with Alzheimer's disease, amyloid B.[116] Several pro-inflammatory cytokines induce BACE1 and amyloid B peptide production, including the inflammasome induced cytokine, IL-18.[117] This suggests that astrocyte melatonin and NAS synthesis may be co-ordinated with BACE1 induction and amyloid B release by the transcription effects of YY1. Factors that downregulate melatonin synthesis, including SNPs in melatoninergic pathways and melatonin receptors, as well as that availability of serotonin, will desynchronise such BACE1/amyloid B production with local melatonin and NAS synthesis. This suggests an interaction with chronic stress levels, given that chronic stress/cortisol increases MAO, [77] in turn decreasing serotonin availability for melatonin and NAS synthesis, thereby allowing chronic stress to drive a desynchronization of melatonin/NAS synthesis and BACE1/amyloid B production. This would then be another route whereby chronic stress may increase the risk of Alzheimer's disease. With melatonin preventing amyloid B toxicity, [118] a relative decrease in YY1 induced melatoninergic pathways coupled to increased BACE1 and amyloid B production would be expected to contribute to the runaway intercellular inflammation triggered by amyloid B.[119]

Similarly, the gene allele most strongly associated with Alzheimer's disease susceptibility, the apolipoprotein (Apo)E4 allele, has been shown to increase levels of melatonin synthesis in C6 glioma cells [120]. These authors show that the ApoE4 induced increase in the melatonergic pathways is driven by decreased MAO-A and MAO-B.[120] This suggests that, as with YY1, chronic stress and factors decreasing serotonin availability will contribute to a relative loss of melatonin protection against the heightened cellular toxicity associated with ApoE4. Generally, data supports this, as the ApoE4 allele is far more associated with Alzheimer's disease, when carriers of the allele report high levels of chronic stress.[121] As well as the effects of stress induced cortisol on MAO, immuno-inflammation and O&NS induced serotonin depletion will also be interacting with the ApoE4 allele, by depriving tryptophan for serotonin synthesis by driving it to TRYCAT production.

Increased amyloid B is also evident in a number of other medical conditions, including Parkinson's disease and the autistic spectrum, suggesting a role for variations in melatonin co-ordination with amyloid B in these conditions also.[122]

Local Melatonin and Parkinson's Disease

It is important to note that the ApoE4 allele increases the susceptibility to dementia in Parkinson's disease patients, suggesting a similar role for synchronized local melatonin and amyloid B production in this disease, as with Alzheimer's disease.[123] This is supported by data showing that melatonin receptors are decreased in the amygdala as well as in the substantia nigra in Parkinson's disease patients.[124] Given that melatonin normally increases its own receptors, this would suggest decreased levels of central melatonin in Parkinson's disease. It is of note that a commonly used medication in the treatment of Parkinson's disease is the MAO inhibitor, selegiline, which inhibits both MAO-A and MAO-B in the CNS.[125]

The MAO gene is also a significant regulator of the course of Parkinson's disease.[126] Such data suggests that a decreased synthesis of local melatonin may well occur in Parkinson's disease, with consequences for its etiology, course and management.

As well as inhibiting amyloid B toxicity in Parkinson's disease, synchronized local melatonin, as well as pineal melatonin, is likely to have wider impacts in this poorly managed condition. Melatonin has significant protective effects on other core changes associated with Parkinson's disease, including decreasing the production of alpha-synuclein and the formation of alpha-synuclein protofibrils, as well as their oligomerization, and secondary structure transitions.[64] Similar to the effects of decreasing amyloid B in Alzheimer's disease, the suppression of alpha-synuclein dampens the inflammatory process *per se*. It has been proposed that alpha-synuclein, as with amyloid B, may initiate something akin to a prion-like disease or a spreading infection, given their impacts on local intercellular inflammatory processes that spread like falling dominoes and involve networks of neurons, microglia and astrocytes.[19, 119] Local melatonin and NAS are proposed to therefore act by inhibiting the initiation, as well as the spread, of such infection-like inflammatory and neurodegenerative processes.

Local Melatonin and MS

MS is an immune mediated disorder, with predominantly white matter, but also grey matter, loss.[127] This neuronal loss is driven by increased T-helper (Th)17 and Th1 T cells. The activation of IDO and TRYCATs is proposed to drive increased depression levels prior to MS disease exacerbations.[128] Such associated decreases in tryptophan availability for serotonin suggests a role for alterations in the melatoninergic pathways in MS. This is indeed the case, with tryptophan hydroxylase-2 and melatonin receptors being MS susceptibility genes, being associated with primary and secondary progressive MS.[43] Melatonin administration to secondary progressive MS patients melatonin decreases lipid peroxidation in red blood cells, whilst increasing the endogenous antioxidants, superoxide dismutase (SOD) and glutathione peroxide (GPx), [129] as well as quality of life of MS patients.[65] Previously we proposed efficacious interactions of vitamin D, melatonin and valproate in the treatment of MS, especially in the case of emergent seizures, [130] which recent data supports.[131] As such, alterations in melatonin are relevant to the etiology, course and treatment of MS.

The ApoE4 allele is also relevant to the course of MS, as with Alzheimer's disease and Parkinson's disease, with the ApoE4 allele increasing the susceptibility to a more severe MS course, [132] with resultant exacerbation of cognitive decline as well as increased grey matter loss.[133, 134] This is in line with the ApoE4 allele being a susceptibility factor for Alzheimer's disease, where it interacts with stress. However, a meta-analysis shows no ApoE4 regulation of MS susceptibility per se by ApoE4, [135] suggesting that the stress and depression interactions with ApoE4 that increase Alzheimer's disease susceptibility may also be operating in MS.[136, 137] Again the enhanced neurotoxicity of the ApoE4 allele may be compensated for by increased astrocyte melatonin and NAS synthesis, which is lost under conditions of stress and depression. Data showing that melatonin increases antioxidants and patient quality of life may be related to this.[65] Astrocytes are dense around myelin, especially at the nodes of Ranvier, where their release of melatonin would afford protection to oligodendrocytes, whilst also increasing remyelination.[138]

Accumulating data shows mitochondrial defects and an energy deficient state in the pathogenesis of MS.[139] Deficits in mitochondrial respiratory chain function are evident in lesion associated oligodendrocytes, hypertrophied astrocytes and axons in MS patients, [140] as well as in MS animal models, [141] suggesting that local melatonin synthesis, by optimizing mitochondrial functioning, will have beneficial effects in a variety of lesion-associated cells [68]. Myelin damage and suboptimal mitochondrial functioning are, at least partly, mediated by increased microglia and macrophage reactivity.[142] Given the role of melatonin in decreasing the reactivity of these cells, local melatonin is likely to contribute to optimizing mitochondrial functioning per se in different lesion-associated cells. Interestingly increased serum TNF-alpha is evident in MS patients coupled to decreased levels of melatonin.[143] This could suggest that increases in TNF-alpha may be contributing to the loss of circadian melatonin, as suggested by the immune-pineal axis model highlighted above.[82] The efficacy of natalizumab treatment is associated with increasing melatonin levels in female relapse-remitting MS patients.[144] Given the role of enhanced immune activity in MS, it is likely that local melatonin plays a significant role via its regulation of immune cell reactivity and phenotype.[2, 104]

Local Melatonin and Glioblastoma

Systemic and local inflammation is a significant contributor to the etiology of many types of tumors, with indicants of systemic immune activation predicting prognosis.[145, 146] As such, factors acting on central inflammatory and O&NS responses will modulate glioma induction, growth and migration. Melatonin decreases glioma survival, growth and migration.[147] Melatonin also decreases glioma initiating cell proliferation as well as decreasing their self-renewal and clonogenic ability, with this being accompanied by a decrease in stem cell marker expression.[148] These authors also showed that melatonin, via the modulation of stem cell properties, induces glioma initiating cell death with autophagy-like ultrastructural features.[148] Melatonin also inhibits hypoxia induced glioblastoma proliferation via its inhibiton of hypoxia-inducing factor-alpha.[149] Such data reinforces melatonin's therapeutic potential in the treatment of malignant glioblastoma by its glioma antiproliferative and pro-apoptotic effects, as well as enhancing chemotherapy efficacy.[150]

Given the above, factors acting to regulate glioma melatonin, as well as that of neighbouring astrocyte and microglia would be expected to impact on glioma initiation and course. In this context, it is important to remember that the dire prognosis associated with glioblastoma arise form their rapid spread along blood vessels and white matter. Given that astrocytes enwrap brain blood vessels and are densely distributed along white matter, increasing astrocyte melatonin synthesis and release could be an important treatment target for the management of this poorly controlled condition. The enhanced synthesis of local astrocyte melatonin would also increase the cytotoxicity of natural killer cells, thereby enhancing glioblastoma death.[151] The paucity of studies investigating glia melatonin synthesis and regulation makes it difficult to link changes in the intracellular processes of neighbouring astrocytes to the intercellular processes occurring in the interaction of astrocytes with glioblastoma. However, some hints may be available from data in other cell types. YY1, which can increase melatonergic pathways in some cells, forms a complex with HMGB2, with this complex repressing genes at chromosome 4q35, which includes the melatonin MT1 receptor.[152] HMGB2 is only expressed in the brain during early development, being mainly evident in adulthood only in glioblastoma. YY1 is expressed in glioblastoma and regulates survival, suggesting that any complex formation with HMGB2 may act to suppress the melatonin MT1 receptor.[153] As such, some of the effects of melatonin may be altered in the tumor microenvironment. Such potential dysregulation of melatonin synthesis and melatonin receptor levels in the cells of the tumor environment requires investigation as to its relevance in vivo, given that melatonin normally induces its own receptors.

Local Melatonin and Gut Permeability

Increased gut permeability has attracted considerable recent attention as to its role in a wide range of medical conditions, including MS, [154] Parkinson's disease, [155] Alzheimer's disease, [156] schizophrenia, [157] and the autistic spectrum, [158] but especially depression.[159-161] Increased gut permeability allows bacteria and small fragments of partially digested food to trigger an immune response, thereby contributing to the immuno-inflammatory environment, with increased immune-inflammation in turn increasing gut permeability. As such, increased gut permeability is intimately associated with a host of medical conditions, with the low level immune-inflammation arising from increased gut permeability being proposed to contribute to how recurrent depressive episodes increase the risk of Alzheimer's disease.[162]

Although expressed in a host of immune cells as well as in the pineal gland, melatonin levels are far higher in the gut than in any other organ, predominantly via release from enterochromaffin cells.[163] Other gut sources of melatonin may also be relevant, including from variations in melatonin synthesis by different gut microbiota. By decreasing gut permeability, gut melatonin is another site for melatonin to influence a range of medical conditions, as well as its pineal release effects on the circadian rhythm and its autocrine and paracrine effects on glia and immune cells.[164] Given the strong association of depression with neurodegenerative, other psychiatric and general medical conditions, as to when increased gut permeability causally contributes to the etiology of depression may have significant implications for the etiology and course of a host of other depression-associated medical conditions, including the neurodegenerative disorders. Diet, stress and alcohol all increase gut permeability and are associated with the etiology and/or course of many neurodegenerative, psychiatric and wider medical conditions.[165-167] As such, many of the associations of these factors with this diverse array of medical conditions may be partly mediated via the regulation of gut permeability, including via the melatonergic pathways.

Treatment Implications

As detailed above, many neurodegenerative, psychiatric and wider medical conditions benefit from treatment with melatonin, both when used as an adjunctive or as the sole treatment [55, 57, 64-66]. This has generally been modeled as a result of melatonin treatment being compensatory to lower levels of pineal circadian melatonin release. However, as detailed above, the many effects of melatonin in cells and tissues, including immune, glia and gut cells, are now being recognized as relevant to the etiology and course of these many medical conditions, suggesting that melatonin may have multiple and parallel sites of action. Upregulating local glia and/or immune cell synthesis of melatonin, in turn decreasing the reactivity of these cells, may prove to be an important pharmaceutical target, with potential wide-ranging benefits across a host of conditions.

CONCLUSION

The melatonergic pathways are significant regulators of a host of key processes that form the biological underpinnings of a wide array of medical conditions. Melatonin is important, in part, via its regulation of the circadian rhythm. However, the targeting of melatonin to glia, immune and gut cells may provide significant treatment benefits, along the way providing a clearer picture of the commonalities in the biological underpinnings of a host of seemingly unrelated medical conditions.

REFERENCES

[1] Anderson G. Neuronal-Immune Interactions in Mediating Stress Effects in the Etiology and Course of Schizophrenia: Role of the Amygdala in Developmental Co-ordination. *Med. Hypotheses*, 2011;76:54-60.

[2] Anderson G., Maes M. Local melatonin regulates inflammation resolution: a common factor in neurodegenerative, psychiatric and systemic inflammatory disorders. *CNS Neurol. Disord. Drug Targets,* 2014;13(5):817-27.

[3] Anderson G., Maes M. Melatonin: an overlooked factor in schizophrenia and in the inhibition of anti-psychotic side effects. *Metab. Brain Dis*., 2012 Jun.; 27(2):113-9.

[4] Masruha M. R., Lin J., de Souza Vieira D. S., Minett T. S., Cipolla-Neto J., Zukerman E., Vilanova L. C., Peres M. F. Urinary 6-sulphatoxymelatonin levels are depressed in chronic migraine and several comorbidities. *Headache,* 2010 Mar.; 50(3):413-9.

[5] Geoffroy P. A., Boudebesse C., Henrion A., Jamain S., Henry C., Leboyer M., Bellivier F., Etain B. An ASMT variant associated with bipolar disorder influences sleep and circadian rhythms: a pilot study. *Genes Brain Behav*., 2014 Mar.; 13(3):299-304.

[6] Anderson G., Maes M. TRYCAT pathways link peripheral inflammation, nicotine, somatization and depression in the etiology and course of Parkinson's disease. *CNS Neurol. Disord. Drug Targets,* 2014 Feb.; 13(1):137-49.

[7] Escribano B. M., Colin-Gonzalez A. L., Santamaria A., Tunez I. The Role of Melatonin in Multiple Sclerosis, Huntington's Disease and Cerebral Ischemia. CNS Neurol. *Disord. Drug Targets,* 2014;13(6):1096-119.

[8] Kalliolia E., Silajdžić E., Nambron R., Hill N. R., Doshi A., Frost C., Watt H., Hindmarsh P., Björkqvist M., Warner T. T. Plasma melatonin is reduced in Huntington's disease. *Mov. Disord.*, In press.

[9] Bolitho S. J., Naismith S. L., Rajaratnam S. M., Grunstein R. R., Hodges J. R., Terpening Z., Rogers N., Lewis S. J. Disturbances in melatonin secretion and circadian sleep-wake regulation in Parkinson disease. *Sleep Med.*, 2014 Mar.; 15(3):342-7.

[10] Basler M., Jetter A., Fink D., Seifert B., Kullak-Ublick G. A., Trojan A. Urinary excretion of melatonin and association with breast cancer: meta-analysis and review of the literature. *Breast Care,* (Basel). 2014 Jul.; 9(3):182-7.

[11] Pernambuco A. P., Schetino L. P., Viana R. S., Carvalho L. S., d'Ávila Reis D. The involvement of melatonin in the clinical status of patients with fibromyalgia syndrome. *Clin. Exp. Rheumatol.,* In press.

[12] Srinivasan V., Spence D. W., Pandi-Perumal S. R., Brown G. M., Cardinali D. P. Melatonin in mitochondrial dysfunction and related disorders. *Int. J. Alzheimer's Dis.,* 2011; 2011:326320.

[13] Anderson G., Maes M. Redox Regulation and the Autistic Spectrum: Role of Tryptophan Catabolites, Immuno-inflammation, Autoimmunity and the Amygdala. *Curr. Neuropharmacol.*, 2014 Mar.; 12(2):148-67.

[14] Ding K., Wang H., Xu J., Li T., Zhang L., Ding Y., Zhu L., He J., Zhou M. Melatonin stimulates antioxidant enzymes and reduces oxidative stress in experimental traumatic brain injury: the Nrf2-ARE signaling pathway as a potential mechanism. *Free Radic. Biol. Med.*, 2014 Aug.; 73:1-11.

[15] Maes M., Kubera M., Obuchowiczwa E., Goehler L., Brzeszcz J. Depression's multiple comorbidities explained by (neuro)inflammatory and oxidative and nitrosative stress pathways. *Neuro. Endocrinol. Lett.,* 2011; 32(1):7-24.

[16] Maccio' A., Madeddu C., Gramignano G., Mulas C., Tanca L., Cherchi M. C., Floris C., Omoto I., Barracca A., Ganz T. The role of inflammation, iron, and nutritional status in cancer-related anemia: results of a large prospective observational study. *Haematologica,* In press.

[17] Aytaç B., Coşkun O., Alioğlu B., Durak Z. E., Büber S., Tapçi E., Ocal R., Inan L. E., Durak I., Yoldaş T. K. Decreased antioxidant status in migraine patients with brain white matter hyperintensities. *Neurol. Sci.*, In press.

[18] Anderson G., Maes M. Oxidative/Nitrosative Stress and Immuno-inflammatory Pathways in Depression: Treatment Implications. *Curr. Pharmaceut. Design.,* 2014; 20(23):3812-47.

[19] Anderson G., Maes M. Neurodegeneration in Parkinson's disease: interactions of oxidative stress, tryptophan catabolites and depression with mitochondria and sirtuins. *Mol. Neurobiol.*, 2014 Apr.; 49(2):771-83.

[20] Maes M., Kubera M., Mihaylova I., Geffard M., Galecki P., Leunis J. C., Berk M. Increased autoimmune responses against auto-epitopes modified by oxidative and nitrosative damage in depression: implications for the pathways to chronic depression and neuroprogression. *J. Affect. Disord.*, 2013 Jul.; 149(1-3):23-9.

[21] Streck E. L., Czapski G. A., Gonçalves da Silva C. Neurodegeneration, mitochondrial dysfunction, and oxidative stress. *Oxid. Med. Cell Longev.*, 2013;2013:826046.

[22] Dodd S., Maes M., Anderson G., Dean O. M., Moylan S., Berk M. Putative neuroprotective agents in neuropsychiatric disorders. *Prog. Neuropsychopharmacol. Biol. Psychiatry*, 2013 Apr. 5;42:135-45.

[23] Crichton G. E., Bryan J., Murphy K. J. Dietary antioxidants, cognitive function and dementia--a systematic review. *Plant Foods Hum. Nutr.,* 2013 Sep.; 68(3):279-92.

[24] Finsterer J. Treatment of central nervous system manifestations in mitochondrial disorders. *Eur. J. Neurol.,* 2011 Jan.; 18(1):28-38.

[25] Iuvone P. M., Boatright J. H., Tosini G., Ye K. N-acetylserotonin: circadian activation of the BDNF receptor and neuroprotection in the retina and brain. *Adv. Exp. Med. Biol.,* 2014; 801:765-71.

[26] Gupta S., Haldar C., Singh S. Daily variations in plasma melatonin and melatonin receptor (MT1), PER1 and CRY1 expression in suprachiasmatic nuclei of tropical squirrel, Funambulus pennanti. *J. Comp. Physiol. A. Neuroethol. Sens. Neural Behav. Physiol.*, 2013 Sep.; 199(9):763-73.

[27] Csaba G. The pineal regulation of the immune system: 40 years since the discovery. *Acta. Microbiol. Immunol. Hung.*, 2013 Jun.; 60(2):77-91.

[28] Sapède D., Cau E. The pineal gland from development to function. *Curr. Top Dev. Biol.,* 2013; 106:171-215.

[29] Pires-Lapa M. A., Tamura E. K., Salustiano E. M., Markus R. P. Melatonin synthesis in human colostrum mononuclear cells enhances dectin-1-mediated phagocytosis by mononuclear cells. *J. Pineal Res.,* 2013 Oct.; 55(3):240-6.

[30] Singh R. P., Hasan S., Sharma S., Nagra S., Yamaguchi D. T., Wong D., Bh H., Hossain A. Th17 cells in inflammation and autoimmunity. *Autoimmun. Rev.,* 2014 Aug. 23.

[31] Jimenéz-Aranda A., Fernández-Vázquez G., Mohammad A-Serrano M., Reiter R. J., Agil A. Melatonin improves mitochondrial function in inguinal white adipose tissue of Zücker diabetic fatty rats. *J. Pineal Res.*, 2014 Aug.; 57(1):103-9.

[32] Hardeland R. Melatonin in aging and disease -multiple consequences of reduced secretion, options and limits of treatment. *Aging Dis.,* 2012 Apr.; 3(2):194-225.

[33] Melamud L., Golan D., Luboshitzky R., Lavi I., Miller A. Melatonin dysregulation, sleep disturbances and fatigue in multiple sclerosis. *J. Neurol. Sci.,* 2012 Mar. 15; 314(1-2):37-40.

[34] Robillard R., Naismith S. L., Rogers N. L., Scott E. M., Ip T. K., Hermens D. F., Hickie I. B. Sleep-wake cycle and melatonin rhythms in adolescents and young adults with mood disorders: comparison of unipolar and bipolar phenotypes. *Eur. Psychiatry*, 2013 Sep.; 28(7):412-6.

[35] Li S. X., Liu L. J., Xu L. Z., Gao L., Wang X. F., Zhang J. T., Lu L. Diurnal alterations in circadian genes and peptides in major depressive disorder before and after escitalopram treatment. *Psychoneuroendocrinology*, 2013 Nov.; 38(11):2789-99.

[36] Mahdi A. A., Fatima G., Das S. K., Verma N. S. Abnormality of circadian rhythm of serum melatonin and other biochemical parameters in fibromyalgia syndrome. *Indian J. Biochem. Biophys.*, 2011 Apr.; 48(2):82-7.

[37] Rana S., Munawar M., Shahid A., Malik M., Ullah H., Fatima W., Mohsin S., Mahmood S. Deregulated expression of circadian clock and clock-controlled cell cycle genes in chronic lymphocytic leukemia. *Mol. Biol. Rep.,* 2014 Jan.; 41(1):95-103.

[38] Cipolla-Neto J., Amaral F. G., Afeche S. C., Tan D. X., Reiter R. J. Melatonin, energy metabolism, and obesity: a review. *J. Pineal Res.,* 2014 May; 56(4):371-81.

[39] Anderson G., Maes M., Berk M. Schizophrenia is primed for an increased expression of depression through activation of immuno-inflammatory, oxidative and nitrosative stress, and tryptophan catabolite pathways. *Prog. Neuropsychopharmacol. Biol. Psychiatry,* 2013 Apr. 5; 42:101-14.

[40] Zhan Y., Liu F., Li C., Gao Q., Liu S., Wang Y. Association between single nucleotide polymorphism of rs4753426 of melatonin receptor 1B gene and gestational diabetes mellitus. *Zhonghua Fu Chan Ke Za Zhi*, 2014 Apr.; 49(4):276-80.

[41] Li C., Qiao B., Zhan Y., Peng W., Chen Z. J., Sun L., Zhang J., Zhao L., Gao Q. Association between genetic variations in MTNR1A and MTNR1B genes and gestational diabetes mellitus in Han Chinese women. *Gynecol. Obstet Invest.,* 2013; 76(4):221-7.

[42] Jonsson L., Anckarsäter H., Zettergren A., Westberg L., Walum H., Lundström S., Larsson H., Lichtenstein P., Melke J. Association between ASMT and autistic-like traits in children from a Swedish nationwide cohort. *Psychiatr. Genet.*, 2014 Feb;24(1):21-7.

[43] Natarajan R., Einarsdottir E., Riutta A., Hagman S., Raunio M., Mononen N., Lehtimäki T., Elovaara I. Melatonin pathway genes are associated with progressive subtypes and disability status in multiple sclerosis among Finnish patients. *J. Neuroimmunol.*, 2012 Sep. 15; 250(1-2):106-10.

[44] Etain B., Dumaine A., Bellivier F., Pagan C., Francelle L., Goubran-Botros H., Moreno S., Deshommes J., Moustafa K., Le Dudal K., Mathieu F., Henry C., Kahn J. P., Launay J. M., Mühleisen T. W., Cichon S., Bourgeron T., Leboyer M., Jamain S. Genetic and functional abnormalities of the melatonin biosynthesis pathway in patients with bipolar disorder. *Hum. Mol. Genet.*, 2012 Sep. 15; 21(18):4030-7.

[45] Park H. J., Park J. K., Kim S. K., Cho A. R., Kim J. W., Yim S. V., Chung J. H. Association of polymorphism in the promoter of the melatonin receptor 1A gene with schizophrenia and with insomnia symptoms in schizophrenia patients. *J. Mol. Neurosci.,* 2011 Oct.; 45(2):304-8.

[46] Deming S. L., Lu W., Beeghly-Fadiel A., Zheng Y., Cai Q., Long J., Shu X. O., Gao Y. T., Zheng W. Melatonin pathway genes and breast cancer risk among Chinese women. *Breast Cancer Res. Treat.,* 2012 Apr.; 132(2):693-9.

[47] Kripke D. F., Nievergelt C. M., Tranah G. J., Murray S. S., McCarthy M. J., Rex K. M., Parimi N., Kelsoe J. R. Polymorphisms in melatonin synthesis pathways: possible influences on depression. *J. Circadian Rhythms*, 2011 Aug 9;9:8.

[48] Gałecka E, Szemraj J, Florkowski A, Gałecki P, Bieńkiewicz M, Karbownik-Lewińska M, Lewiński A. Single nucleotide polymorphisms and mRNA expression for melatonin MT(2) receptor in depression. Psychiatry Res. 2011 Oct. 30; 189(3):472-4.

[49] Samimi-Fard S., Abreu-Gonzalez P., Dominguez-Rodriguez A., Jimenez-Sosa A. A case-control study of melatonin receptor type 1A polymorphism and acute myocardial infarction in a Spanish population. *J. Pineal Res.*, 2011 Nov.; 51(4):400-4.

[50] Delavest M., Even C., Benjemaa N., Poirier M. F., Jockers R., Krebs M. O. Association of the intronic rs2072621 polymorphism of the X-linked GPR50 gene with affective disorder with seasonal pattern. *Eur. Psychiatry*, 2012 Jul.; 27(5):369-71.

[51] Romo-Nava F., Alvarez-Icaza González D., Fresán-Orellana A., Saracco Alvarez R., Becerra-Palars C., Moreno J., Ontiveros Uribe M. P., Berlanga C., Heinze G., Buijs R. M. Melatonin attenuates antipsychotic metabolic effects: an eight-week randomized, double-blind, parallel-group, placebo-controlled clinical trial. *Bipolar. Disord.*, 2014 Jun.; 16(4):410-21.

[52] Modabbernia A., Heidari P., Soleimani R., Sobhani A., Roshan Z. A., Taslimi S., Ashrafi M., Modabbernia M. J.. Melatonin for prevention of metabolic side-effects of

olanzapine in patients with first-episode schizophrenia: randomized double-blind placebo-controlled study. *J. Psychiatr. Res.*, 2014 Jun.; 53:133-40.

[53] Davis J., Moylan S., Harvey B. H., Maes M., Berk M. Neuroprogression in schizophrenia: Pathways underpinning clinical staging and therapeutic corollaries. *Aust. N Z J. Psychiatry,* 2014 May 6; 48(6):512-529.

[54] Berk M., Berk L., Dodd S., Cotton S., Macneil C., Daglas R., Conus P., Bechdolf A., Moylan S., Malhi G. S. Stage managing bipolar disorder. *Bipolar. Disord.*, 2014 Aug.; 16(5):471-7.

[55] Wade A. G., Farmer M., Harari G., Fund N., Laudon M., Nir T., Frydman-Marom A., Zisapel N. Add-on prolonged-release melatonin for cognitive function and sleep in mild to moderate Alzheimer's disease: a 6-month, randomized, placebo-controlled, multicenter trial. *Clin. Interv. Aging*, 2014 Jun. 18; 9:947-61.

[56] Brusco L. I., Márquez M., Cardinali D. P. Melatonin treatment stabilizes chronobiologic and cognitive symptoms in Alzheimer's disease. *Neuro Endocrinol. Lett.,* 2000; 21(1):39–42.

[57] Cardinali D. P., Vigo D. E., Olivar N., Vidal M. F., Furio A. M., Brusco L. I. Therapeutic application of melatonin in mild cognitive impairment. *Am. J. Neurodegener Dis.,* 2012; 1(3):280–291.

[58] Datieva V. K., Rosinskaia A. V., Levin O. S. The use of melatonin in the treatment of chronic fatigue syndrome and circadian rhythm disorders in Parkinson's disease. *Zh. Nevrol. Psikhiatr. Im. S. S. Korsakova,* 2013; 113(7 Pt 2):77-81.

[59] Hansen M. V., Danielsen A. K., Hageman I., Rosenberg J., Gögenur I. The therapeutic or prophylactic effect of exogenous melatonin against depression and depressive symptoms: A systematic review and meta-analysis. *Eur. Neuropsychopharmacol.*, In press.

[60] Fallah R., Shoroki F. F., Ferdosian F. Safety and Efficacy of Melatonin in Pediatric Migraine Prophylaxis. *Curr. Drug Saf.,* In press.

[61] Stubbe C. E., Valero M. Complementary strategies for the management of radiation therapy side effects. *J. Adv. Pract. Oncol.*, 2013 Jul.; 4(4):219-31.

[62] Glenister R., McDaniel K., Francis H., Venter J., Jensen K., Dusio G., Gaudio E., Glaser S., Meng F., Alpini G. Therapeutic actions of melatonin on gastrointestinal cancer development and progression. *Transl. Gastrointest. Cancer,* 2013 Jan. 1; 2(1).

[63] de Zanette S. A., Vercelino R., Laste G., Rozisky J. R., Schwertner A,. Machado C. B., Xavier F., de Souza I. C., Deitos A., Torres I. L., Caumo W. Melatonin analgesia is associated with improvement of the descending endogenous pain-modulating system in fibromyalgia: a phase II, randomized, double-dummy, controlled trial. *BMC Pharmacol. Toxicol.*, 2014 Jul. 23; 15:40.

[64] Ono K., Mochizuki H., Ikeda T., Nihira T., Takasaki J., Teplow D. B., Yamada M. Effect of melatonin on α-synuclein self-assembly and cytotoxicity. *Neurobiol. Aging*, 2012 Sep;33(9):2172-85.

[65] Adamczyk-Sowa M., Pierzchala K., Sowa P., Polaniak R., Kukla M., Hartel M. Influence of melatonin supplementation on serum antioxidative properties and impact of the quality of life in multiple sclerosis patients. *J. Physiol. Pharmacol.*, 2014 Aug.; 65(4):543-50.

[66] Adamczyk-Sowa M., Pierzchala K., Sowa P., Mucha S., Sadowska-Bartosz I., Adamczyk J., Hartel M. Melatonin acts as antioxidant and improves sleep in MS patients. *Neurochem. Res.*, 2014b Aug.; 39(8):1585-93.

[67] Martín M., Macías M., León J., Escames G., Khaldy H., Acuña-Castroviejo D. Melatonin increases the activity of the oxidative phosphorylation enzymes and the production of ATP in rat brain and liver mitochondria. *Int. J. Biochem. Cell Biol.,* 2002 Apr.; 34(4):348-57.

[68] Kashani I. R., Rajabi Z., Akbari M., Hassanzadeh G., Mohseni A., Eramsadati M. K., Rafiee K., Beyer C., Kipp M., Zendedel A. Protective effects of melatonin against mitochondrial injury in a mouse model of multiple sclerosis. *Exp. Brain Res.,* 2014 Sep.; 232(9):2835-46.

[69] Ledderose C., Bao Y., Lidicky M., Zipperle J., Li L., Strasser K., Shapiro N. I., Junger W. G. Mitochondria Are Gate-keepers of T Cell Function by Producing the ATP That Drives Purinergic Signaling. *J. Biol. Chem.*, 2014 Sep. 12;289(37):25936-45.

[70] Beauquis J., Vinuesa A., Pomilio C., Pavía P., Saravia F. Hippocampal and cognitive alterations precede amyloid deposition in a mouse model for Alzheimer's disease. *Medicina (B Aires),* 2014; 74(4):282-286.

[71] Regensburger M., Prots I., Winner B. Adult hippocampal neurogenesis in Parkinson's disease: impact on neuronal survival and plasticity. *Neural. Plast.*, 2014; 2014:454696.

[72] Huehnchen P., Prozorovski T., Klaissle P., Lesemann A., Ingwersen J., Wolf S. A., Kupsch A., Aktas O., Steiner B. Modulation of adult hippocampal neurogenesis during myelin-directed autoimmune neuroinflammation. *Glia,* 2011 Jan.; 59(1):132-42.

[73] Miller B. R., Hen R. The current state of the neurogenic theory of depression and anxiety. *Curr. Opin. Neurobiol.*, 2014 Sep. 17; 30C:51-58.

[74] Tocharus C., Puriboriboon Y., Junmanee T., Tocharus J., Ekthuwapranee K., Govitrapong P. Melatonin enhances adult rat hippocampal progenitor cell proliferation via ERK signaling pathway through melatonin receptor. *Neuroscience,* 2014 Sep. 5; 275: 314-21.

[75] Wang B., Jin K. Current perspectives on the link between neuroinflammation and neurogenesis. *Metab. Brain Dis.*, In press.

[76] Soliman A., Udemgba C., Fan I., Xu X., Miler L., Rusjan P., Houle S., Wilson A. A., Pruessner J., Ou X. M., Meyer J. H. Convergent effects of acute stress and glucocorticoid exposure upon MAO-A in humans. *J. Neurosci.,* 2012, 32(48), 17120-17127.

[77] Ou X. M., Chen K., Shih J. C. Glucocorticoid and androgen activation of monoamine oxidase A is regulated differently by R1 and Sp1. *J. Biol. Chem.,* 2006 Jul. 28; 281 (30): 21512-25.

[78] Ruksee N., Tongjaroenbuangam W., Mahanam T., Govitrapong P. Melatonin pretreatment prevented the effect of dexamethasone negative alterations on behavior and hippocampal neurogenesis in the mouse brain. *J. Steroid. Biochem. Mol. Biol.*, 2014 Sep.; 143:72-80.

[79] Anderson G., Maes M. Reconceptualizing adult neurogenesis: role for sphingosine-1-phosphate and fibroblast growth factor-1 in co-ordinating astrocyte-neuronal precursor interactions. *CNS Neurol. Disord. Drug Targets*, 2014 Feb.; 13(1):126-36.

[80] Tiberi L., van den Ameele J., Dimidschstein J., Piccirilli J., Gall D., Herpoel A., Bilheu A., Bonnefont J., Iacovino M., Kyba M., Bouschet T., Vanderhaeghen P. BCL6

controls neurogenesis through Sirt1-dependent epigenetic repression of selective Notch targets. *Nat. Neurosci.,* 2012 Dec.; 15(12):1627-35.
[81] da Silveira Cruz-Machado S., Pinato L., Tamura E. K., Carvalho-Sousa C. E., Markus R. P. Glia-pinealocyte network: the paracrine modulation of melatonin synthesis by tumor necrosis factor (TNF). *PloS One,* 2012;7(7);e40142.
[82] Pontes G. N., Cardoso E. C., Cameiro-Sampaio M. M., Markus R. P. Pineal melatonin and the innate immune response: the TNF-alpha increase after cesarean sections suppresses nocturnal melatonin production. *J. Pineal Res.*, 2007; 43: 365-71.
[83] Swardfager W., Lanctot K., Rothenburg L., Wong A., Capell J., Hermann N. A. Meta-analysis of cytokines in Alzheimer's disease. *Biol. Psychiatry,* 2010; 68: 930-941.
[84] Khemka V. K., Ganguly A., Bagchi D., Ghosh A., Bir A., Biswas A., Chattopadhyay S., Chakrabarti S. Raised serum proinflammatory cytokines in Alzheimer's disease with depression. *Aging Dis.,* 2014 Jun. 1; 5(3):170-6.
[85] Uysal S., Meriç Yilmaz F., Boğdaycioğlu N., Mungan Öztürk S., Ak F. Increased serum levels of some inflammatory markers in patients with multiple sclerosis. *Minerva Med.,* 2014 Jun ; 105(3):229-35.
[86] Koziorowski D., Tomasiuk R., Szlufik S., Friedman A. Inflammatory cytokines and NT-proCNP in Parkinson's disease patients. *Cytokine,* 2012 Dec;60(3):762-6.
[87] Munkholm K., Braüner J. V., Kessing L. V., Vinberg M. Cytokines in bipolar disorder vs. healthy control subjects: a systematic review and meta-analysis. *J. Psychiatr. Res.,* 2013 Sep.; 47(9):1119-33.
[88] Fatima G., Mahdi A. A., Das S. K., Anjum B., Verma N. S., Kumar P., Shrivastava R. Lack of Circadian Pattern of Serum TNF-α and IL-6 in Patients with Fibromyalgia Syndrome. *Indian J. Clin. Biochem.*, 2012 Oct.; 27(4):340-3.
[89] Young J. J., Bruno D., Pomara N. A review of the relationship between proinflammatory cytokines and major depressive disorder. *J. Affect Disord.*, 2014 Dec. 1;169C:15-20.
[90] Tian L., Tan Y., Chen D., Lv M., Tan S., Soares J. C., Zhang X. Y. Reduced serum TNF alpha level in chronic schizophrenia patients with or without tardive dyskinesia. *Prog. Neuropsychopharmacol. Biol. Psychiatry*, 2014 Oct 3;54:259-64.
[91] Rozen T., Swidan S. Z. Elevation of CSF tumor necrosis factor alpha levels in new daily persistent headache and treatment refractory chronic migraine. *Headache,* 2007 Jul.-Aug.; 47(7):1050-5.
[92] Cheng W., Li Y., Hou X., Zhang N., Ma J., Ding F., Li F., Miao Z., Zhang Y., Qi Q., Li G., Shen Y., Liu J., Huang W., Wang Y. HSP60 is involved in the neuroprotective effects of naloxone. *Mol. Med. Rep.*, 2014 Oct.; 10(4):2172-6.
[93] Jacobsen J. H., Watkins L. R., Hutchinson M. R. Discovery of a Novel Site of Opioid Action at the Innate Immune Pattern-Recognition Receptor TLR4 and its Role in Addiction. *Int. Rev. Neurobiol.*, 2014; 118:129-63.
[94] Fang H., Wu Y., Huang X., Wang W., Ang B., Cao X., Wan T. Toll-like receptor 4 (TLR4) is essential for Hsp70-like protein 1 (HSP70L1) to activate dendritic cells and induce Th1 response. *J. Biol. Chem.*, 2011 Sep. 2;286(35):30393-400.
[95] Merighi S., Gessi S., Varani K., Fazzi D., Stefanelli A., Borea P. A. Morphine mediates a proinflammatory phenotype via μ-opioid receptor-PKCε-Akt-ERK1/2 signaling

pathway in activated microglial cells. *Biochem. Pharmacol.*, 2013 Aug. 15; 86(4): 487-96.

[96] Chuchuen U., Ebadi M., Govitrapong P. The stimulatory effect of mu- and delta-opioid receptors on bovine pinealocyte melatonin synthesis. *J. Pineal Res.*, 2004 Nov.; 37(4): 223-9.

[97] Chetsawang B., Govitrapong P. Stimulatory effect of morphine on rat pineal melatonin synthesis via a cyclic AMP-dependent transcription pathway. *Neurosci. Lett.,* 2005 Nov. 25; 389(1):57-60.

[98] Chong, R. Y., Oswald, L., Yang, X., Uhart, M., Lin, P. I., and Wand, G. S. The mu-opioid receptor polymorphism A118G predicts cortisol responses to naloxone and stress. *Neuropsychopharmacology,* 2006; 31(1), 204–211.

[99] Ruzicka B. B., Akil H. The interleukin-1beta-mediated regulation of proenkephalin and opioid receptor messenger RNA in primary astrocyte-enriched cultures. *Neuroscience,* 1997 Jul.; 79(2):517-24.

[100] Anderson G., Maes M., Berk M. Inflammation-related disorders in the tryptophan catabolite (TRYCAT) pathway in depression and somatization. *Adv. Protein Chemistry Structural Biol.*, 2012;88:27-48.

[101] Mathieu-Kia A. M., Fan L. Q., Kreek M. J., Simon E. J., Hiller J. M. Mu-, delta- and kappa-opioid receptor populations are differentially altered in distinct areas of postmortem brains of Alzheimer's disease patients. *Brain Res.,* 2001 Mar. 2; 893(1-2): 121-34.

[102] Slavich G. M., Tartter M. A., Brennan P. A., Hammen C. Endogenous opioid system influences depressive reactions to socially painful targeted rejection life events. *Psychoneuroendocrinology*, 2014 Nov.; 49:141-9.

[103] Hriscu M., Saulea G., Ostriceanu S., Baciu I. Circadian phagocytic activity in rats under light-dark and constant light regimens. *Rom. J. Physiol.*, 2002-2003; 39-40: 17-26.

[104] Muxel S. M., Pires-Lapa M. A., Monteiro A. W., Cecon E., Tamura E. K., Floeter-Winter L. M., Markus R. P. NF-κB drives the synthesis of melatonin in RAW 264.7 macrophages by inducing the transcription of the arylalkylamine N acetyltransferase (AA-NAT) gene. *PLoS One,* 2012;7(12):e52010.

[105] Lardone P. J., Guerrero J. M., Fernández-Santos J. M., Rubio A., Martín-Lacave I., Carrillo-Vico A. Melatonin synthesized by T lymphocytes as a ligand of the retinoic acid-related orphan receptor. *J. Pineal Res.*, 2011 Nov.; 51(4):454-62.

[106] Sakaguchi K., Itoh M. T., Takahashi N., Tarumi W., Ishizuka B. The rat oocyte synthesises melatonin. *Reprod. Fertil Dev.*, 2013;25(4):674-82.

[107] Liu Y. J., Zhuang J., Zhu H. Y., Shen Y. X., Tan Z. L., Zhou J. N. Cultured rat cortical astrocytes synthesize melatonin: absence of a diurnal rhythm. *J. Pineal Res.,* 2007 Oct.; 43(3):232-8.

[108] Liu Y. J., Meng F. T., Wu L., Zhou J. N. Serotoninergic and melatoninergic systems are expressed in mouse embryonic fibroblasts NIH3T3 cells. *Neuro. Endocrinol. Lett.,* 2013; 34(3):236-40.

[109] Al-Ghoul W. M., Abu-Shaqra S., Park B. G., Fazal N. Melatonin plays a protective role in postburn rodent gut pathophysiology. *Int. J. Biol. Sci.*, 2010 May 17; 6(3):282-93.

[110] Tan D. X., Manchester L. C., Liu X., Rosales-Corral S. A., Acuna-Castroviejo D., Reiter R. J. Mitochondria and chloroplasts as the original sites of melatonin synthesis: a

hypothesis related to melatonin's primary function and evolution in eukaryotes. *J. Pineal Res.*, 2013; 54: 127-38.

[111] Kerenyi N. A., Sotonyi P., Somogyi E. Localizing acetylserotonin transferase by electron microscopy. *Histochem.*, 1975; 46:77.

[112] Bernard M., Voisin P. Photoreceptor-specific expression, lightdependent localization, and transcriptional targets of the zinc-finger protein Yin Yang 1 in the chicken retina. *J. Neurochem.*, 2008; 105(3): 595-604.

[113] Nowak K., Lange-Dohna C., Zeitschel U., Günther A., Lüscher B., Robitzki A., Perez-Polo R., Rossner S. The transcription factor Yin Yang 1 is an activator of BACE1 expression. *J. Neurochem.*, 2006; 96(6): 1696-707.

[114] Rylski M., Amborska R., Zybura K., Konopacki F. A., Wilczynski G. M., Kaczmarek L. Yin Yang 1 expression in the adult rodent brain. *Neurochem. Res.*, 2008; 33(12): 2556-64.

[115] Siednienko J., Maratha A., Yang S., Mitkiewicz M., Miggin S. M., Moynagh P. N. Nuclear factor κB subunits RelB and cRel negatively regulate Toll-like receptor 3-mediated β-interferon production via induction of transcriptional repressor protein YY1. *J. Biol. Chem.*, 2011 Dec. 30; 286(52):44750-63.

[116] Lange-Dohna C., Zeitschel U., Gaunitz F., Perez-Polo J. R., Bigl V., Rossner S. Cloning and expression of the rat BACE1 promoter. *J. Neurosci. Res.*, 2003 Jul. 1; 73(1): 73-80.

[117] Sutinen E. M., Pirttilä T., Anderson G., Salminen A., Ojala J. O. Proinflammatory interleukin-18 increases Alzheimer's disease associated amyloid-B production in human neuron-like cells. *J. Neuroinflamm.*, 2012; 9: 199.

[118] Gutierrez-Cuesta J., Tajes M., Jiménez A., Coto-Montes A., Camins A., Pallàs M. Evaluation of potential pro-survival pathways regulated by melatonin in a murine senescence model. *J. Pineal Res.*, 2008 Nov.; 45(4):497-505.

[119] Anderson G., Ojala J. O. Alzheimer's and Seizures: Interleukin-18: Indoleamine 2:3-Dioxygenase and Quinolinic Acid. *Int. J. Trytophan Res.*, 2010; 3: 169-73.

[120] Liu Y. J., Meng F. T., Wang L. L., Zhang L. F., Cheng X. P., Zhou J. N. Apolipoprotein E influences melatonin biosynthesis by regulating NAT and MAOA expression in C6 cells. *J. Pineal Res.*, 2012; 52(4): 397-402.

[121] Lee B. K., Glass T. A., Wand G. S., McAtee M. J., Bandeen-Roche K., Bolla K. I., Schwartz B. S. Apolipoprotein e genotype, cortisol, and cognitive function in community-dwelling older adults. *Am. J. Psychiatry*, 2008 Nov.; 165(11):1456-64.

[122] Lahiri D. K., Sokol D. K., Erickson C., Ray B., Ho C. Y., Maloney B. Autism as early neurodevelopmental disorder: evidence for an sAPPα-mediated anabolic pathway. *Front. Cell Neurosci.*, 2013 Jun. 21; 7:94.

[123] Monsell S. E., Besser L. M., Heller K. B., Checkoway H., Litvan I., Kukull W. A. Clinical and pathologic presentation in Parkinson's disease by apolipoprotein e4 allele status. *Parkinsonism Relat. Disord.*, 2014 May; 20(5):503-7.

[124] Adi N., Mash D. C., Ali Y., Singer C., Shehadeh L., Papapetropoulos S. Melatonin MT1 and MT2 receptor expression in Parkinson's disease. *Med. Sci. Monit.*, 2010; 16: BR61–BR67.

[125] Fowler J. S., Logan J., Volkow N. D., Shumay E., McCall-Perez F., Jayne M., Wang G. J., Alexoff D. L., Apelskog-Torres K., Hubbard B., Carter P., King P., Fahn S., Telang F., Shea C., Xu Y., Muench L. Evidence That Formulations of the Selective MAO-B

Inhibitor, Selegiline, Which Bypass First-Pass Metabolism, Also Inhibit MAO-A in the Human Brain. *Neuropsychopharmacology*, In press.

[126] Hao H., Shao M., An J., Chen C., Feng X., Xie S., Gu Z., Chan P.; Chinese Parkinson Study Group. Association of Catechol-O-Methyltransferase and monoamine oxidase B gene polymorphisms with motor complications in parkinson's disease in a Chinese population. *Parkinsonism Relat. Disord.,* In press.

[127] Wootla B., Watzlawik J. O., Denic A., Rodriguez M. The road to remyelination in demyelinating diseases: current status and prospects for clinical treatment. *Expert. Rev. Clin. Immunol.*, 2013 Jun.; 9(6):535-49.

[128] Akpinar Z., Tokgöz S., Gökbel H., Okudan N., Uğuz F., Yilmaz G. The association of nocturnal serum melatonin levels with major depression in patients with acute multiple sclerosis. *Psychiatry Res.,* 2008 Nov. 30; 161(2):253-7.

[129] Miller E., Walczak A., Majsterek I., Kędziora J. Melatonin reduces oxidative stress in the erythrocytes of multiple sclerosis patients with secondary progressive clinical course. *J. Neuroimmunol.,* 2013 Apr. 15; 257(1-2):97-101.

[130] Anderson G., Rodriguez M. Multiple Sclerosis, Seizures and Anti-Epileptics: Role of IL-18, IDO and Melatonin. *Eur. J. Neurology*, 2011 May; 18(5):680-5.

[131] Golan D., Staun-Ram E., Glass-Marmor L., Lavi I., Rozenberg O., Dishon S., Barak M., Ish-Shalom S., Miller A. The influence of vitamin D supplementation on melatonin status in patients with multiple sclerosis. *Brain Behav. Immun.,* 2013 Aug.; 32:180-5.

[132] Tamam Y., Tasdemir N., Yalman M., Tamam B. Association of apolipoprotein E genotypes with prognosis in multiple sclerosis. *Eur. Rev. Med. Pharmacol. Sci.,* 2011 Oct.; 15(10):1122-30.

[133] Shi J., Tu J., Gale S. D., Baxter L., Vollmer T. L., Campagnolo D. I., Tyry T. M., Zhuang Y., Kuniyoshi S. M. APOE ε4 is associated with exacerbation of cognitive decline in patients with multiple sclerosis. *Cogn. Behav. Neurol.*, 2011 Sep.; 24(3): 128-33.

[134] Horáková D., Kýr M., Havrdová E., Doležal O., Lelková P., Pospíšilová L., Bergsland N., Dwycr M. G., Cox J. L., Husscin S., Seidl Z., Vaněčková M., Krásenský J., Zivadinov R. Apolipoprotein E ε4-positive multiple sclerosis patients develop more gray-matter and whole-brain atrophy: a 15-year disease history model based on a 4-year longitudinal study. *Folia Biol.,* (Praha). 2010; 56(6):242-51.

[135] Xuan C., Zhang B. B., Li M., Deng K. F., Yang T., Zhang X. E. No association between APOE ε 4 allele and multiple sclerosis susceptibility: a meta-analysis from 5472 cases and 4727 controls. *J. Neurol. Sci.,* 2011 Sep. 15; 308(1-2):110-6.

[136] Metti A. L., Cauley J. A., Newman A. B., Ayonayon H. N., Barry L. C., Kuller L. M., Satterfield S., Simonsick E. M., Yaffe K. Plasma beta amyloid level and depression in older adults. *J. Gerontol. A. Biol. Sci. Med. Sci.*, 2013 Jan.; 68(1):74-9.

[137] Kim J. M., Stewart R., Kim S. Y., Kim S. W., Bae K. Y., Yang S. J., Kim S. W., Bae K. Y., Yang S. J. Synergistic associations of depression and apolipoprotein E genotype with incidence of dementia. *Int. J. Geriatr. Psychiatry*, 2011 Sep.; 26(9):893-8.

[138] Villapol S., Fau S., Renolleau S., Biran V., Charriaut-Marlangue C., Baud O. Melatonin promotes myelination by decreasing white matter inflammation after neonatal stroke. *Pediatr. Res.*, 2011 Jan.; 69(1):51-5.

[139] Campbell G. R., Ohno N., Turnbull D. M., Mahad D. J. Mitochondrial changes within axons in multiple sclerosis: an update. *Curr. Opin. Neurol.*, 2012 Jun.; 25(3): 221-30.

[140] Mahad D., Ziabreva I., Lassmann H., Turnbull D. Mitochondrial defects in acute multiple sclerosis lesions. *Brain,* 2008 Jul.; 131(Pt 7):1722-35.

[141] Sathornsumetee S., McGavern D. B., Ure D. R., Rodriguez M. Quantitative Ultrastructural Analysis of a Single Spinal Cord Demyelinated Lesion Predicts Total Lesion Load, Axonal Loss, and Neurological Dysfunction in a Murine Model of Multiple Sclerosis. *Am. J. Pathol.*, 2000; 157:1365-76.

[142] Fischer M. T., Sharma R., Lim J. L., Haider L., Frischer J. M., Drexhage J., Fischer M. T., Sharma R., Lim J. L. NADPH oxidase expression in active multiple sclerosis lesions in relation to oxidative tissue damage and mitochondrial injury. *Brain,* 2012 Mar.; 135 (Pt 3):886-99.

[143] Farhadi N., Oryan S., Nabiuni M. Serum levels of melatonin and cytokines in multiple sclerosis. *Biomed. J.*, 2014 Mar.-Apr.; 37(2):90-2.

[144] Bahamonde C., Conde C., Agüera E., Lillo R., Luque E., Gascón F., Feijóo M., Cruz A. H., Sánchez-López F., Túnez I. Elevated melatonin levels in natalizumab-treated female patients with relapsing-remitting multiple sclerosis: relationship to oxidative stress. *Eur. J. Pharmacol.*, 2014 May 5; 730:26-30.

[145] Brenner D. R., Scherer D., Muir K., Schildkraut J., Boffetta P., Spitz M. R., Le Marchand L., Chan A. T., Goode E. L., Ulrich C. M., Hung R. J. A Review of the Application of Inflammatory Biomarkers in Epidemiologic Cancer Research. *Cancer Epidemiol. Biomarkers Prev.*, 2014 Sep.; 23(9):1729-1751.

[146] Hu B., Yang X. R., Xu Y., Sun Y., Sun C., Guo W., Zhang X., Wang W. M., Qiu S. J., Zhou J., Fan J. Systemic immune-inflammation index (SII) predicts prognosis of patients after curative resection for hepatocellular carcinoma. *Clin. Cancer Re*s., In press.

[147] Wang J., Hao H., Yao L., Zhang X., Zhao S., Ling E. A., Hao A., Li G. Melatonin suppresses migration and invasion via inhibition of oxidative stress pathway in glioma cells. *J. Pineal Res.*, 2012 Sep.; 53(2):180-7.

[148] Martín V., Sanchez-Sanchez A. M., Puente-Moncada N., Gomez-Lobo M., Alvarez-Vega M. A., Antolín I., Rodriguez C. Involvement of autophagy in melatonin-induced cytotoxicity in glioma-initiating cells. *J. Pineal Res.,* 2014 Oct.; 57(3):308-16.

[149] Zhang Y., Liu Q., Wang F., Ling E. A., Liu S., Wang L., Yang Y., Yao L., Chen X., Wang F., Shi W., Gao M., Hao A. Melatonin antagonizes hypoxia-mediated glioblastoma cell migration and invasion via inhibition of HIF-1α. *J. Pineal Res.,* 2013 Sep.; 55(2): 121-30.

[150] Martín V., Sanchez-Sanchez A. M., Herrera F., Gomez-Manzano C., Fueyo J., Alvarez-Vega M. A., Antolín I., Rodriguez C. Melatonin-induced methylation of the ABCG2/BCRP promoter as a novel mechanism to overcome multidrug resistance in brain tumour stem cells. *Br. J. Cancer*, 2013 May 28; 108(10):2005-12.

[151] Calvo J. R., González-Yanes C., Maldonado M. D. The role of melatonin in the cells of the innate immunity: a review. *J. Pineal Res.,* 2013 Sep.; 55(2):103-20.

[152] Gabellini D., Green M. R., Tupler R. Inappropriate gene activation in FSHD: a repressor complex binds a chromosomal repeat deleted in dystrophic muscle. *Cell,* 2002 Aug. 9; 110(3):339-48.

[153] Ishii H., Hulett M. D., Li J. M., Santiago F. S., Parish C. R., Khachigian L. M. Yin Yang-1 inhibits tumor cell growth and inhibits p21WAF1/Cip1 complex formation with cdk4 and cyclin D1. *Int. J. Oncol.,* 2012 May; 40(5):1575-80.

[154] Joscelyn J., Kasper L. H. Digesting the emerging role for the gut microbiome in central nervous system demyelination. *Mult. Scler.*, In press.

[155] Forsyth C. B., Shannon K. M., Kordower J. H., Voigt R. M., Shaikh M., Jaglin J. A., Estes J. D., Dodiya H. B., Keshavarzian A. Increased intestinal permeability correlates with sigmoid mucosa alpha-synuclein staining and endotoxin exposure markers in early Parkinson's disease. *PLoS One,* 2011; 6(12):e28032.

[156] Bekkering P., Jafri I., van Overveld F. J., Rijkers G. T. The intricate association between gut microbiota and development of type 1, type 2 and type 3 diabetes. *Expert Rev. Clin. Immunol.,* 2013 Nov.; 9(11):1031-41.

[157] Severance E. G., Yolken R. H., Eaton W. W. Autoimmune diseases, gastrointestinal disorders and the microbiome in schizophrenia: more than a gut feeling. *Schizophr. Res.,* 2014 Jul. 14.

[158] Samsam M., Ahangari R., Naser S. A. Pathophysiology of autism spectrum disorders: revisiting gastrointestinal involvement and immune imbalance. *World J. Gastroenterol.*, 2014 Aug. 7; 20(29):9942-51.

[159] Anderson G., Maes M. The Gut-Brain Axis: The Role of Melatonin in Linking Psychiatric, Inflammatory and Neurodegenerative Conditions. *Adv. Integr. Med.*, In press.

[160] Berk M., Williams L. J., Jacka F. N., O'Neil A., Pasco J. A., Moylan S., Allen N. B., Stuart A. L., Hayley A. C., Byrne M. L., Maes M. So depression is an inflammatory disease, but where does the inflammation come from? *BMC Med.*, 2013 Sep. 12; 11:200.

[161] Maes M., Kubera M., Leunis J. C. The gut-brain barrier in major depression: intestinal mucosal dysfunction with an increased translocation of LPS from gram negative enterobacteria (leaky gut) plays a role in the inflammatory pathophysiology of depression. *Neuro. Endocrinol. Lett.,* 2008 Feb.; 29(1):117-24.

[162] Dotson V. M., Beydoun M. A., Zonderman A. B. Recurrent depressive symptoms and the incidence of dementia and mild cognitive impairment. *Neurology,* 2010 Jul. 6; 75 (1): 27-34.

[163] Raikhlin N. T., Kvetnoy I. M. Melatonin and enterochromaffine cells. *Acta Histochem.,* 1976; 55(1):19-24.

[164] Trivedi P. P., Jena G. B. Melatonin reduces ulcerative colitis-associated local and systemic damage in mice: investigation on possible mechanisms. *Dig. Dis. Sci.,* 2013 Dec.; 58(12):3460-74.

[165] Mariadason J. M., Catto-Smith A., Gibson P. R. Modulation of distal colonic epithelial barrier function by dietary fibre in normal rats. *Gut,* 1999 Mar.; 44(3):394-9.

[166] Vanuytsel T., van Wanrooy S., Vanheel H., Vanormelingen C., Verschueren S., Houben E., Salim Rasoel S., Tóth J., Holvoet L., Farré R., Van Oudenhove L., Boeckxstaens G., Verbeke K., Tack J. Psychological stress and corticotropin-releasing hormone increase intestinal permeability in humans by a mast cell-dependent mechanism. *Gut,* 2014 Aug.; 63(8):1293-9.

[167] Sommansson A., Saudi W. S., Nylander O., Sjöblom M. Melatonin inhibits alcohol-induced increases in duodenal mucosal permeability in rats in vivo. *Am. J. Physiol. Gastrointest. Liver Physiol.*, 2013 Jul. 1; 305(1):G95-G105.

In: Indoleamines
Editor: Ángel Catalá

ISBN: 978-1-63482-097-4

Chapter 9

TRYPTOPHAN, INDOLAMINES AND SLEEP: FROM HUMAN MILK TO TRYPTOPHAN-ENRICHED DIETS

***R. Bravo*[1], *L. Franco*[1], *C. Galán*[1], *L. Ugartemendia*[1], *J. M. Tena*[2], *J. Cubero*[3], *A. B. Rodríguez*[1] and *C. Barriga*[1]**

[1]Chrononutrition Laboratory. Department of Physiology. Faculty of Science. University of Extremadura. Badajoz, Spain

[2]Anesthesiology & Reanimation Area. Hospital University Complex. Badajoz (Spain)

[3]Health Education Laboratory. Science Education Area. University of Extremadura. Badajoz, Spain

ABSTRACT

Chrononutrition is the field of Chronobiology that establishes the principle of consuming foodstuffs at times of the day when they are more useful for health, improving, therefore, biological rhythms and physical performance. Tryptophan is an essential amino acid which has to be taken from diet and its ingestion has been widely studied by our research group. Previous studies carried out in laboratory animals reported that after tryptophan ingestion, serotonin and melatonin levels increase in plasma. The next step was to study tryptophan oscillations in human milk and its possible effects on sleep/wake circadian rhythm, urinary indolamine metabolites in newborns and its possible role as sleep/wake synchronizer. Taking into account tryptophan circadian rhythm in breastfeeding milk and its association with sleep conditions, our research was focused in creating dissociated formula diets and to enhance both indolamines levels and sleep conditions in other ages (children with sleep problems and elderly people) and children with neurological disorders. These different studies reported similar conclusions: enhancing indolamine levels through tryptophan supplements or tryptophan-enriched diets sleep conditions are able to improve both in animals and humans.

INTRODUCTION

Although it is well-known that tryptophan (TRP) is an essential amino acid which has to be taken in from the diet, not so many years ago, in 1974, Gessa et al. [1] was reported that TRP is used for the synthesis of both the neurotransmitter serotonin and the indole melatonin. These experiments were carried out through acute TRP depletion in rats. Since then, TRP-free diets started to be used in experiments, and psychological effects, sleep problems, and anxiety disorders, among others, linked with low levels of serotonin and melatonin were reported [2]. Our research group has focused our objectives on knowing how this essential amino acid may influence circadian rhythms in animals and humans.

Once TRP is ingested, it is transported and absorbed through the walls of the intestine, and later, 80-90 % binds to albumin and 10-20 % remains free. After TRP is transported across the intestine, it will be carried by the blood into the central nervous system (CNS) where it will be converted into serotonin (5-HT) [3,4]. The indole melatonin is mainly produced during dark hours by the pineal gland from serotonin, and entrains circadian rhythms. Specialized photo-receptive cells in the retina detect light, and this information goes directly to the suprachiasmatic nuclei through the retinohypothalamic tract, and indirectly through the geniculohypothalamic tract [5,6].

The synthesis of 5-HT begins with the action of the enzyme tryptophan hydroxylase, the limiting enzyme in this process, which converts the amino acid tryptophan into 5-hydroxytryptophan. This is then decarboxylated by the action of the enzyme 5-hydroxytryptophan decarboxylase resulting in the formation of 5-HT [7,8]. The availability of TRP for the synthesis of 5-HT is dependent on the amount of free or total amino acids on the ration between TRP and free amino acids. Thus, the concentration of 5-HT in the brain is directly proportional to the tryptophan concentration in brain and plasma, and the dietary intake of tryptophan directly influences the amount of 5-HT in the plasma, the brain, and indeed throughout the body [9].

5-HT is a neurotransmitter involved in many of the brain's functions. It participates in the synchronization of the circadian clock located in the suprachiasmatic nucleus (SCN), and in the regulation of the sleep/wake cycle, as well as having a fundamental role in the biosynthetic pathway of melatonin, the pineal indole that also carries out regulatory functions in circadian rhythms (such as sleep induction) [8,10].

Melatonin is considered to be the principal component of the internal biological clock in the SCN. It is now clear that melatonin production is subject to circadian rhythms, with its levels being high at night and low by day, and also to seasonal variations throughout the year. It is synthesized at night mainly in the pinealocytes present in the pineal gland in response to an endogenous internal clock regulated by environmental photoperiodic stimuli [11,12,13].

Melatonin is an extremely interesting indole, and its physiological effects and pharmacological properties have been widely studied. Melatonin regulates sleep cycles and readjusts altered circadian rhythms, hence its use in the treatment of jet-lag. It enhances the efficacy of cancer treatments by counteracting the adverse effects of chemotherapy, combats cell-level ageing by acting as a free-radical scavenger, and it is even known to synchronize circannual behaviours in some animals. But until several years ago, little was known about the effect of the administration of the precursor of its synthesis - the essential amino acid TRP [12].

Tryptophan and Circadian Rhythms in the Rat and the Ring Dove: The First Steps to Chrononutrition

The first studies carried out by our research group were focused on the rat for molecular studies and on the ring dove both for molecular studies too and for activity/inactivity research, since rats have a polyphasic but ring dove a monophasic sleep circadian rhythm.

Our first studies in rats revealed that 5-HT indeed shows a circadian rhythm in plasma, but also in different parts of the brain such as the pineal gland, hypothalamus, cerebellum, hippocampus, and striatum [4]. In this way, we were able to confirm that melatonin's immediate precursor also follows a circadian rhythm when it acts as a neurotransmitter in the CNS.

We then decided to work with the essential amino acid TRP as an indolamine precursor. Oral doses of 125 mg TRP / 1 mL NaCl saline solution per animal/day were administered for one week to young rats an hour before lights off. Another group of control rats received only 1 mL of saline solution following the same schedule as the TRP-treated rodents. This experiment produced an increase in 5-HT levels with respect to the control rats during the 24-hour period in plasma and brain [5]. When the same trial was repeated to study the effects on melatonin synthesis, similar results were obtained, i.e., an increase in the melatonin rhythm was found after TRP administration [14]

Taking into account that previous studies reported that exogenous melatonin shows positive effects on sleep/wake circadian rhythm, our next objective was to examine whether TRP could improve activity/inactivity circadian rhythm (an indirect method to get conclusions about sleep conditions) in birds with similar sleep patterns to humans. The animals chosen were young and old ring dove (old animals show lower serotonin and melatonin values than young animals) since they belong to a diurnally active and monophasic species. After different TRP doses administered for one week, birds showed significant increases in both serotonin and melatonin levels with respect to a young control group and an old control group. Moreover, old TRP-treated animals showed similar concentrations in serotonin and melatonin as young animals, but also both young treated animals and old treated animals improved their sleep conditions as the activity records showed during the treatment week [15].

Tryptophan in Breast Milk: Chrononutrition Begins

It is well known that circadian rhythms are still untrained in the first stages of life [16]. Several studies carried out on infants have shown that, until week 12 of life, the circadian rhythm of melatonin has yet to be established, and that the subsequent circadian consolidation of the pineal hormone is clearly reflected in coordination of the sleep/wake cycle [17].

It has been reported that, in several stages of life, oral TRP administration is able to increase the levels of both the neurotransmitter serotonin and the indole melatonin [8]. Our research group took into account circadian oscillations in biological fluids, and we considered the hypothesis that breast milk could have variations in its composition. Consequently, the aim of our next trial was to relate, over 24 hours, the levels of the amino acid tryptophan in breast milk with the levels of 6-sulfatoxymelatonin —the metabolite of melatonin excreted in the urine—in infants who had been exclusively breast-fed for three months, and to study the effects on the sleep of these infants compared with formula-fed infants.

In our study participated 2 groups composed of 8 infants per group with their mothers: the first group of infants was breast-fed and the other one was fed a formula milk (Blemil 1 Plus, Ordesa S.L., Spain). Both groups were provided with their respective diet every 4 hours. With respect to sample collections, urine was collected over a period of 24 h both in mothers (in a urine bottle) and babies (in adhesive bags); in the case of breast-fed infants samples of 2 mL of breast milk were collected from mothers. Our results showed that the rhythm of 6-sulfatoxymelatonin excretion seems to be entrained and influenced by the rhythm of TRP in the mother's milk. Although there appeared a lag in infants' 6-sulfatoxymelatonin rhythm by 2h 44min behind the rhythm of TRP in the breast milk, some authors have proposed that this fact may be due to the formation of melatonin from the ingested TRP (estimated time of 60 minutes) and the catabolism of melatonin to 6-sulfatoxymelatonin which is then excreted with an elimination half-life of between 30-50 minutes. With respect to sleep, our research group logged sleep data through actimeters (*Actiwatch*®, Cambridge Neurotechology, U.K.) which infants wore on their ankles. Sleep patterns in dark cycles were recorded and later analysed with the software *Sleep Analysis*® (Cambridge Neurotechnology, U.K.) for *Windows*® environment. Sleep study revealed that breast-fed infants had better sleep conditions due to the assumed sleep, actual sleep time, and sleep efficiency being greater ($p<0.05$ for every variable) with respect to formula-fed infants.

Following this research, our group studied more amino acids in breast milk – methionine, aspartic acid, histidine, phenylalanine, and tyrosine. But the only amino acid which follows a circadian rhythm in colostrum is TRP. The rest of the amino acids acquire a circadian rhythm in transition milk (methionine for wakefulness) and in mature milk (the rest).

These findings support the declaration by the World Health Organization (WHO) and the United Nations Children's Fund (UNICEF) that breast milk is the optimal food for infants and can never be equalled by artificial substitutes. In this affirmation, one can include the fact that breast milk entrains the circadian rhythm with positive effects on babies' sleep and wakefulness.

DISSOCIATED FORMULA MILKS: APPLIED CHRONONUTRITION

Some 30% of infants present sleep problems during the first year of life [18]. This represents a serious disorder in that it may affect brain development since sleep in the newborn is decisive for correct brain maturation and for the correct synthesis of different hormones, such as growth hormone [19]. In addition, infant sleep problems constitute a major source of distress in many families. It is known that the quality and quantity of sleep can be influenced through diet. Feeding-induced sleep modulation probably results from an alteration in the availability of various neuromodulators. Among the components of food, amino acids serve as precursors for the synthesis of different neurotransmitters. The perfect food for a baby is breast milk. This presents circadian changes in some of its components, in particular those involved in sleep/wakefulness. Thus, the levels of tryptophan in breast milk have been observed to change during the course of the day. The acrophase (time of maximum concentration) occurs at night, with tryptophan possibly acting as a promoting substance in the anabolism of the biogenic amines serotonin and melatonin – both sleep inducers as has been mentioned above.

Our next objective was to dissociate chronobiologically an infant formula into its macro- and micro-nutrients with the aim of facilitating the consolidation of the sleep/wake circadian rhythm. Therefore, the nutritional components of the commercial formula Blemil Plus 1 Forte (Ordesa S.L) were separated into two different preparations according to whether their nutritional components facilitated sleep or wakefulness. To work towards this aim, we decided to perform a double-blind trial with babies less than 5 months old who suffered from sleep problems (more than three nocturnal arousals) and were fed formula milks. In this study infants (n=30) consumed different formula milk combinations over a period of three weeks. Three diets were considered in this study: Diet A, consisting of standard milk (Blemil plus 1 Forte, Ordesa S.L.) both day and night for a week; Diet B, consisting of standard milk Blemil Plus 1 Forte during the day (06:00-18:00) and Blemil Plus 1 Night (formula dissociated into its nutritional components to consolidate sleep) during the night (18:00-6:00) for another week; Diet C, infants were given Blemil Plus 1 Day (formula dissociated into its nutritional components that consolidate wakefulness) during the day (06:00-18:00), and Blemil Plus 1 Night (Ordesa S.L.) during the night (18:00-06:00). Every participant had a different diet combination through the 3 weeks of the study. The analysis of the actigraphy data (*Actiwatch* ®) indicated an improvement in the infants' nocturnal sleep in both the week with Diet B (Control/Night) and that with Diet C (dissociated Night/Day milk). In particular, Diet C showed increased hours of Actual Sleep (7.68±0.54h vs 6.77±0.12h for the Diet A control; $p<0.05$) and improved Sleep Latency (0.44±0.04h vs 0.60±0.08h for the Diet A control; $p<0.05$). The same infants receiving the Diet B in another different week showed an improvement in Sleep Efficiency (76.43±3.4% vs the Diet A control 69.86±0.94%; $p<0.05$) and Sleep Latency (0.45±0.04h vs the Diet A control 0.60±0.08h; $p<0.05$). Thus, the use of infant milk formulas with chronobiologically adapted nutritional components improved the consolidation of the sleep/wake cycle in bottle-fed infants.

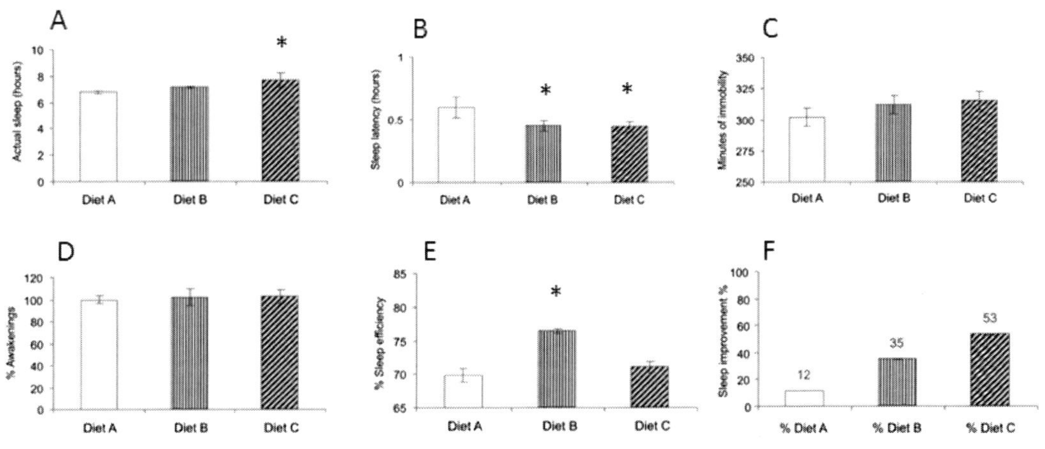

Figure 1. Different sleep parameters (Mean±SD) of infants 4-20 weeks of age, fed for 3 weeks with three different formula diets: Diet A (control): Normal Initation Milk; Diet B: of 06:00-18:00 Normal Initation Milk and 18:00-06:00 Night; Diet C: Day/Night formulas in the scheduled time previosuly. A: Actual sleep time; B: Sleep latency; C: Minutes of immobility; D: Percentage of nocturnal arousals; E: Sleep Effiiciency; F: Percentage of sleep improvement. *: statistically significant with respect to the week with the diet A (p<0,05).

Table 1. Nutrient composition of the Blemil Plus 1 Night and Blemil Plus 1 Day, both of which satisfy the 1996/49/CE and 2003/14/CE directives for infant milk formulas (Patent Nº 057846149-2114, Europe)

NUTRIENTS (per 100 g milk powder)	BLEMIL 1 PLUS NIGHT (3.4 g tryptophan/100 g protein)	BLEMIL 1 PLUS DAY (1.5 g tryptophan/ 100 g protein)
Macronutrients		
Proteins	10.7 g	12 g
Tryptophan total	0.40 g	0.18 g
Fats	26 g	26 g
Vegetable	16.4 g (63%)	25.7 g (98.75%)
MCT	9.6 g (37%)	
Formulaid		0.3 g (1.25%)
Carbohydrates	59.3 g	58 g
Lactose	44.8 g	44.8 g
Maltodextrin	14.5 g	13.2 g
Taurine	32 mg	32 mg
L-carnitine	17 mg	17 mg
Minerals		
Minerals	2.5 g	2.5 g
Sodium	175 mg	175 mg
Potassium	535 mg	535 mg
Chlorine	290 mg	290 mg
Calcium	420 mg	420 mg
Phosphorus	230 mg	230 mg
Iron	6 mg	6 mg
Magnesium	42 mg	42 mg
Zinc	4.4 mg	4.4 mg
Copper	300 mcg	300 mcg
Iodine	70 mcg	70 mcg
Manganese	50 mcg	50 mcg
Selenium	10.7 mcg	10.7 mcg
Ca/P ratio	1.8	1.8
Vitamins		
Vitamin A	500 mcg/1756 IU	640 mcg/2133 IU
Vitamin D	10.3 mcg/412 IU	10.3 mcg/412 IU
Vitamin E	12.2 mg	25 mg
Vitamin K	42 mcg	42 mcg
Vitamin B1	520 mcg	520 mcg
Vitamin B2	620 mcg	620 mcg
Vitamin B6	825 mcg	825 mcg
Vitamin B8	16 mcg	16 mcg
Vitamin B12	2 mcg	2.5 mcg
Vitamin 9	42 mcg	42 mcg
Vitamin C	50 mg	60 mg
Calcium pantothenate	3.2 mg	3.2 mg
Nicotinamide	6 mg	6 mg
Nucleotides		
Cytidine 5´monophosphate		7.9 mg
Uridine 5´monophosphate	5.3 mg	
Adenosine 5´monophosphate	2.7 mg	
Guanosine 5´monophosphate		1.6 mg
Inosine 5´monophosphate		1.6 mg

Although infant food matrices try to be as similar as possible to breast milk, complete equality is not possible. As we reported, there are marked circadian variations in the composition of human milk that must have a key functional importance in the development of pre-weaning infants. But also, recent studies have shown that even the time of breastfeeding is important.

TRYPTOPHAN-ENRICHED CEREALS IN NEUROLOGICAL DISORDERS

Cerebral palsy is defined as a group of disorders of movement and posture causing activity limitations that can be attributed to non-progressive disturbances that occur in the developing fetal or infant brain. The motor disorders of cerebral palsy are often accompanied by disturbances of sensation, cognition, communication, perception, behaviour, and/or a concurrent seizure disorder [20].

One hypothesis relates to an intrinsic anomaly in sleep regulation, involving an endogenous dysfunction in hormonal release necessary for the maintenance of the circadian rhythm [21]. A few studies have shown abnormal melatonin secretion in children with mental retardation or intellectual disability which frequently coexists in the context of cerebral palsy [21,22]. To analyse the melatonin effect in cerebral palsy, Braam et al. conducted in 2008 [23] a randomized placebo-controlled trial of melatonin treatment in individuals with intellectual disability and chronic insomnia, nine of them with cerebral palsy. In this study, individuals were given a 4-week trial with either melatonin or placebo following 1 week of baseline sleep observation. Parental observation was used to measure sleep in children. In particular, following administration of melatonin children with paralysis had a significant improvement in sleep latency, number of night-time awakenings, and total sleep time [23]. Another randomized placebo-controlled study of melatonin treatment in children with neuro-developmental disabilities was published in the same year. The study had a crossover design in which the patients were receiving a 10-day course of melatonin or placebo followed by 3 to 5 days of washout period and then 10 days of the alternative treatment. A significant improvement in total night-time sleep and sleep latency was documented both by direct observation and by actigraphy recordings [24].

Children with ASD involving difficulties with both the initiation and maintenance of sleep may have reduced levels of 6-sulfatoxymelatonin, which is the urinary breakdown product of melatonin [28]. In fact, an overall diminished pattern of nocturnal melatonin secretion or a delay in its secretion at night is considered likely in ASD [24]. Given the endogenous melatonin deficiency in a significant percentage of children with ASD, supplementation with 1-6 mg [29,30,31] of this hormone is recommended to improve the difficulties that appear in these children during the development of sleep. Subsequently, another study was conducted lasting 14 weeks in 24 children with ASD: supplemental melatonin was provided at a dose rate of 1-3 mg, and the result reported in this trial was that sleep latency decreases as measured by actigraphy [32].

Our next objective was to analyse whether a TRP-enriched diet might improve sleep conditions in this kind of population. To this end, 8 cerebral palsy and 7 children who suffered from ASD participated in our study for three experimental weeks: during one week children ingested TRP-enriched cereals for breakfast, in a second week TRP-enriched cereals were ingested only for dinner, and in a third week the children consumed TRP-enriched cereals both for breakfast and dinner. These three experimental weeks were each separated by a control week, so that the trial took 5 weeks. During these five weeks, the children wore a wrist actimeter on their non-dominant hand during all the trial. At the end of the study, the results in the case of cerebral palsy were that the classical actigraphy sleep variables improved with the TRP-enriched diets, but, in the case of the ASD children, there was only a trend for the total activity score to decrease, although this parameter should be taken into account due to its relationship with interday stability which was the only variable which showed an improvement with this population.

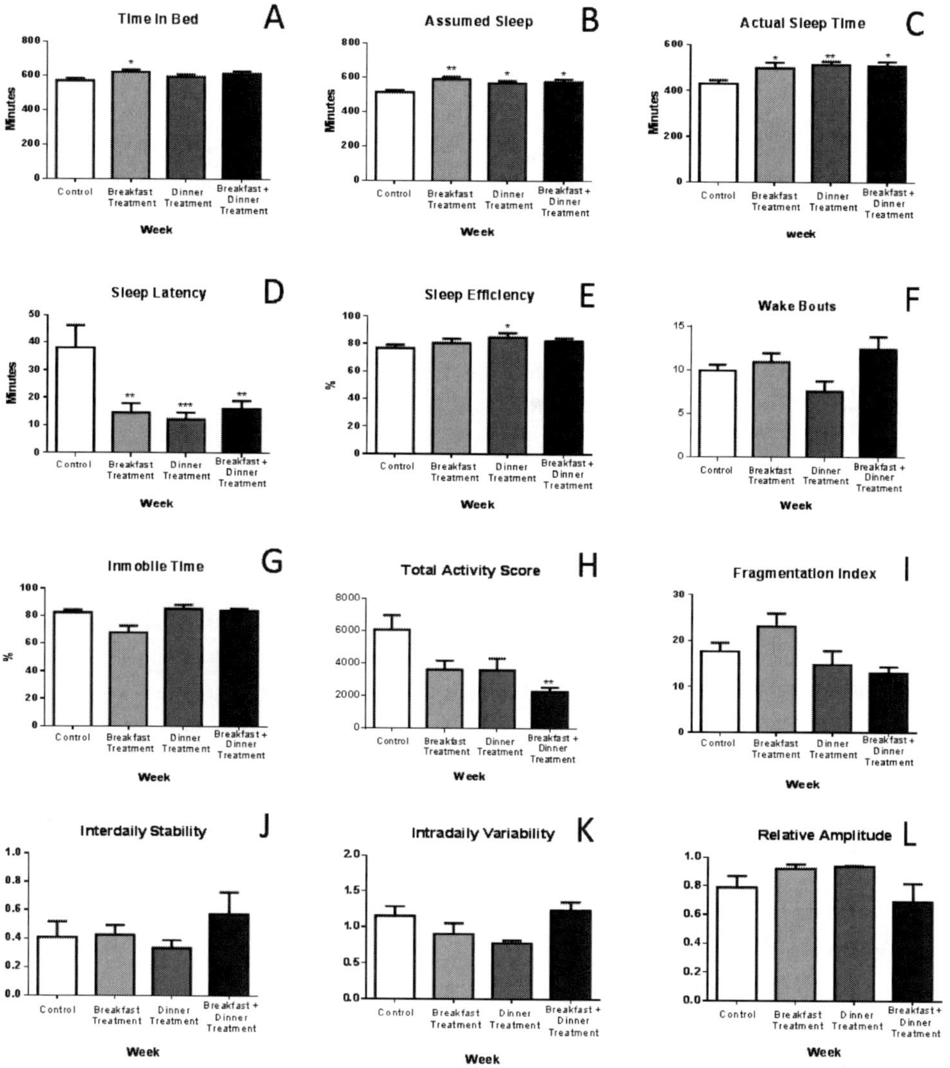

Figure 2. Graphs A to I represent different tryptophan-enriched cereals treatments and its influence on several sleep parameters in palsy cerebral children (N=7); in graphs J to L are represented non-parametric circadian rhythm variables. Each value represents mean ± SEM with respect to control. *p<0.05;**p<0.005; ***<0.001.

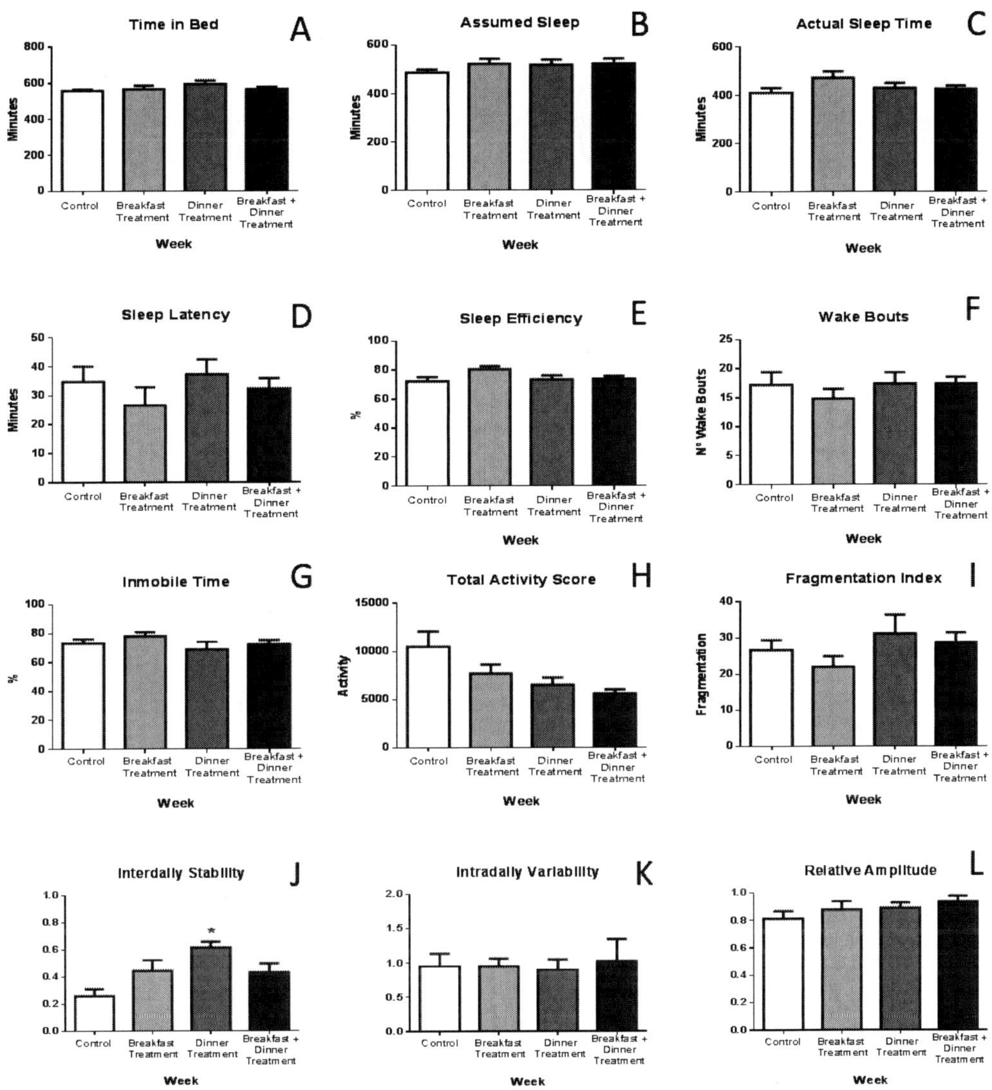

Figure 3. Graphs A to I represent different tryptophan-enriched cereals treatments and its influence on several sleep parameters in autistic children (N=7); in graphs J to L are represented non-parametric circadian rhythm variables. Each value represents mean ± SEM with respect to control. *p<0.05.

TRYPTOPHAN-ENRICHED CEREALS FOR AGEING: CHRONONUTRITION HELPING TO ACHIEVE A YOUNGER METABOLISM

Sleep disturbance in elderly people is a topic which has been extensively reviewed. In fact, nearly 30% of people over 50 years old suffer from sleep problems due to ageing [33], and the oldest groups are especially prone to impaired sleep. Elderly people trend to experience a greater number of nocturnal arousals, wake-up times, an earlier bedtime, or increased daytime napping. But also, chronobiologically, the most prominent age-related changes in circadian functioning are reduced amplitude and an advanced phase. These changes are particularly relevant in ageing. Indeed, many studies have reported that sleep/wake circadian rhythm disruption due to ageing is caused by alterations in indolamine metabolism. The loss of melatonin in advanced age leads to disturbances in circadian pacemakers which are directly involved in chronodisruption, inducing both chronopathologies and health impairments [8,13,34].

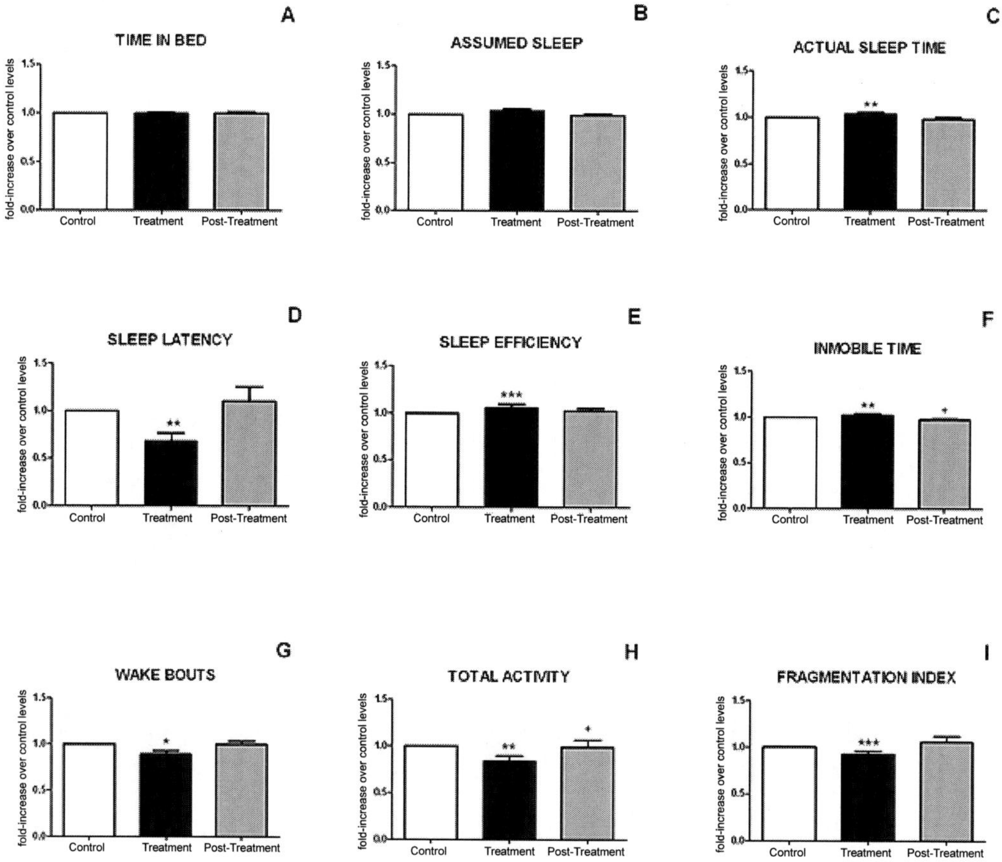

Figure 4. Effect of tryptophan-enriched cereal intake in elderly people on time in bed (A), assumed sleep (B), actual sleep time (C), sleep latency (D), sleep efficiency (D), immobile time (F), number of wake bouts (G), total activity score (H), and fragmentation index (I). Results are expressed as fold-increase over control levels (normalized and expressed as 1). Each value represents the mean ± SEM. Control: first week (22.5 mg of tryptophan was consumed in breakfast and dinner); treatment: second week (60 mg of tryptophan was ingested in breakfast and dinner); posttreatment week: habitual diet. *$p<0.05$ control vs. treatment; **$p<0.01$ control vs. treatment; ***$p<0.001$ control vs. treatment; +$p<0.05$ treatment vs. posttreatment; Kruskal–Wallis test and Dunn's multiple comparison test (posttest) were performed. N=35.

This study was carried out on 35 elderly volunteers aged 55-75 years who suffered from sleep problems (mainly sleep onset and sleep fragmentation problems) and non-reported sleep pathologies or depression during the trial. Three experimental weeks with a different diet per week were considered: First week (control week), participants consumed control cereals (22.5 mg tryptophan in 30 g cereals per dose) at breakfast and dinner; second week (treatment week), they consumed 30 g of TRP-enriched cereals (containing 60 mg tryptophan) both at breakfast and dinner; the third week (post-treatment week), they consumed their habitual diet. Both the control cereals and the TRP-enriched cereals were manufactured by Ordesa S.L. To collect data, we decided to measure sleep activity through actimetry (*Actiwatch* ®) due to our group having had excellent results with this technique in previous studies. Serotonin and melatonin urinary metabolites (5-hydroxyindoleacetic acid and 6-sulfatoxymelatonin) were measured using ELISA kits.

Our results showed that during the treatment week most of the sleep parameters were enhanced as compared to both the control week and the post-treatment week. In fact, Actual Sleep Time (3.5%; $p<0.01$), Sleep Efficiency (5.7%, $p<0.001$), and Immobile Time (2.8%, $p<0.01$) increased with respect to the control levels. Sleep Latency (31.6%, $p<0.05$), number of Wake Bouts (11.1%;$p<0.5$), Nocturnal Total Activity (16.2%, $p<0.01$), and Sleep Fragmentation (7.8%, $p<0.001$) decreased. Also, quantification of serotonin and melatonin urinary metabolites showed that serotonin and melatonin were at higher levels during the treatment week. It was notable the increase in serotonin levels (the metabolite with the greatest increase) compared with sleep latency (sleep variable with the greatest improvement), showing the relationship between serotonin and sleep induction, as other researchers have shown previously. On the other hand, the melatonin levels did not increase in the way that did the serotonin levels, and consequently the sleep variables related with melatonin (sleep maintenance) improved less than sleep latency.

The consumption of those diets which are able to have a restorative effect on serotonin and melatonin levels not only improves age-related circadian rhythm disorders both in animals and humans, but in fact, some studies have shown that melatonin may exert anti-inflammatory effects when contained in vegetable matrices with high levels of TRP, serotonin, and melatonin. Moreover, it may increase immunoglobulins (especially immunoglobulin M) in aged animals and exert other effects such as improving bactericidal activity.

Tryptophan-Enriched Diets: A Healthy Lifestyle

Every trial carried out by our research group in the chrononutrition field has demonstrated that some nutrients may influence some circadian functions, but they may also be positive for health in ageing and some neurological disorders.

But not only ageing or neurological diseases make us suffer from chronodisruption. In fact, the two main causes of chronodisruption are chronic changes in timing of light and behavioural activity (physical activity, rotating shift work, meal times, etc.). When the sleep/wake circadian rhythm shows chronodisruption symptoms and becomes uncoupled with the endogenous timing system, normal day/night variations in many hormones and neurotransmitters are altered and may have adverse health consequences [35].

TRP-enriched diets may achieve beneficial effects on serotonin or melatonin levels with their consequent positive influence on health. In this way, our research group has searched scientific databases to elaborate tables of foods whose composition shows high levels of TRP, serotonin, or melatonin. We recommend these tables for the design of diets to improve sleep disorders, those neurological diseases related with impairments in indolamine levels, or psychological disorders like anxiety or depression.

Moreover, to design a TRP-enriched diet it is important to take into account that meals must have at least some glycæmic effect in the blood. TRP ingested with glucides is better transported into the blood. This is provoked by insulin secretion which induces the entry of large neutral amino acids (LNAA: leucine, isoleucine, tyrosine, phenylalaline and valine) into muscle cells [3]. This fact is particularly relevant due to TRP being transported bound to albumin, such as LNAA; in this way a glycæmic meal with high levels of TRP should be better at increasing indolamine levels in the blood.

Table 2. Food with high levels of TRP. Sources: United States Department of Agriculture (USDA), USDA Nutrient Database for Standard Reference and reference 36

Food	*Tryptophan content*
Seeds:	
Winged beans	0.762 g/100gr
Soybeans	0.591 g/100 g
Roasted pumpkin	0.569 g/100 g
Sesame	0.371 g/100g
Walnuts	0.32 g/100g
Parsley	0.475 g/100g
Cheese:	
Mozzarella	0.571 g/100g
Chedar	0.558 g/100g
Parmesan	0.482 g/100g
Edam	0.352 g/100g
Gouda	0.352 g/100g
Cod	0.704 g/100g
Tuna	0.34 g/100g
Elk	0.545 g/100g
Pork:	
Bacon	0.464 g/100g
Loin	0.386 g/100g
Ham	0.374 g/100g
Sirloin	0.37 g/100g
Chops	0.369 g/100g
Liver	0.366 g/100g
Rabbit:	
Wild	0.436 g/100g
Domesticated	0.401 g/100g
Goat	0.403 g/100g
Beef:	
Round	0.402 g/100g
Liver	0.368 g/100g
Porterhouse steak	0.355 g/100g
Chicken	0.386 g/100g
Wild boar	0.38 g/100g
Veal:	
Liver	0.361 g/100g
Quail breast	0.354 g/100g
Horse	0.349 g/100g

Food	*Tryptophan content*
Spiny lobster	0.368 g/100g
Cuttlefish	0.364 g/100g
Octopus	0.33 g/100g
Cherry: Navalinda Van Pico limón Bourlat Ambrunés Pico Negro Pico colorado	 82.65 ± 4.29 ppm 68.58 ± 3.28 ppm 62.69 ± 5.61 ppm 61.36 ± 5.81 ppm 57.24 ± 7.11 ppm 37.76 ± 2.43 ppm 36.53 ± 9.84 ppm

Table 3. Food with indoleamine serotonin high levels. Sources: references 37 and 38

Food	*Serotonin content*
Tomato	221.9 µg/g
Cherry tomato	156.1 µg/g
Chinese cabbage	110.9 ± 22.5 µg/g
Spinach	34.4 ± 2.4 µg/g
Plantain	30 ± 7.5 µg/g
Cherry: Navalinda Pico colorado Burlat Pico Negro	 30.7 ng/100g 36.6 ng/100g 12.6 ng/100g 2.8 ng/100g
Hot pepper	17.9 µg/g
Chicory	8.5 ± 3.2 µg/g
Green onion	8 ± 0.8 µg/g
Kiwi fruit	5.9 ± 0.9 µg/g
Plums	4.7 ± 0.8 µg/g
Strawberry	3.77 ± 0.66 µg/g
Lettuce	3.3 ± 0.6 µg/g
Paprika	1.8 µg/g
Egg plant	1.5 – 12 µg/g
Wild maracuja	1.4 - 3.5 µg/g
Papaya	1.1 - 2.1 µg/g
Avocados: Haas Fuerte Booth	 1.6 ± 0.40 µg/g 1.5 ± 0.21 µg/g 0.2 ± 0.004 µg/g
Pineapple	1.5 µg/g
Grapefruit	0.9 µg/g
Honeydew melon	0.6 µg/g
Olives	0.2 µg/g

Table 4. Food with indole melatonin high levels. Sources: references 39 and 40

Food	Melatonin content
White mustard	189 ng/g
Black mustard	129 ng/g
Almond	39 ng/g
Fennel	28 ng/g
Cherry	15-18 ng/g
Green cardamom	15 ng/g
Anise	7 ng/g
Oat	1.8 ng/g
Indian corn	1.3 ng/g
Rice	1 ng/g
Radish	0.6 ng/g
Tomato	0.5 ng/g
Banana	0.5 ng/g
Ginger	0.5 ng/g
Cabbage	107.4 ± 7.3 pg/g
Carrot	55.3 ± 11.9 pg/g
Pineapple	36.2 ± 8.4 pg/g
Onion	31.5 ± 4.8 pg/g
Cucumber	24.6 ± 3.5 pg/g
Kiwi fruit	24.4 ± 1.7 pg/g
Strawberry	12.4 ± 3.1 pg/g
Asparagus	9.5 ± 3.2 pg/g

Tryptophan Deficiency and Disorders Related to Its Metabolites

Acute TRP depletion provokes an immediate sharp decrease in plasma TRP concentrations, and consequently lower 5-HT production as has been indicated previously. 5-HT plays an important role in many physiological functions, like mood, anxiety, aggressiveness, sleep, and appetite, among others, although the mechanisms have not been clearly established in every aspect [41].

Different studies have suggested that the serotonergic system may be implicated in Attention Deficit/Hyperactivity Disorders (ADHD). ADHD frequently shows an aggressive behaviour, a symptom which has been associated with impairments in serotonergic functions. It has been reported that molecular alterations in 5-HT synthesis and its receptors may be possible causes of these disorders [42]. Moreover, disorders in serotonergic and dopaminergic functions which have been linked with other behaviour alterations related to riskiness and impulsiveness are also related to low TRP concentrations in the CNS [43]. Also, it has been reported that acute TRP depletion may cause patients who have suffered from depression and related disorders to have a relapse. With respect to memory and cognition, there have been reported several forms of impairment linked to low TRP levels. This is especially so for long term memory whose consolidation is made difficult when serotonin levels are inappropriate [44,45,46}. These memory impairments which are related to reduced functioning of the serotonergic system may have even worse consequences such as dementia. Indeed, impaired serotonergic system functioning has also been related to cognitive effects in Alzheimer's disease. Also, neuroleptic treatment of schizophrenic patients who had low levels of TRP improved their cognitive status [47].

It has been reported that TRP participates too in food behaviour. Studies on non-human animals have shown a linkage between a decrease in serotonergic function and weight increase due to impairments in satiety. Indeed, bulimia nervosa patients with impaired serotonergic function show greater weight gain and impulsive food intake than those in whom this function is normal [48].

In the field of oncology, TRP catabolism (kynurenine metabolism) is ever more becoming regarded as an important micro-environmental factor which suppresses the capacity of tumours to survive, and, in vitro, it has shown apoptotic effects together with melatonin in tumour cells. In the last few years, these findings have opened up new paths for research into TRP metabolism and its effect on cancer [51].

Finally, with regard to cardiovascular diseases, there have been important advances involving the essential amino acid TRP. It has been observed that indolamine-2,3-dioxygenase disorders in macrophages, dendritic cells, and other cell types provoke a decrease in TRP plasma levels. Connected to this is the recent linkage found between increased risk for cardiovascular disease and an imbalance in the kynurenine pathway [52].

To summarize, TRP and its metabolites are involved in numerous aspects of health. Either through nutrition or through pharmacology, increased TRP, serotonin, and melatonin levels may be considered to be healthy. In particular, for public health, there is a need to make it generally well-known that TRP-enriched diets or melatonin consumption may be beneficial for many health problems or disorders in today's society.

References

[1] Gessa, GL; Biggio, G; Fadda, F; Corsini, GV; Tagliamonte, A. Effect of oral administration of tryptophan-free amino acid mixtures on serum tryptophan, brain tryptophan and serotonin metabolism. *Journal of Neurochemistry* 1974, 22: 869-870.

[2] Harada, T; Nakade, M; Wada, K; Akimitsu, O; Noji, T; Krejci, M; Takeuchi, H. Tryptophan and sleep: breakfast tryptophan content and sleep. In Predy VR, Patel VB, Le L. Handbook of nutrition, diet and sleep. Wageningen (Netherlands): Wageningen Academic Publishers, 2013.

[3] Silber, BY; Schmitt, JAJ. Effect of tryptophan loading on human cognition, mood, and sleep. Neuroscience and Biobehavioral Reviews 2009, doi:10.1016/j.neubiorev.2009.08.005

[4] Sánchez, S; Sánchez, CL; Paredes, SD; Cubero, J; Rodríguez, AB; Barriga, C. Circadian variations of serotonin in plasma and different brain regions of rats. *Mollecular and Cellular Biochemistry*, 2008, 317: 105-111.

[5] Sánchez, S; Sánchez, CL; Paredes, SD; Rodríguez, AB; Barriga, C. Circadian levels of serotonin in plasma and brain after oral administration of tryptophan in rats. *Basic & Clinical Pharmacology & Toxicology*, 2009, 104(1): 52-59.

[6] Toker, L; Amar, S; Bersudsky, Y; Benjamin, J; Klein, E; Agam, G. The biology of tryptophan depletion and mood disorders. *The Israel Journal of Pshychiatry and Related Sciences*, 2010, 47(1): 46-54.

[7] Hussain, AM & Mitra, AK. Effect of aging on tryptophan hydroxilase in rats brain: implications on serotonin level. *The American Society for Pharmacology and Experimental Therapeutics*, 2000, 28(9): 1038-1042.

[8] Bravo, R; Matito, S; Cubero, J; Paredes, SD; Franco, L; Rivero, M; Rodríguez, AB; Barriga, C. Tryptophan-enriched cereals intake improves nocturnal sleep, melatonin, serotonin, total antioxidant capacity levels and mood in elderly humans. *AGE*, 35:1277-1285.

[9] Markus, CR; Jonkman, LM; Lammers, JHCM; Deutz, NEP; Messer, MH; Rigtering, N. Evening intake of α-lactalbumin increases plasma tryptophan availability and improves morning alertness and brain measures of attention. *The American Journal of Clinical Nutrition*, 2005, 81: 1026-1033.

[10] Delgado, J; Terrón, MP; Garrido, M; Pariente, JA; Barriga, C; Rodríguez, AB; Paredes, SD. A cherry nutraceutical modulates melatonin, serotonin, and total antioxidant capacity: effects on ageing and chronotype. *Journal of Applied Biomedicine*. 2011, 10:109-117.

[11] Cubero, J; Narciso, D; Terrón, MP; Rial, RV; Esteban, S; Rivero, M; Parvez, H; Rodríguez, AB; Barriga, C. Chronunutrition applied to formula milks to consolidate infants´ sleep/wake cycle. *Neuroendocrinology Letters*, 2008, 28(4): 360-366.

[12] Kotronoulas, G; Stamatakis, A; Stylianopoulou, F. Hormones, hormonal agents, and neuropeptides involved in the neuroendocrine regulation of sleep in humans. *Hormones*, 2009, 8(4): 232-248.

[13] Garrido, M; Rodríguez, AB; Barriga, C; Paredes, SD. Cherry-enriched diets improve sleep from young to elderly populations. In Predy VR, Patel VB, Le L. Handbook of nutrition, diet and sleep. Wageningen (Netherlands): Wageningen Academic Publishers, 2013.

[14] Sánchez, S; Sánchez, CL; Paredes, SD; Rodríguez, AB; Barriga, C. The effect of tryptophan administration on the circadian rhythms of melatonin in plasma and the pineal gland of rats. *Journal of Applied Biomedicine*, 2008, 6:177-186.

[15] Paredes, SD; Marchena, AM; Bejarano, I; Espino, J; Barriga, C; Rial, RV; Reiter, RJ; Rodríguez, AB. Melatonin and tryptophan affect the activity-rest rhythm, core and peripheral temperatures, and interleukin levels in the ringdove: changes with age. The journals of Gerontology. *Series A, Biological sciences and medical sciences*, 64(3): 340-350.

[16] Ardura, J; Gutiérrez, R; Andrés, J; Agapito, T. Emergences and evolution of the circadian rhythm of melatonin in children. Hormone Research, 2003; 59(2):66-72.

[17] Cubero, J; Valero, V; Sánchez, J; Rivero, M; Párvez, H; Rodríguez, AB; Barriga, C. The circadian rhythm of tryptophan in breast milk affects the rhythms of 6-sulfatoxymelatonin and sleep in newborn. *Neuroendocrinology Letters*, 2005, 26(6): 657-661.

[18] Estivill, E. Behaviour of insomniacs and implication for their management. *Sleep Medicine Reviews*, 2002, 6: 3-6.

[19] Cubero, J; Chanclón, B; Sánchez, S; Rivero, M; Rodríguez, AB; Barriga, C. Improving the quality of infant sleep thorugh the inclusion at supper of cereals enriched with

tryptophan, adenosine-5´-phosphate, and uridine-5´-phosphate. *Nutritional Neuroscience*, 2009, 12(6): 272-280.

[20] Bax, M; Goldstein, M; Rosenbaum, P. Executive committee for the definition of cerebral palsy. Proposed definition and classficiation of cerebral palsy. *Developmental Medicine & Child Neurology,* 2005, 47: 571,576.

[21] Zucconi, M; Bruni O. Sleep disorders in children with neurologic diseases. *Semininars in Pediatric Neurololgy*, 2001, 8: 258-275.

[22] Pillar, G; Etzioni, A; Shahar, E; Lavie, P. Melatonin treatment in an institutionalised child with psychomotor retardation and an irregular sleep-wake pattern. *Archives of Disease in Childhood*. 1998, 79: 63-64.

[23] Braam, W; Didden, R; Smits, M; Curfs, L. Melatonin treatment in individuals with intellectual disability and chronic insomnia: a randomized placebo-controlled study. *Journal of Intellectual Disabilility Research*, 2008, 52: 256-264.

[24] Wasdell, MB; Jan, JE; Bomben MM. A randomized, placebo-controlled trial of controlled release melatonin treatment of delayed sleep phase syndrome and impaired sleep maintenance in children with neurodevelopmental disabilities. *Journal of Pineal Research*, 2008, 44:57e64.

[25] Van Waelvelde, H; Oostra, A; Dewitte, G; Van Den Broeck, C; Jongmans MJ. Stability of motor problems in young children with or at risk of autism spectrum disorders, ADHD, and or developmental coordination disorder. *Developmental Medicine & Child Neurology*. 2010, 52(8):e174–8.

[26] Johnson, KP; Giannotti, F; Cortesi, F. Sleep patterns in autism spectrum disorders. *Child & Adolescent Psychiatric Clinics in North America*, 2009, 18:917e28.

[27] Kotagal, S; Broomall, E. Sleep in children with autism spectrum disorder. *Pediatric Neurology,* 2012, 47(4):242-51.

[28] Nicholas, B; Rudrasingham, V; Nash, S; Kirov, G; Owen, MJ; Wimpory, DC. Association of PER1 and NPAS2 with autistic disorder: Support for the clock genes/social timing hypothesis. *Molecular Psychiatry*. 2007, 12:581e92.

[29] Thirumalai, SS; Shubin, RA; Robinson, R. Rapid eye movement sleep behavior disorder in children with autism. *Journal of Child Neurology,* 2002, 17:173e8.

[30] Johnson, KP; Malow B.A. Assessment and pharmacologic treatment of sleep disturbance in autism. *Child & Adolescent Psychiatric Clinics in North America*. 2008, 17:773e85.

[31] Lloyd, R; Tippmann-Pieker, M; Slocumb, N; Kotagal, S. Characteristics of REM sleep behavior disorder in children. *Journal of Clinical Sleep Medicine*. 2012, 8:127e31.

[32] Malow B., Adkins K.W., McGrew S.G., et al. Melatonin for sleep in children with autism: A controlled trial examining dose, tolerability, and outcomes. *Journal of Autism and Developmental Disorders*. 2012, Aug; 42(8):1729-37.

[33] Zhdanova, IV; Wurtman, RJ; Regan, MM; Taylor, JA; Shi, JP; Leclair, OU. Melatonin treatment for age-related insomnia. 2001, *The Journal of Clinical Endocrinology and Metabolism* 86(10):4727–4730.

[34] Van Someren, EJW. Circadian and sleep disturbances in the elderly. *Experimental Gerontology*, 2000, 35:1229–1237.

[35] McGowan, NM; Coogan, AN. Circadian and behavioural responses to shift work-like schedules of light/dark in the mouse. *Journal of Molecular Psychiatry*, 2013, 1(1): 7.

[36] Cubero, J; Toribio, F; Garrido, M; Hernández, M; Maynar, J; Barriga, C; Rodríguez, AB. Assays of the amino acid trytpophan in cherries by HPLC-fluorescence. *Food Analytical Methods*, 2010, 3: 36-39.

[37] Ramakrishna, A; Giridhar, P; Aswathanarayana, G. Phytoserotonin: A review. *Plant signaling & Behavior*, 2011, 6(6): 800-809.

[38] Feldman, JM; Lee, EM. (1985) Serotonin content of foods: effect on urinary excretion of 5-hydroxyindoleacetic acid. *The American Society for Clinical Nutrition*, 1985, 42(4):639-643.

[39] Hardeland, R; Pandi-Perumal, SR. Melatonin, a potent agent in antioxidative defense: Actions as a natural food constituent, gastrointestinal factor, drug and prodrug. *Nutrition & Metabolism*, 2005, 2:22.

[40] Paredes, SD; Barriga, C; Reiter, RJ; Rodríguez AB. Assessment of the potential role of Tryptophan as the precursor of Serotonin and Melatonin for the aged sleep-wake cycle and immune function: Streptopelia risoria as a model. *International Journal of Tryptophan Research*, 2009 2: 23-36.

[41] Bell, C; Abrams, J; Nutt, D. Tryptophan deplection and its implications for psychiatry. *British Journal of Psychiatry*, 2001, 178, 399-405.

[42] Tan, G; Ren, D; Xin, R; Qian, Y; Wang, D, Jiang, S.Lack of association between the tryptophan hydroxylase gene A218C polymorphism and attention-deficit hyperactivity disorder in Chinese Han population. *American Journal of Medical Genetics*, 2001, 105(6):485-8.

[43] Rogers, RD; Blackshaw, AJ; Middleton, HC; Mathews, K; Hawtin, K; Crowley, C. Tryptophan depletion impairs stimulus-reward learning while methylphenidate disrupts attentional control in healthy young adults: implications for the monoaminergic basis of impulsive behaviour. *Psycopharmacology*, 1999, 146(4): 482-491.

[44] Smith, KA; Morris, JS; Friston, KJ; Cowen, PJ; Dolan, RJ. Brain mechanisms associated with *depressive relapse and associated cognitive impairment following acute-tryptophan depletion. The British Journal of Phsychiatry*, 1999, 174: 525-529.

[45] Schmitt, J; Jorissen, B; Sobczak, S. Tryptophan impairs memory consolidation but improves focussed attention in healthy young volunteers. *Journal of Psychopharmacology*, 2000, 14: 21-29.

[46] Porter, LH; Matison, RT; McDougle, CJ, et al. Neurobiology of tryptophan deplection in depression: effects of m-clorophenylpiperazine (mCPP). *Neuropsychopharmacology*, 17,342-350.

[47] Golightly, K; Lloyd, J; Hobson, J; Effect of tryptophan deplection on cognition in schizophernia. *Journal of Neurology, Neurogurgery and Psychiatry*, 1999, 23: 56-62.

[48] Kaye, WH; Gendall, KA; Fernstrom, MH et al. Effects of acute tryptophan deplection on mood in bulimia nervosa. *Biological Psychiatry*, 2000, 47, 151-157.

[49] Cubero, J; Valero, V; Narciso, D; Sánchez, J; Rodríguez, AB; Barriga, C. Applications of the oral administration of the amino acid L-Triptophan as a posible antioxidant precursor. *Acta Alimentaria Hungarica,* 2007, 36 (4): 419-424.

[50] Platten, M; Wick Wolfgang; van den Eynde, BJ. Tryptophan catabolism in cancer: beyond IDO and tryptophan deplection. *Cancer Research*, 2012, 72: 5435-5440.

[51] Bejarano, I; Redondo, PC; Espino, J; Rosado, JA; Paredes, SD; Barriga, C; Reiter, RJ; Pariente JA; Rodríguez AB. Melatonin induces mitochondrial-mediated apoptosis in human myeloid HL-60 cells. *Journal of Pineal Research*, 2009, 46(4): 392-400.

[52] Mangge, H; Stelzer, I; Rininghaus, EZ; Weghuber, D; Postolache, TT; Fuchs, D. Distruberd tryptophan metabolism in cardiovascular disease. *Current Medicinal Chemistry*, 2014, 21(17): 1931-1937.

In: Indoleamines
Editor: Ángel Catalá
ISBN: 978-1-63482-097-4

Chapter 10

FUNCTION OF INDOLEAMINES IN BIOLOGICAL PROCESSES WITH EMPHASIS ON LIPID PEROXIDATION IN RETINA

Ángel Catalá*

Instituto de Investigaciones Fisicoquímicas Teóricas y Aplicadas, (INIFTA-CCT La Plata-CONICET), Facultad de Ciencias Exactas, Universidad Nacional de La Plata, Argentina

ABSTRACT

The three main actors involved in the panorama described are: indoleamines, lipid peroxidation and retina. First, a brief introduction about the history of these players is presented.

Keywords: Lipid peroxidation, retina, Membranes, PUFAs, melatonin, serotonin

ABBREVIATIONS

DPPH	2,2-diphenyl-2-picrylhydrazyl
HHE	4-hydroxy-2-hexenal
HNE	4-hydroxy-2-nonenal
5HO-TRP	5-HO-tryptophan
5MTP	5-methoxytryptamine
20:4 n-6	Arachidonic acid
BHT	Butylated hydroxytoluene
22:6 n-3	Docosahexaenoic acid
22:6 n-3	Docosahexaenoic acid
22:5 n-3 (EPA)	Eicosapentaenoic acid

* Corresponding author: Ángel Catalá. E-mail: catala@inifta.unlp.edu.ar.

GC–MS	Gas chromatography–mass spectrometry
HIOMT	Hydroxyindole-O-methyltransferase
18:2 n-6	Linoleic acid
MLT	Melatonin (N-acetyl-5-methoxytryptamine)
MAO	Monoamine oxidase
NAS	N-acetylserotonin
PC	Phosphatidylcholine
PC	Phosphatidylcholine
PE	Phosphatidylethanolamine
PS	Phosphatidylserine
PUFAs	Polyunsaturated fatty acids
ROS	Reactive oxygen species
TBARS	thiobarbituric reactive substances

Introduction

1. Brief Description of Pineal Indols

Melatonin and serotonin are mammalian neurohormones that have been implicated in the regulation of hormonal and neurological processes and responses to environmental stimuli in a wide range of species. Melatonin is a methoxyindole derivative produced in the pineal gland from the essential amino acid tryptophan in a four-step process, with the enzyme N-acetyltransferase being the regulator of melatonin synthesis (Reiter, 1980).

Subsequent studies have demonstrated that melatonin is also produced in other tissues such as retina, bone marrow, gastrointestinal tract, gonads and immune system (Bubenik, 1980; Tan et al., 1999; Guerrero and Reiter, 2002).

Although melatonin was initially considered a modulator of circadian rhythms and seasonal reproduction in several mammals, extensive research has verified the antioxidant action of melatonin and its beneficial potential in preserving the nervous system from oxidative damage (Reiter, 1998). Melatonin efficiently protects biomolecules, including lipids, proteins and nucleic acids (Reiter et al., 2008) and preserves membrane fluidity (García et al., 1997).

One of the main functions of indoleamines in biological processes is the protection of lipid peroxidation. Peroxidation of PUFAs in lipid bilayer membranes causes loss of fluidity, a fall in membrane potential, increased permeability to protons and calcium ions, and eventually, breakdown of cell membranes because of cellular deformities. The structural and functional integrity of the cell membranes is necessary for signal transduction, molecular recognition and transport, cellular metabolism, etc. The damage inflicted upon biological systems by reactive oxygen species have been implicated in numerous disease processes including inflammation, degenerative diseases, and tumor formation and involved in physiological phenomena such as aging (Kohen and Nyska, 2002). Initiation is the most important phase of lipid peroxidation especially in a cellular context; preventive therapy of lipid peroxidation-associated disease would target the initiation process (Dix and Aikens, 1993).

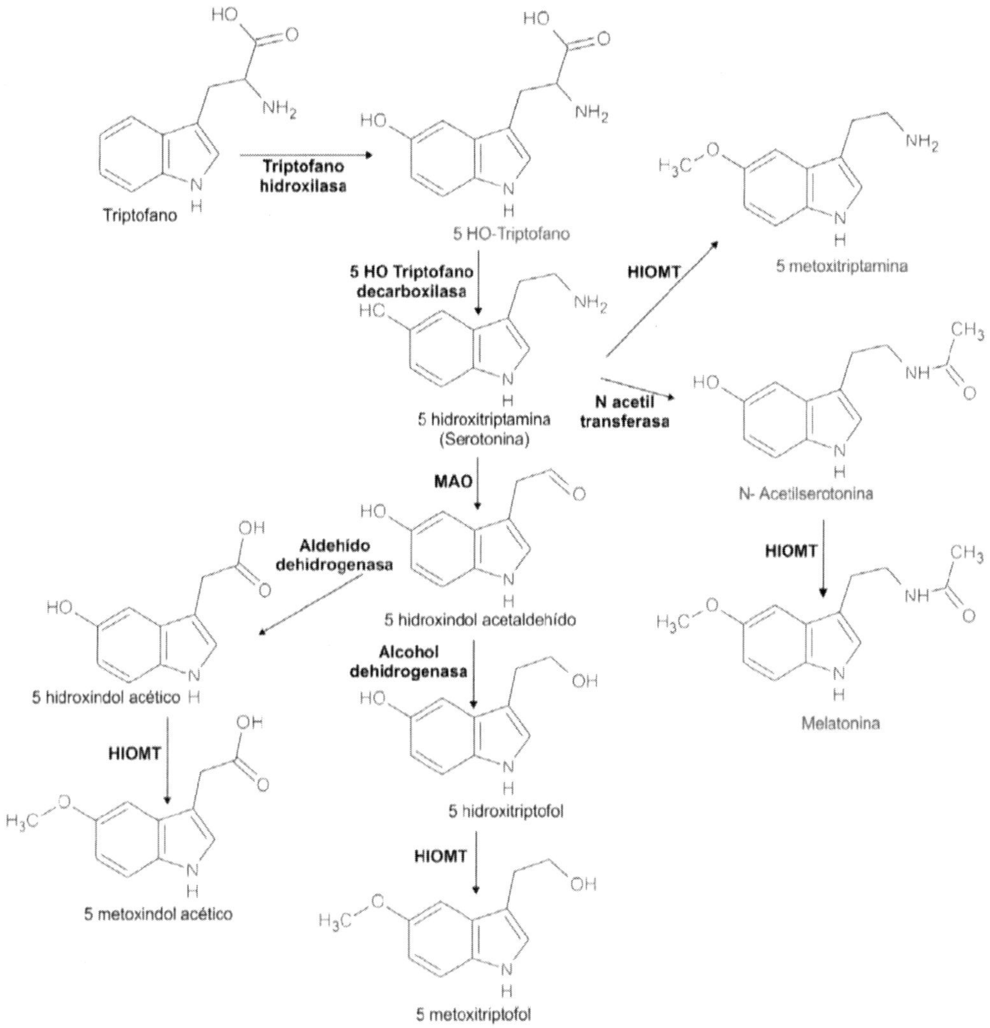

Enzymes: HIOMT, hydroxyindole-O-methyltransferase; MAO, monoamine oxidase.

Figure 1. Schematic representation of the metabolism of pineal indols. Indols more studied on lipid peroxidation are highlighted in red.

Indeed, many ocular disorders including glaucoma, cataracts, diabetic retinopathy and retinal degeneration have been attributed to lipid peroxidation processes (Ueda et al., 1996). Because of intense exposure to light and oxygen and their high PUFA content which is prone to lipid peroxidation, the retina is highly susceptible to oxidative stress (Stone et al., 1979; Winkler et al., 1999).

2. Lipid Peroxidation: Fundamental Concepts

During the last four decades the interest in polyunsaturated fatty acids has increased significantly and the number of published studies is rising each year. The current impetus for this interest has been mainly the observation that polyunsaturated fatty acids (PUFAs) and their metabolites have a diversity of physiological roles including: energy provision, membrane structure, cell signaling and regulation of gene expression. In addition the observation that PUFAs are targets of lipid peroxidation opens a new important area of investigation. Lipid peroxidation is one of the major outcomes of free radical-mediated injury to tissue. Peroxidation of fatty acyl groups occurs mostly in membrane phospholipids. Peroxidation of lipids can greatly alter the physicochemical properties of membrane lipid bilayers, resulting in severe cellular dysfunction.

In addition, a variety of lipid by products is produced as a consequence of lipid peroxidation. Lipids containing polyunsaturated fatty acids are subject to free radical–initiated oxidation and can contribute in chain reactions that amplify damage to biomolecules as described previously (Catalá, 2009, 2011, 2012). Lipid peroxidation often occurs in response to oxidative stress, and a great diversity of aldehydes is formed when lipid hydroperoxides break down in biological systems.

Some of these aldehydes are highly reactive and may be considered as second toxic messengers which disseminate and augment initial free radical events. The aldehydes most intensively studied up to now are 4-hydroxy-2-nonenal, 4-hydroxy-2-hexenal, and malondialdehyde. 4-hydroxy-2-nonenal (HNE) is known to be the main aldehyde formed during lipid peroxidation of n6 polyunsaturated fatty acids, such as linoleic acid C18:2 n-6 and arachidonic acid C20:4n-6 (Catalá, 2009).

On the other hand lipid peroxidation of n3 polyunsaturated fatty acids such as γ-linolenic acid C18:3 n-3 and docosahexaenoic acid C22:6 n-3 generates a closely related compound, 4-hydroxy-2-hexenal (HHE), which is a potential mediator of mitochondrial permeability transition. 4-hydroxy-2-alkenals represent the most prominent aldehyde substances generated during lipid peroxidation (Figure 2).

3. The Retina Contains Very High Levels of 22:6n-3 Which Is Very Sensitive to Lipid Peroxidation

The retina contains very high levels of 22:6n-3 representing the highest concentration of PUFAs of any vertebrate tissue (Stinson et al., 1991).

In fact, 50% of all acyl chains in the outer segments of photoreceptors phospholipids (both sn-1 and sn-2) are 22:6n-3 (in PC, PE, and PS).

Minor phospholipids, like phosphatidylinositol and phosphatidic acid, contain predominantly 20:4n-6 (Daemen, 1973).

Thus, rod outer segments represent an excellent model to define the role of 22:6n-3 in membrane structure and function. It have been suggested that polyunsaturation alters membrane properties that are critical for activity of integral receptor proteins (Cantor, 1999).

Therefore, exploring the structure of polyunsaturated bilayers is a prerequisite for understanding how neural membranes function. It has been demonstrated that distribution of saturated and polyunsaturated hydrocarbon chains differs significantly, supporting the hypothesis that DHA-containing membranes are under considerable elastic stress that may influence the function of integral membrane proteins (Mihailescu and Gawrisch, 2006).

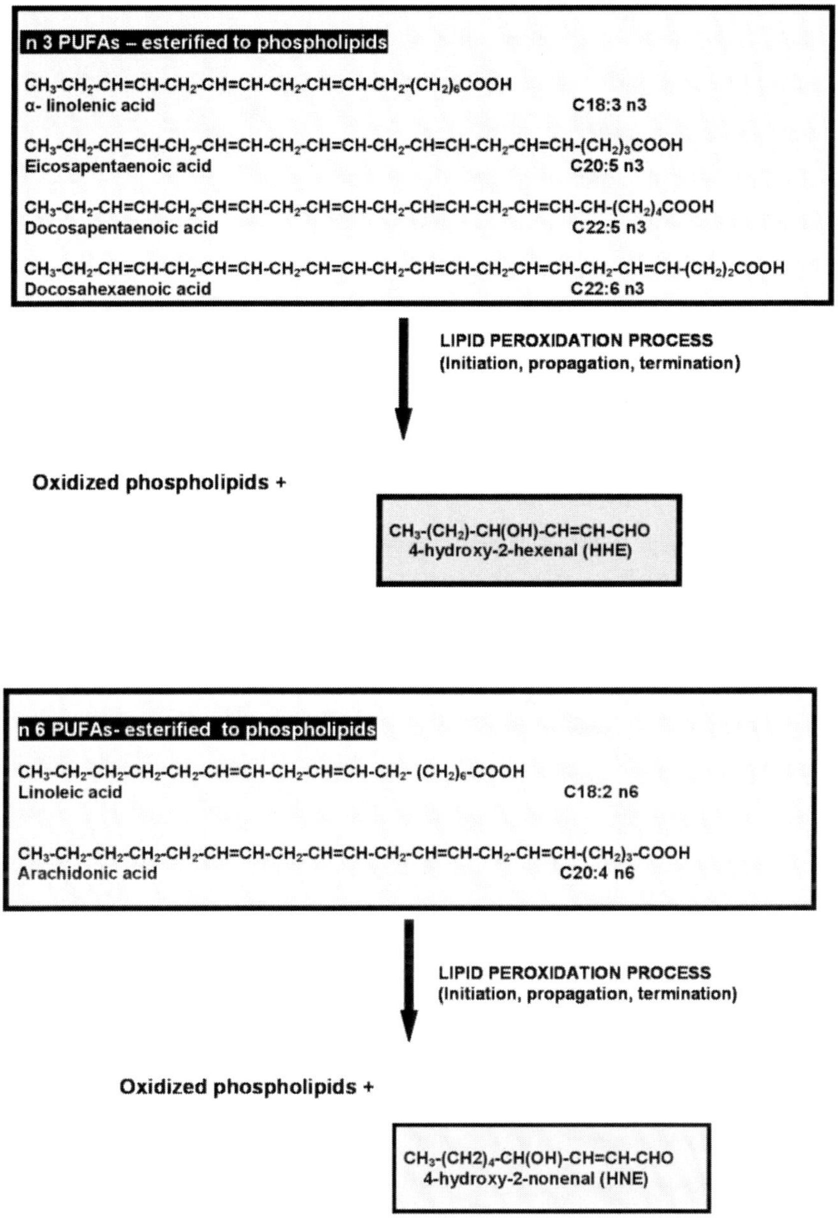

Figure 2. Schematic diagram of reactive hydroxy-alkenals produced during lipid peroxidation of n-3 and n-6 polyunsaturated fatty acids.

4. Studies *In Vivo* and *In Vitro*

Reactive oxygen species are continuously formed during cellular metabolism and are removed by antioxidants defenses. The reactive oxygen species from endogenous and exogenous sources results in continuous and accumulative oxidative damage to cellular components and alters many cellular functions. The most vulnerable molecules to oxidative damage are proteins, lipids and DNA (Kohen and Nyska, 2002; Catalá, 2011).

In mammalian retina, free radicals and lipoperoxides seem to play important roles in the evolution of different retinopathies including glaucoma, cataractogenesis, diabetic retinopathy, ocular inflammation and retinal degeneration (Ueda et al., 1996; De La Paz and Anderson, 1992). The retina is the neurosensorial tissue of the eye, it is very rich in membranes and therefore, in polyunsaturated fatty acids (PUFAs). Because free radicals production induces the lipid peroxyl radical formation, known as secondary free radicals, this chain reaction of lipid peroxidation can damage the retina, especially the membranes that play important roles in visual function (Catalá, 2006).

Melatonin (N-acetyl-5-methoxytryptamine) is an indoleamine hormone synthesized from tryptophan in pinealocytes, and it acts as an endocrine hormone to influence various circadian activities in many target organs. Two pineal biochemical functions lipoxygenation and melatonin synthesis, appear to be synergistically regulated by the status of n-3 essential fatty acids (Catalá, 2010). Melatonin is also synthesized by retinal photoreceptors, but apparently does not normally enter the circulation, so it acts more likely as a paracrine signaling molecule within the eye.

Hundred of studies have indicated that melatonin may act as a scavenger of reactive oxygen species (ROS) such as hydroxyl radical, alkoxyl radical, hypochlorous acid and singlet oxygen. Melatonin has the capacity to scavenge hydrogen peroxide in a doses dependent manner; however it has no effect as a scavenger of ion superoxide and peroxyl radicals (Allegra et al., 2003). A number of in vitro and in vivo studies have reported that exogenously administered melatonin provides protection against oxidative stress (Karbownik et al., 2001, Khaldy et al., 2000, Sinha et al., 2001 and Giusti et al., 1997). This may be not only due to the ability of melatonin to directly scavenge $HO^{\bullet}$ radical and other ROS but also by up regulation of endogenous antioxidant defense mechanisms. For example, melatonin increases the activity of glutathione peroxidase (Barlow-Walden et al., 1995) and superoxide dismutase (Antolín et al., 1996).

However, other studies suggest that melatonin is a relatively ineffective antioxidant (Wang et al., 2001, Wölfler et al., 1999 and Duell et al., 1998), or may even have prooxidant properties (Fagali and Catalá, 2007 and Wölfler et al., 2001). It has been suggested that the protective role of N-acetylserotonin and melatonin in preserving optimal levels of fluidity of the biological membranes may be related to its ability to reduce lipid peroxidation (García et al., 1997, 2001). Furthermore the protection of long chain polyunsaturated fatty acids in biological systems by melatonin and related indoleamines during lipid peroxidation (ascorbate-Fe^{2+} dependent) has been demonstrated (Leaden et al., 2002, Leaden and Catalá, 2005; Guajardo et al., 2003).

5. The Antioxidant Behaviour of Indoleamines during Lipid Peroxidation: A Comparative Study

In a previous study (Fagali and Catalá, 2012) we have analyzed the antioxidant activity of melatonin (MLT), N-acetylserotonin (NAS), 5-HO-tryptophan (5HO-TRP) and 5-methoxytryptamine (5MTP) in two different lipid systems with high content of polyunsaturated fatty acids (PUFAs): triglycerides (rich in 20:5 n-3, 22:6 n-3) dissolved in chloroform and sonicated liposomes made of retinal lipids (rich in 22:6 n-3).

In the triglyceride-chloroform-system the peroxidation reaction was initiated by cumene hydroperoxide (CHP) whereas liposomes were peroxidized with Fe^{2+}. The techniques employed at that work were: (1) TBARS production, (2) DPPH assay, (3) determination of conjugated dienes production and (4) analysis of fatty acid profile by GC-MS. Butylated hydroxytoluene (BHT) was employed as a reference because of its well known antioxidant capacity. Our results showed that MLT and 5MTP were unable to protect PUFAs against lipid peroxidation in both systems, whereas NAS and 5HO-TRP were better antioxidants that BHT in the triglyceride-system but ineffective in the liposome-system. We concluded that the antioxidant behaviour of pineal indoles depends not only on their functional groups but also on the assay system and could be explained by the polar paradox theory.

6. The Antioxidant Behaviour of Indoleamines Using Triglycerides (Rich in 20:5 N-3, 22:6 N-3) Dissolved in Chloroform

The precursors of MLT, 5HO-TRP and NAS, were the most efficient radical scavengers in experiments carried out in triglycerides dissolved in chloroform. In this regard in a recent paper it has been demonstrated that 5-HO-TRP is a more potent in vitro hydroxyl radical scavenger than MLT or vitamin C (Keithahn and Lerchl, 2005).

As our results indicate, 5HO-TRP showed the highest capacity to inhibit the TBARS production and this situation agrees with its high activity against DPPH radical. These results lead to the conclusion that this compound has a strong trend to give a hydrogen atom ($H^{\bullet}$) to free radicals generated during lipid peroxidation process. In consequence, the chain of lipid peroxidation is cut and the PUFAs are protected. Other 5-HO- derivative, NAS, also was able to inhibit TBARS production and was an efficient DPPH radical scavenger.

Wherever, its effectiveness was lesser than that of 5HO-TRP. MLT and 5MTP (5-methoxy derivatives) were not efficient $H^{\bullet}$ donors as they were analyzed by DPPH assay. These results agree with the lack of inhibition of both compounds on TBARS production (indeed, they presented pro-oxidant effect). Therefore, MLT and 5MTP were not good antioxidants in the system of triglycerides in chloroform.

The comparison between structural analogues revealed a critical importance of functional groups. The existence of a methoxy residue in position 5, as in 5MTP and MLT, strongly reduced the capacity of radical scavenging. On the other hand, the presence of the hydroxyl group in position 5 of the indole ring, as in 5HO-TRP and NAS, established high free-radical scavenging activity and efficient antioxidant activity during the lipid peroxidation of Tg PUFAs n-3. However, the mechanism by which the indole ring interacts with free radicals is still only partially understood.

Several evidences are on line with our results (i) Teixeira et al. have demonstrated that MLT shows low reactivity with DPPH (Teixeira et al., 2003), (ii) other study showed that MLT exerts only limited direct antioxidant activities (Marshall et al., 1996), (iii) the antioxidant activity of MLT in soybean PC liposomes is much lower than that of alpha-tocopherol, under comparable assay conditions (Livrea et al., 1997), and (iv) the ability of MLT to scavenge the lipid peroxyl radical ($LOO^{\bullet}$) is debated (Reiter et al., 2001).

7. The Antioxidant Behaviour of Indoleamines Using Sonicated Liposomes Made of Retinal Lipids (Rich in 22:6 N-3)

Peroxidation of liposomes made of retinal lipids using Fe^{2+} as initiator was assayed by detection of conjugated dienes, TBARS and analysis of fatty acid profiles. NAS, MLT, 5HO-TRP and 5MTP did no present antioxidant effect when it was determined by TBARS and conjugated dienes techniques; moreover, 5HO-TRP presented significant pro-oxidant effect. Analyzing fatty acid profile, we can see that PUFAs present in liposomes were affected by peroxidation to almost disappear. BHT (5 μM) was able to protect them while MLT and related indoleamines, at the same concentrations, were not. We think the PUFAs loss is a conclusive result, since the lack of protective effect becomes apparent and this technique is a direct way to evaluate antioxidant power of indoleamines.

We have performed studies using a simple liposomal model that facilitate evaluation of the kinetic of the lipid peroxidation process without the interference produced by retinal proteins. Liposomes, whose phospholipid composition, structure and dynamics can be controlled, are frequently accepted as an appropriate model for in vitro studies of membrane structures and properties. They are composed by lipid bilayers, structurally similar to cell membrane lipidic environment. Conjugated dienes formed from oxidized PUFAs, and TBARS products derived from the breakdown of these fatty acids located in phospholipids could be analyzed during peroxidation of liposomes made of retinal lipids using Fe^{2+} as initiator. Through both techniques it was noticeable the powerful antioxidant effect of BHT, while MLT, 5MTP and NAS did not show antioxidant effect when they were added to the reaction mixture, even when we assayed higher concentrations of MLT (10 and 20 μM). 5HO-TRP not only did not show antioxidant activity but it is possible that it have prooxidant effect as we observed with TBARS and conjugated dienes assays.

Our results do not allow us to conclude about the reasons why MLT and 5MTP stimulated TBARS production in the triglycerides-system and 5HO-TRP stimulated TBARS and conjugated dienes in liposomes-system, but we can speculate about the importance of theirs functional groups on this effect.

The differences found in the antioxidant behaviour of NAS and 5HO-TRP (both hydrophilic) in both sets of experiments may be explained by the known theory of the Polar Paradox (Shahidi and Zhong, 2011). This theory would also explain the fact that these compounds showed more efficient antioxidant activity than BHT (lipophilic) in the pure lipid system containing triglycerides dissolved in chloroform. This theory interprets the apparent contradiction of the fact that the hydrophilic antioxidants are more effective in purely lipid media while lipophilic antioxidants are better in aqueous media. It is known that, together with its innate power, the effectiveness of an antioxidant is also affected by its interfacial properties and in the medium partition. Early studies of pure lipid oxidation were based on the statement that peroxidation occurs in a homogeneous medium. The air–lipid was considered the start site of peroxidation propagating then into oil. Under this postulation, partially soluble antioxidants would be directed at the air–oil interface where peroxidation occurs and, hence, would protect the system from oxidative changes. However, the distribution of polar antioxidants in this interface was questioned because the air is even less polar than the oil. Thus, we supposed that micro or nanoambients affect the chemistry of lipid peroxidation and antioxidant altering the physical position of lipid substrates and antioxidants. For example, different types or micellar or lamellar structures (in the presence of traces of water) may be formed by self-assembly of lipid components (such as phospholipids) or peroxidation products (hydroperoxides, aldehydes and ketones). Currently there is insufficient evidence to support the hypothesis that these partnership structures are the sites where lipid peroxidation occurs. Polar antioxidants, instead of being located in the air–oil as previously believed, are preferentially located at the interface of these colloidal structures (e.g., oil–water interface) and are thus more efficient in inhibiting the peroxidation, than nonpolar that is dissolved in the lipid phase. This theory was supported by the fact that polar antioxidants are unable to decrease the surface tension, they succeeded in reducing the interfacial tension (Chaiyasit et al., 2007). On the basis of these considerations, more comprehensive studies are required to better understand the behaviour of indoleamines during lipid peroxidation in different media.

Conclusion

Melatonin and serotonin are mammalian neurohormones that have been implicated in the regulation of hormonal and neurological processes and responses to environmental stimuli in a wide range of species.

One of the main functions of indoleamines in biological processes is the protection of lipid peroxidation. Peroxidation of PUFAs in lipid bilayer membranes causes loss of fluidity, a fall in membrane potential, increased permeability to protons and calcium ions, and eventually, breakdown of cell membranes because of cellular deformities.

The retina contains very high levels of 22:6n-3 representing the highest concentration of PUFAs of any vertebrate tissue. 50% of all acyl chains in the outer segments of photoreceptors phospholipids (both sn-1 and sn-2) are 22:6n-3 (in PC, PE, and PS).

We have analyzed the antioxidant activity of melatonin (MLT), N-acetylserotonin (NAS), 5-HO-tryptophan (5HO-TRP) and 5-methoxytryptamine (5MTP) in two different lipid systems with high content of polyunsaturated fatty acids (PUFAs): triglycerides (rich in 20:5 n-3, 22:6 n-3) dissolved in chloroform and sonicated liposomes made of retinal lipids (rich in 22:6 n-3).

Our results showed that MLT and 5MTP were unable to protect PUFAs against lipid peroxidation in both systems, whereas NAS and 5HO-TRP were better antioxidants that BHT in the triglyceride-system but ineffective in the liposome-system. We concluded that the antioxidant behaviour of pineal indoles depends not only on their functional groups but also on the assay system and could be explained by the polar paradox theory.

Acknowledgments

Studies in the author laboratory were supported by Consejo Nacional de Investigaciones Científicas y Técnicas (CONICET), PIP 0157.

The author is Member of Carrera del Investigador Científico, Consejo Nacional de Investigaciones Científicas y Técnicas (CONICET), Argentina.

REFERENCES

Allegra, M., Reiter, R. J., Tan, D. X., Gentile, C., Tesoriere, L., Livrea, M. A. The chemistry of melatonin's interaction with reactive species. *J. Pineal Res.* 34 (2003)1-10.

Antolín, I., Rodríguez, C., Saínz, R. M., Mayo, J. C., Uría, H., Kotler, M. L., Rodríguez-Colunga, M. J., Tolivia, D., Menéndez-Peláez, A. Neurohormone melatonin prevents cell damage: effect on gene expression for antioxidant enzymes. *FASEB J.* 10 (1996) 882-890.

Barlow-Walden, L. R., Reiter, R. J., Abe, M., Pablos, M., Menendez-Pelaez, A., Chen, L. D., Poeggeler, B. Melatonin stimulates brain glutathione peroxidase activity. *Neurochem. Int.* 5 (1995) 497-502.

Bubenik, G. A. Localization of melatonin in the digestive tract of the rat. Effect of maturation, diurnal variation, melatonin treatment and pinealectomy. *Horm. Res.* 12 (1980) 313-323.

Cantor, R. S. Lipid composition and the lateral pressure profile in bilayers. *Biophys. J.* 76 (1999) 2625-2639.

Catalá, A. (Eds.), *Open Access book project: "Lipid Peroxidation"* ISBN 980-953307 143 0 InTech – Open Access Publisher 2012.

Catalá, A. Lipid peroxidation of membrane phospholipids generates hydroxyl-alkenals and oxidized phospholipids active in physiological and/or pathological conditions, *Chem. Phys. Lipids* 157 (2009) 1-11.

Catalá, A. Lipid peroxidation. In: Pantopoulos, K., Schipper, H., editors. *Principles of Free Radical Biomedicine.* Chapter 7. Vol. 1. New York, NY, US: McGill University, Nova Science; 2011. pp. 137-160.

Catalá, A. Lipid peroxidation of membrane phospholipids in the vertebrate retina. *Front. Biosci.* (Schol Ed.). 3 (2011) 52-60.

Catalá, A. (Ed.), *Lipid Peroxidation: Biological Implications*, Transworld Research Network Kerala, India (2011) pp. 1-173.

Chaiyasit, W., Elias, R. J., McClements, D. J., Decker, E. A. Role of physical structures in bulk oils on lipid oxidation. *Crit. Rev. Food Sci. Nutr.* 47(2007) 299-317.

Daemen, F. J. M. Vertebrate rod outer segment membranes. *Biochim. Biophys. Acta* 300 (1973) 255-288.

Dix, T. A., Aikens, J. Mechanisms and biological relevance of lipid peroxidation initiation. *Chem. Res. Toxicol.* 6 (1993) 2-18.

Duell, P. B., Wheaton, D. L., Shultz, A., Nguyen, H. Inhibition of LDL oxidation by melatonin requires supraphysiologic concentrations. *Clin. Chem.* 44 (1998)1931-1936.

Fagali, N., Catalá, A. The effect of melatonin and structural analogues on the lipid peroxidation of triglycerides enriched in omega-3 polyunsaturated fatty acids. *Life Sci.* 81 (2007) 299-305.

Fagali, N., Catalá, A. The antioxidant behaviour of melatonin and structural analogues during lipid peroxidation depends not only on their functional groups but also on the assay system. *Biochem. Biophys. Res. Commun.* 423 (2012) 873 877.

García, J. J., Reiter, R. J., Guerrero, J. M., Escames, G., Yu, B. P., Oh, C. S., Muñoz-Hoyos, A. Melatonin prevents changes in microsomal membrane fluidity during induced lipid peroxidation. *FEBS Lett.* 408 (1997) 297-300.

García, J. J., Reiter, R. J., Karbownik, M., Calvo, J. R., Ortiz, G. G., Tan, D. X., Martínez-Ballarín, E., Acuña-Castroviejo, D. N-acetylserotonin suppresses hepatic microsomal membrane rigidity associated with lipid peroxidation. *Eur. J. Pharmacol.* 428 (2001) 169-175.

Giusti, P., Lipartiti, M., Gusella, M., Floreani, M., Manev, H. In vitro and in vivo protective effects of melatonin against glutamate oxidative stress and neurotoxicity. *Ann. N. Y. Acad. Sci.* 825 (1997) 79-84.

Guajardo, M. H., Terrasa, A. M., Catalá, A. Protective effect of indoleamines on in vitro ascorbate-Fe2+ dependent lipid peroxidation of rod outer segment membranes of bovine retina. *J. Pineal Res.* 35 (2003) 276-282.

Guerrero, J. M., Reiter, R. J. Melatonin-immune system relationships. *Curr. Top. Med. Chem.* 2 (2002) 167-179.

Khaldy, H., Escames, G., León, J., Vives, F., Luna, J. D., Acuña-Castroviejo, D. Comparative effects of melatonin, L-deprenyl, Trolox and ascorbate in the suppression of hydroxyl radical formation during dopamine autoxidation in vitro. *J. Pineal Res.* 29 (2000)100-107.

Karbownik, M., Lewinski, A., Reiter, R. J. Anticarcinogenic actions of melatonin which involve antioxidative processes: comparison with other antioxidants. *Int. J. Biochem. Cell Biol.* 33 (2001) 735-53.

Keithahn, C., Lerchl, A. 5-Hydroxytryptophan is a more potent in vitro hydroxyl radical scavenger than melatonin or vitamin C. *J. Pineal Res.* 38 (2005) 62-66.

Kohen, R., Nyska, A. Oxidation of biological systems: oxidative stress phenomena, antioxidants, redox reactions, and methods for their quantification. *Toxicol. Pathol.* 30 (2002) 620-650.

Leaden, P., Barrionuevo, J., Catalá, A. The protection of long chain polyunsaturated fatty acids by melatonin during nonenzymatic lipid peroxidation of rat liver microsomes. *J. Pineal Res.* 32 (2002) 129-134.

Leaden, P. J., Catalá, A. Protective effect of melatonin on ascorbate-Fe2+ lipid peroxidation of polyunsaturated fatty acids in rat liver, kidney and brain microsomes: a chemiluminescence study. *J. Pineal Res.* 39 (2005)164-169.

Livrea, M. A., Tesoriere, L., D'Arpa, D., Morreale, M. Reaction of melatonin with lipoperoxyl radicals in phospholipid bilayers. *Free Radic. Biol. Med.* 23 (1997) 706-711.

Marshall, K. A., Reiter, R. J., Poeggeler, B., Aruoma, O. I., Halliwell, B. Evaluation of the antioxidant activity of melatonin in vitro. *Free Radic. Biol. Med.* 21(1996) 307-315.

Mihailescu, M., Gawrisch, K. The structure of polyunsaturated lipid bilayers important for rhodopsin functions: a neutron diffraction study. *Biophys. J.* 90 (2006) L04-L06.

Reiter, R. J. Oxidative damage in the central nervous system: Protection by melatonin. *Prog. Neurobiol.* 56 (1998) 359-384.

Reiter, R. J. The pineal and its hormones in the control of reproduction in mammals. *Endocr. Rev.* 1(1980) 109-131.

Reiter, R. J., Tan, D. X., Manchester, L. C., Qi, W. Biochemical reactivity of melatonin with reactive oxygen and nitrogen species: a review of the evidence. *Cell Biochem. Biophys.* 34 (2001) 237-256.

Reiter, R. J., Korkmaz, A., Paredes, S. D., Manchester, L. C., Tan, D. X. Melatonin reduces oxidative/nitrosative stress due to drugs, toxins, metals, and herbicides. *Neuro Endocrinol. Lett.* 29 (2008) 609-613.

Shahidi, F., Zhong, Y. Revisiting the polar paradox theory: a critical overview. *J. Agric. Food Chem.* 59 (2011) 3499-3504.

Sinha, K., Degaonkar, M. N., Jagannathan, N. R., Gupta, Y. K. Effect of melatonin on ischemia reperfusion injury induced by middle cerebral artery occlusion in rats. *Eur. J. Pharmacol.* 428 (2001) 185-92.

Stinson, A. M., Wiegand, R. D., Anderson, R. E. Fatty acid and molecular species compositions of phospholipids and diacylglycerols from rat retinal membrane. *Exp. Eye Res.* 52 (1991) 213-218.

Stone, W. L., Farnsworth, C. C., Dratz, E. A. A reinvestigation of the fatty acid content of bovine, rat and frog retinal rod outer segments. *Exp. Eye Res.* 28 (1979) 387-397.

Tan, D. X., Manchester, L. C., Reiter, R. J., Qi, W. B., Zhang, M., Weintraub, S. T., Cabrera, J., Sainz, R. M., Mayo, J. C. Identification of highly elevated levels of melatonin in bone marrow: Its origin and significance. *Biochim. Biophys. Acta.* 1472 (1999) 206-214.

Teixeira, A., Morfim, M. P., de Cordova, C. A., Charão, C. C., de Lima, V. R., Creczynski-Pasa, T. B. Melatonin protects against pro-oxidant enzymes and reduces lipid peroxidation in distinct membranes induced by the hydroxyl and ascorbyl radicals and by peroxynitrite. *J. Pineal Res.* 35 (2003) 262-268.

Ueda, T., Ueda, T., Armstrong, D. Preventive effect of natural and synthetic antioxidants on lipid peroxidation in the mammalian eye. *Ophthalmic Res.* 28 (1996) 184-192.

Wang, H. X., Liu, F., Ng, T. B. Examination of pineal indoles and 6-methoxy-2-benzoxazolinone for antioxidant and antimicrobial effects. *Comp. Biochem. Physiol. C Toxicol. Pharmacol.* 130 (2001) 379-388.

Winkler, B. S., Boulton, M. E., Gottsch, J. D., et al. Oxidative damage and age-related macular degeneration. *Mol. Vis.* 5 (1999) 1-32.

Wölfler, A., Abuja, P. M., Schauenstein, K., Liebmann, P. M. N-acetylserotonin is a better extra- and intracellular antioxidant than melatonin. *FEBS Lett.* 449 (1999) 206-210.

Wölfler, A., Caluba, H. C., Abuja, P. M., Dohr, G., Schauenstein, K., Liebmann, P. M. Prooxidant activity of melatonin promotes fas-induced cell death in human leukemic Jurkat cells. *FEBS Lett.* 502 (2001)127-131.

In: Indoleamines
Editor: Ángel Catalá

ISBN: 978-1-63482-097-4

Chapter 11

INDOLEAMINES IN EDIBLE PLANTS: ROLE IN HUMAN HEALTH EFFECTS

Akula Ramakrishna*
India Vegetable R & D, Vegetable Quality analytics Lab, Monsanto Crop Breeding Station, Kallinayakanahalli, Chikkaballapur, Karnataka, India

ABSTRACT

Melatonin (MEL) and Serotonin (SER), also known as [5-Hydroxytryptamine (5-HT)] are reported to be in a range of plant types which are edible. Their occurrence in plants species appears to be ubiquitous. Their presence in high quantities in plants assumes significance owing to their physiological effects upon consumption by human beings. MEL is a well-know animal hormone mainly released by the pineal gland known to influence circadian rhythm, sleep, apart from immune enhancement. Similarly, SER is a well-known neurotransmitter that regulates mood, sleep and anxiety in mammals. It is implicated in memory, behavioral changes, scavenging reactive oxygen species, antipsychotic, etc. Similarly Role of SER and MEL in plant morphogenesis, and various physiological processes through intense research is beginning to unfold. These molecules are in common foods viz banana, pineapple, plum, nuts, milk, grape wine. N- Feruloyl serotonin and p-coumaroyl serotonin found in certain seeds are found to possess antioxidant, anti-inflammatory, antitumor, antibacterial and anti-stress potential apart from reducing depression and anxiety. MEL is found in Mediterranean diets, nuts, cherries, tomato berries, and olive products. Consumption of foods rich in MEL is known to increase blood MEL levels which has been implicated in protective effect against cardiovascular damage, cancer initiation and growth. MEL is also found in wines, green tea, beer, olive oil etc. Moreover, presence of SER and MEL in Coffee beans (green and roasted beans) and decoction has been reported. In this chapter occurrence of indoleamines in edible plants and their implications in human health are presented.

* Email:ramapcbt@gmail.com.

Introduction

SER and MEL potent neurotransmitters in animal systems, are present in various plant species, structurally and functionally [1, 2]. The definite function of these compounds in plants is yet to be understood. SER, a monoamine neurotransmitter, also called 5-Hydroxytryptamine (5-HT), is found in plants as well as animals [2]. SER is a physiologically active amine that regulates mood, sleep, and anxiety in mammals [3]. SER has also been reported in wide range of plant species. The precise function of SER at physiological level particularly, growth regulation, flowering, xylem sap exudation, ion permeability, and plant morphogenesis etc. [4, 5].

In vertebrates, MEL (*N*-acetyl-5-methoxytryptamine), a well-known animal hormone, is released by the pineal gland although it is formed in numerous other organs. As a hormone, it plays a key role in various physiological processes such as circadian rhythmicity, sleep and seasonal photoperiod regulation, sexual and reproductive behaviour apart from immunological enhancement [6, 7]. MEL has been studied not only for its hormonal activity but also role in scavenging different types of reactive oxygen and nitrogen species [8]. SER and MEL synthesized from Tryptophan. SER is synthesized from tryptophan by two enzymes, *tryptophan decarboxylase* (TDC) and tryptamine *5-hydroxylase* (T5H) [5]. SER is converted in to MEL by two enzymes, *serotonin N-acetyl transferase* and *Hydroxy-o-methyl transferase* [9]. MEL is occurred as secondary metabolite in plants and, in fact, it has been found in many edible products broadly consumed such as tomatoes, strawberries, cherries, olive oil and coffee [10-14]. Consumption of edible plant tissues that contain the SER and MEL would be useful as their anti-oxidant activity has a bearing their as medicinal value for treatment of human diseases such as cardiovascular damage [15], Alzheimer's disease [16] and cancer [17].

Free biogenic amine levels in some arabica and robusta coffee samples have been used as chemical markers for the two coffee species [18]. Intake of tryptophan rich foods can result in elevated MEL levels both in tissue and in blood [19-21]. Over the course of a lifetime, the regular consumption of melatonin-rich foods may improve the general health of consumers [22-23]. This chapter describes the occurrence of indoleamines in edible plants and their health benefits.

Occurrence of Serotonin in Edible Plants

It is well established that the levels of SER vary in different plant parts and also varied in respective tissues. In leaves, SER is reported to be 0.007 μg/g fresh weight (FW), while seeds harbored 2,000 μg/g in West Africa leguminous plant *Griffonia simplicifolia* [24]. Fruits, vegetables, and seeds are the major tissues in which SER occurs abundantly **[25]**. SER occurs in fruits (e.g. pineapple, bananas and tomatoes), avocado, eggplant, walnuts and also in the sting of the nettle *Urtica dioica* [26] and in the pods of cowhage [27] in which SER is suggested to play a protective role against predators. In the fruits of pineapple, SER distribution in tissues is not uniform, with a large amount in the outer skin of fruit, considerably less in inner skin and pulp [28]. SER levels are known to increase as the fruits ripen in many species, including tomato, although the inverse is true of the fruit of pineapple

(*Ananas comosus*) [29]. SER has been analyzed in the fruits of *Juglans regia,* a temperate species where its amount is found to be lowest in spring but sharply increases during autumn which levels is more than that found in animal tissues [29], coinciding with seed dispersal period [30].

The presence of SER in coffee beans might have beneficial influence on consumer health [18]. Even *Coffea* seed wax contain 4.1 mg g^{-1} SER [31] and also polished coffee called Astra, contains 20-40 mg SER/100g [32]. In our study we were able to identify MEL & SER both in green and roasted beans and also in decoction (brew). Da Silveira *et al* (2007) [33] reported that during roasting of coffee, there is a significant reduction in the levels of amines. Similarly, roasting temperature influenced indoleamine profile. However, coffee beans submitted to a light roast contained low levels of SER [18, 34]. In regular coffee, SER contributed most to total levels, followed by cadaverine, spermidine and tryptamine [33]. According to Lima *el al* (1999) [35], intake of four to five 50 mL cups of coffee beverage provide 24.5 μg tryptamine and 39 μg SER daily.

The presence of indoleamines in roasted beans which pass into coffee brew might be useful for consumers. For the first time we were able to identify the presence of significant levels of MEL in coffee beans [13]. Recent studies have shown that sweet cherries contain substantial amounts of MEL and SER, and may have several health benefits of relevance in a healthy diet [36].

Pulp of underripe and ripe yellow banana contains SER at concentrations of 31.4 and 18.5 ng/g, respectively [37,38]. Higher concentrations of SER in the peel (150,000 ng g-1) of the banana compared to the pulp 24,000 ng g-1 [39]. During ripening, concentrations of SER increased significantly in both the inner and outer banana peel. However, in French plantain (*Musa paradisica* var. *sapientum*) though very high SER concentrations were found during ripening (from 49,900 to 56,700 ng·g-1), but during ripening it decrease (12,000 ng·g-1) [40]. In Prata bananas (*M. acuminate* × *M. balbisiana*) during ripening, SER levels were between 15,000 and 20,000 ng·g^{-1} until the fourteenth day of storage [41].

Health Benefits of Phytoserotonin

SER is involved in the regulation of a number of important functions in humans, including sleeping, hunger, thirst, mood and sexual activity [46]. Some of the foods (banana, pineapple, plum, nuts and milk) rich in SER and tryptophan may elevate mood by raising brain SER levels [47].

The presence of SER in coffee beans might have beneficial influence on consumer health [18]. Sweet cherries contain significant amount of SER and may have a great number of health benefits if incorporated in a healthy diet [36]. Moreover, SER is also found in grape and wine which supports the health benefits due to their bioactive chemical diversity [48, 49]. *N*-ferulyl serotonin (FS) and *p*-coumaroyl serotonin (CS) have been isolated from safflower seeds and possess antioxidant, anti-inflamatory activities [50], anti-tumor, anti-bacterial [51] and anti-stress potential apart from being involved in reducing depression and anxiety [52]. SER has been administered to boost its levels in human brain in treating SER deficiency syndrome [53]. Some of the extensive uses of SER include treating patients suffering from the parkinson-like symptoms [53] and controlling obesity [54]. SER action of hallucinogenic drugs is known from the early 1950s to the present day. There is now converging evidence from biochemical, electro- physiological, and behavioral studies reported that the two major classes of psychedelic hallucinogens, (e.g. LSD) and the phenethylamines (e.g. mescaline) have a common site of action as partial agonists at 5-HT2A and other 5-HT2 receptors in the central nervous system [55].

Occurrence of Melatonin in Edible Plants

Occurrence of MEL in plants has been published by various researchers [8, 56-65]. To date, numerous publications have reported the presence of MEL in plants and plant products with a wide range of concentrations from picograms to micrograms per gram of tissue [66-71]. MEL has been identified in the most popular beverages of the world such as coffee [13], tea, wine, and beer [72-77]. Melatonin concentration in these crops varies from several nano grams per gram to several thousand picograms per gram. As in the most common beverages, melatonin has been discovered in widely consumed crops including corn, rice, wheat, barley, and oats [78]. Detailed studies have shown that even in the different varieties of the same species, substantial differences in melatonin concentrations exist, for example in different varieties of corn and rice [78].

Table 1. Occurrence of serotonin in edible plants

Plant	Serotonin content	Reference
Strawberry	3.77 ± 0.66 μg/g fw	Ly et al 2008
Sweet cherry		
Cultivar		[36]
Burlat	12.6 ng/100g fw	
Navalinda	30.7 ng/100g fw	
Van	10.6 ng/100g fw	
Pico Limon Negro	27.1 ng/100g fw	
Sweet heart	19.2 ng/100g fw	
Pico Negro	2.8 ng/100g fw	
Ambrunes	37.6 ng/100g fw	
Pico Colorado	36.6 ng/100g fw	
Instant coffee bevarege	0.04-1.80 mg/100g	[33]
Tomato	221.9 μg/g fw	[42]
Cherry tomato	156.1 μg/g fw	[42]
Hot pepper	17.9 μg/g fw	
Paprika	1.8 μg/g fw	[42]
Egg plant	1.5-12 μg/g fw	[43]
Green onion	8 ± 0.8 μg/g fw	[42]
Chicory	8.5± 3.2 μg/g fw	[42]
Pineapple	Fruit 1.5 μg/g fw	[44]
Papaya	Fruit 1.1-2.1 μg/g fw	[45]

HEALTH BENEFITS OF PHYTOMELATONIN

MEL effectively scavenges a variety of reactive oxygen species (ROS) and reactive nitrogen species (RNS), and protects cells, tissues and organisms from oxidative stress [79-82]. Most prominent foods rich in MEL in Mediterranean diets are nuts, grapes, cherries, tomato berries, species, olive-based products (black or green olive oil) and a variety of grains. Currently, it is believed that some of the health benefits from Mediterranean diets are derived from their phytochemicals including phenylpropanoids (or phenolic compounds), isoprenoids, alkaloids and MEL [75]. Putative effect of MEL in functional food (walnut) [19] and in officinal plant (St.John's wort) has been reported. It is reported that consumption of walnuts, increases blood MEL concentrations. An increase in circulating MEL in human subjects after consuming coffee has been reported [83].

The rise in blood MEL levels correlates with an increased antioxidative capacity of this fluid. This increment may also be protective against cardiovascular damage and cancer initiation or growth [19]. It is well documented that MEL strongly suppresses human prostate tumour cell proliferation in culture, as well as in animal models and in humans [84-87]. A human study has shown that individuals who consume vegetables rich in MEL increase their urinary MEL metabolite excretion, which is consistent with an elevated level of MEL in the blood [88]. Consumption of edible plant tissues that contain the SER and MEL will be useful as their anti-oxidant activity is linked to their medicinal value for treatment of human diseases [19].

Table 2. Occurrence of melatonin in edible plants

Plant	Melatonin (Pg/g FW)	Reference
Beet root	2	[77]
Cabbage	107.4	[70]
Pineapple	36.2	[70]
Cucuber fruit	24.6	[70]
Rice seed	1,006	[70]
Barley seed	378.1	[70]
Sweet corn	1,366	[70]
Banana	0.46	[67]
Apple	47.6	[70]
Strawberry	12.4	[70]
Sweet cherries	6-224	[36]
Carrot	55.3	[70]
Sunflower seed	29,000 dw	[68]
Tomato fruit	32.2	[70]
Cardmom seed	15,000	[68]
Curcuma	120,000	[68]
Ginger	583.7	[70]
Onion	31.5	[70]
Walnuts	3-4 ng	[19]

The presence of MEL in plant products and its consumption is reported to elevate MEL profiles of blood leading to improved antioxidant activity [19,75] and amphipathic potential [89]. Similarly, red wine contains significant amounts of MEL and its consumption increases the blood levels of this indoleamine in humans [90]. Phytomelatonin in walnuts is reported to contribute to enhancement of indoles in offering protection against cardiovascular damage [19]. Hardman *et al* (2008) [91] have reported the suppression of cancer growth by dietary walnut.

CONCLUSION

In view of the antioxidant power of SER and MEL, they may be of nutraceutical value apart from their role as neurotransmitters. Intake of melatonin-rich plants seems to have a promising future as a healthy photochemical. It would, therefore, seem a worthwhile task to search for plants with high levels of endogenous MEL and SER that could be used as a natural source of nutraceuticals. The consumption of these foods by humans or animals could have health benefits considering the important functions of melatonin as a potent free radical scavenger and antioxidant.

Table 3. Melatonin content in processed foods

Food	Melatonin content	Reference
Italian wines	0.4–0.5 ng/mL	[79]
Chinese green tea	2.12 μg/g Fw	[72]
Olive oil	53-119 pg/mL	[12]
Spanish wines	50–80 pg/mL	[12]
Red wine	50-80 pg/mL	[90]
Beer	52–170 pg/mL	[12]
Extra virgin olive oil	71–119 pg/mL	[69]

REFERENCES

[1] Tretyn A, Kendrick RE (1991) Acetylcholine in plants: presence, metabolism and mechanism of action. *Bot. Rev.*, 57:33–73.

[2] Roshchina VV (2001) Neurotransmitters in plant life. Science Publishers, Enfield, 4–81

[3] Veenstra-Vander Weele J, Anderson GM, Cook EH (2000) Pharmacogenetics and the serotonin system: initial studies and future directions. *Eur. J. Pharm.*, 410:165–81.

[4] Ramakrishna A, Giridhar P, Ravishankar GA (2011) Phytoserotonin: a review. *Plant Signal & Behav.*, 6:800–809.

[5] Kang S, Kang K, Lee K, Back K (2007) Characterization of tryptamine 5-hydroxylase and serotonin synthesis in rice plants. *Plant Cell Rep.*, 26:2009–2015.

[6] Foulkes NS, Borjigin J, Snyder SH, Sassone-Corsi P (1997) Rhythmic transcription: the molecular basis of circadian melatonin synthesis. *Trends Neurosci.*, 20:487–492.

[7] Reiter RJ, Peoggeler B, Tan DX, Chen LD, Manchester LC, Guerrero JM (1993) Antioxidant capacity of melatonin, a novel action not requiring a receptor. *Neuroendocrinol Lett.,* 15:103–116.

[8] Reiter RJ, Tan D-X, Burkhardt S, Manchester LC (2001) Melatonin in plants. *Nutr. Rev.,* 59:286–290 .

[9] Murch SJ, Campbell SSB, Saxena PK (2001) The role of serotonin and melatonin in plant morphogenesis: regulation of auxin induced root organogenesis in *in vitro*-cultured explants of St. John's wort (*Hypericum perforatum* L.). *In Vitro Cell Dev. Biol. Plant,* 37:786–793.

[10] Kirakosyan, A., Seymour, E. M., Llanes, D. E. U., Kaufman, P. B., & Bolling, S. F. (2009) Chemical profile and antioxidant capacities of tart cherry products. *Food Chem.,* 115: 20–25.

[11] Stürtz, M., Cerezo, A. B., Cantos-Villar, E., & Garcia-Parrilla, M. C. (2011) Determination of the melatonin content of different varieties of tomatoes (*Lycopersicon esculentum*) and strawberries (*Fragaria ananassa*). *Food Chem.,* 127: 1329–1334.

[12] Garcia-Parrilla, Cantos E, Troncoso AM (2009) Analysis of melatonin in foods. *J. Food Comp. Ana.,* 22:177–183.

[13] Ramakrishna A, Giridhar P, Udaya Sankar K and Ravishankar GA. Melatonin and serotonin profiles in beans of *Coffea* sps. *J. Pineal Res.,* (2012b) 52:470-476.

[14] Arnao MB (2014) Phytomelatonin: Discovery, Content, and Role in Plants. Adv Bot, 1-11, http://dx.doi.org/10.1155/2014/815769.

[15] Reiter RJ, Tan D-X, Mayo JC, Sainz RM, Leon J, Czarnocki Z (2003) Melatonin as an antioxidant: biochemical mechanisms and pathophysiological implications in humans. *Acta Biochim. Pol.,* 50:1129–1146.

[16] Pappolla MA, Chyan YJ, Poeggeler B, Frangione B, Wilson G, Ghiso J, Reiter RJ (2000) An assessment of the antioxidant and antiamyloidogenic properties of melatonin: implications for Alzheimer's disease. *J. Neural. Transmis,* 107:203–231.

[17] Sauer LA, Dauchy RT, Blask DE (2001) Polyunsaturated fatty acids, melatonin and cancer prevention. *Biochem. Pharmacol.,* 61:1455–1462.

[18] Casal S, Mendes E, Aalves RM, Alves RC, Ferreira MA (2004) Free and conjugated biogenic amines in green and roasted coffee beans. *J. Agric Food Chem.,* 52:6188–6192.

[19] Reiter RJ, Manchester LC, Tan DX (2005) Melatonin in walnuts: influence on levels of melatonin and total antioxidant capacity of blood. *Nutrition,* 21:920–924.

[20] Hardeland R, Pandi-Perumal SR (2005) Melatonin, a potent agent in ant oxidative defense: actions as a natural food constituent, gastrointestinal factor, drug and prodrug. *Nut. Met.,* 2:1–15.

[21] Bubenik GA (2002) Review: gastrointestinal melatonin: localization, function, and clinical relevance. *Diges Dis. Sci.,* 47:2336–2348.

[22] Janas KM, Posmyk MM. Melatonin, an underestimated natural substance with great potential for agricultural application, *Acta Physiol. Plant,* (2013) 35:3285–3292.

[23] Tan DX, Hardeland R, Manchester LC, Korkmaz A, Ma S, Rosales-Corral S, Reiter RJ (2012) Functional roles of melatonin in plants, and perspectives in nutritional and agricultural science. *J. Exp. Bot.,* 63:577–597.

[24] Fellows LE, Bell EA (1971) Indole metabolism in *Piptadenia peregrina.* Phytochem 10:2083–2091.

[25] Kema IP, de Vries E, Muskiet F (2000) Clinical chemistry of serotonin and metabolites. *J. Chromatogra B,* 747:33–38.

[26] Collier HOJ, Chesher GB (1956) Identification of 5-HT in the sting of the nettle (*Utrica dioica*). *Brit. J. Pharmacol. Chemother,* 11:186–189.

[27] Bowden K, Brown BG, Batty JE (1954) 5-hydroxytryptamine: Its occurrence in cowhage (*Mucuna prurience*). *Nature,* 17: 925–926.

[28] Udenfriend S, Weissbach H, Clark CT (1955) The estimation of 5-hydroxytryptamine (serotonin) in biological tissues. *J. Biol. Chem.,* 215: 337–344.

[29] Erspamer V (1961) Recent research in the field of 5-Hydroxy tryptamine and related indolealkylamines. In: Jucker E ed. Progress in Drug research, Basel, Stuttgart: Birkhaser Verlag, 151–67.

[30] Bergmann L, Grosse W, Ruppel HG (1970) Serotonin in *Juglans regia* L. Planta 94:47–59.

[31] Kele M, Ochmacht R (1996) Determination of serotonin released from coffee wax by Liquid Chromatography. *J. Chromatogra A,* 730:59–62.

[32] Strance A (1993) Astra – a natural coffee with a reduced irritant content. *Przemysl. Spozywezy,* 47:76–77.

[33] da Silveira TML, Tavares E, Gloria MBA (2007) Profiles and levels of bioactive amines in instant coffee. *J. Food Comp. Anal.,* 20:451–457.

[34] Cirilo MPG, Coelho AFS, Araujo CM, Goncalves FRB, Nogueira FD, Gloria MBA (2003) Profile and levels of bioactive amines in green and roasted coffee. *Food Chem.,* 82:397–402.

[35] Lima AS, Gloria MBA (1999) Aminas bioativas em alimentos. *Bol. Soc. Brasil Cien. Tecnol. Aliment, 33*:70–79.

[36] Gonzalez-Gomez D, Lozano M, Fernandez-Leon MF, Ayuso MC, Bernalte MJ, Odriguez, AB (2009) Detection and quantifcation of melatonin and serotonin in eight sweet Cherry cultivars (*Prunus avium* L.) *Eur Food Res. Technol.,* 229:223–229.

[37] Badria FA (2002) Melatonin, serotonin, and tryptamine in some Egyptian food and medicinal plants. *J. Med. Food,* 5:153–157.

[38] Vetorazzi G (1974) 5-hydroxytryptamine content of bananas and banana products. *Food Cosm. Toxicol.,* 12:107–113.

[39] Udenfriend S, Lovenberg W, Sjoerdsma A (1959) Physiologically active amines in common fruits and vegetables. *Arch. Biochem. Biophy.,* 85:487–490.

[40] Foy JM, Parratt JR (1962) Noradrenaline and 5-hydroxytryptamine in plantain (Musa sapientum, var. paradisiaca*). J. Pharm. Pharmaco.,* 113:360–364.

[41] Waalkes TP, Sjoerdsma A, Creveling CR, Weissbach H, Ddenfriend S (1958) Serotonin, norepinephrine and related compounds in banana. Science 127:648–650.

[42] Ly D, Kang K, Choi JY, Ishihara A, Back K, Lee SG (2008) HPLC analysis of serotonin, tryptamine, tyramine, and the hydroxycinnamic acid amides of serotonin and tyramine in food vegetables. *J. Med. Food,* 11:385–389.

[43] Udenfriend S, Weissbach H, Clark CT (1955) The estimation of 5-hydroxytryptamine (serotonin) in biological tissues. *J. Biol. Chem.,* 215: 337–44.

[44] West GB (1960) Carcinoid tumors and pineapples. J Pharm Pharmacol 12:768–769

[45] Erspamer V. Recent research in the field of 5-Hydroxy tryptamine and related indolealkylamines. In: Jucker E ed. Progress in Drug research, Basel, Stuttgart: Birkhaser Verlag, 1961; 151–67.

[46] Coutts RT, Baker GB, Pasutto FM (1986) Food stuffs as sources of psychoactive amines and their precursors: content, significance and identification. *Adv. Drug Res.,* 15:169–232.
[47] Spring B, Chiodo J, Bowen DJ (1987) Carbohydrates, tryptophan and behavior: a methodological review. *Psych. Bull.,* 102:234–256.
[48] Iriti M, Rossoni M, Faoro F (2006) Melatonin content in grape: myth or panacea?. *J. Sci. Food Agric.,* 86:1432–1438.
[49] Iriti M, Faoro F (2006) Grape phytochemicals: A bouquet of old and new nutraceuticals for human health. *Med. Hyp.,* 67:833–838.
[50] Hotta Y, Nagatsu A, Liu W, Muto T, Narumiya C, Lu X, Yajima M, Ishikawa N, Miyazeki K, Kawai N, Mizukami H, Sakakibara J (2002) Protective effects of antioxidative serotonin derivatives isolated from safflower against postischemic myocardial dysfunction. *Mole Cell Biochem.,* 238:151–162.
[51] Kumarasamya Y, Middletona M, Reida RG, Naharb L, Sarkera SD (2003) Biological activity of serotonin conjugates from the seeds of Centaurea nigra. *Fitoterapia,* 74:609–612.
[52] Yamamotova A, Pometlova M, Harmatha J, Raskova H, Rokyta R (2007) The selective effect of N-feruloylserotonins isolated from Leuzea carthamoides on nociception and anxiety in rats. *J. Ethno pharm.,*112: 368–374.
[53] Bell EA, Janzen DH (1996) Medical and ecological considerations of L-DOPA and 5-HTP in seeds. *Nature,* 210:529.
[54] Cangiano C, Ceci F, Cascino A, Del Ben M, Laviano A, Muscaritoli M, Antonucci F, Rossi-Fanelli F (1992) Eating behavior and adherence to dietary prescriptions in obese adult subjects treated with 5-hydroxytryptophan. *Ame. J. Clin. Nut.,* 56:863–867.
[55] Aghajanian GK, Marek GJ (1991) Serotonin, via 5-HT2A receptors, increases EPSCs in layer V pyramidal cells of prefrontal cortex by an asynchronous mode of glutamate release. *Brain Res.,* 825:161–171.
[56] Murch SJ, Simmons CB, Saxena PK (1997) Melatonin in feverfew and other medicinal plants. *Lancet,* 350:1598–1599.
[57] Harumi T, Matsushima S (2000) Separation and assay methods for melatonin and its precursors. *J. Chromatogra B,* 747:95–110.
[58] Van Tassel DL, Oneill SD (2001) Putative regulatory molecules in plants: Evaluating melatonin. *J. Pineal. Res.,* 31:1–7.
[59] Van Tassel DL, Roberts N, Lewy A, O'Neill SD (2001a) Melatonin in plant organs. *J. Pineal. Res.,* 31:8–15.
[60] Hardeland R, Poeggeler B (2003) Non-vertebrate melatonin. J Pineal Res 34:233–241
[61] Kolar J, Machackova I (2005) Melatonin in higher plants: occurrence and possible functions. *J. Pineal. Res.,* 39:333–341
[62] Posmyk MM, Janas KM (2009) Melatonin in plants. *Acta Phy. Plant,* 31:1–11.
[63] Ramakrishna A, Giridhar P, Ravishankar GA (2009) Indoleamines and calcium channels influence morphogenesis in *in vitro* cultures of *Mimosa pudica* L. *Plant Signal & Behav,* 12:1136–1141.
[64] Ramakrishna A, Giridhar P, Jobin M, Paulose CS and Ravishankar GA. Indoleamines and calcium enhance somatic embryogenesis in cultured tissues of *Coffea canephora* P ex Fr. *Plant Cell Tiss Org. Cul.,* (2012) 108: 267–278.

[65] Ramakrishna A, Giridhar P, Udaya Sankar K and Ravishankar GA. Endogenous profiles of indoleamines: serotonin and melatonin in different tissues of *Coffea canephora* P ex Fr. as analyzed by HPLC and LC-MS-ESI. *Acta Physiol. Plant,* (2012a) 34: 393–396.

[66] Tettamanti C, Cerabolini B, Gerola P, Conti A (2000) Melatonin identification in medicinal plants. *Acta Phytotherapeutica,* 3:137–144.

[67] Dubbels R, Reiter RJ, Klenke E, Goebel A, Schnakenberg E, Ehlers C, Schiwara HW, Schloot W (1995) Melatonin in edible plants identified by radioimmunoassay and by high performance liquid chromatography-mass spectrometry. *J. Pineal. Res.,* 18:28–31.

[68] Manchester LC, Tan D-X, Reiter RJ, Park W, Monis K, Qi WB (2000) High levels of melatonin in the seeds of edible plants - Possible function in germ tissue protection. *Life Sci.,* 67:3023–3029.

[69] Paredes SD, Korkmaz A, Manchester LC, Tan DX, Reiter RJ (2009) Phytomelatonin: a review. *J. Exp. Botany,* 60:57–69.

[70] Hattori A, Migitaka H, Iigo M, Itoh M, Yamamoto K, Ohtani-Kaneko R, Hara M, Suzuki T, Reiter RJ (1995) Identification of melatonin in plants and its effects on plasma melatonin levels and binding to melatonin receptors in vertebrates. *Biochem. Mole Biology Int.,* 35:627–634.

[71] Hernandez-Ruiz A, Cano A, Arnao MB (2005) Melatonin acts as a growth-stimulating compound in some monocot species. *J. Pineal. Res.,* 39:137–142.

[72] Chen G, Huo Y, Tan DX, Liang Z, Zhang W, Zhang Y (2003) Melatonin in Chinese medicinal herbs. *Life Sci.,* 73:19–26.

[73] Mercolini L, Addolorata Saracino M, Bugamelli F, Ferranti A, Malaguti M, Hrelia S, Raggi MA (2008) HPLC-F analysis of melatonin and resveratrol isomers in wine using an SPE procedure. *J. Separation. Sci.,* 31:1007–1014.

[74] Iriti M (2009) Melatonin in grape, not just a myth, may be a panacea. *J. Pineal. Res.,* 46:353.

[75] Iriti M, Varoni EM, Vitalini S (2010) Melatonin in traditional Mediterranean diets. *J. Pineal. Res.,* 49:101–105.

[76] Maldonado MD, Moreno H, Calvo JR (2009) Melatonin present in beer contributes to increase the levels of melatonin and antioxidant capacity of the human serum. *Clinical Nutrition,* 28:188–191.

[77] Stege PW, Sombra LL, Messina G, Martinez LD, Silva MF (2010) Determination of melatonin in wine and plant extracts by capillary electrochromatography with immobilized carboxylic multi-walled carbon nanotubes as stationary phase. *Electrophoresis,* 31:2242–2248.

[78] Wang J, Liang C, Li S, Zheng J (2009) Study on analysis method of melatonin and melatonin content in corn and rice seeds. *Chinese Agricultural Science Bulletin,* 25: 20–24.

[79] Tan DX, Chen LD, Poeggeler B, Manchester LC, Reiter RJ (1993) Melatonin: a potent, endogenous hydroxyl radical scavenger. *Endocrine J.,* 1:57–60.

[80] Kaur C, Sivakumar V, Ling EA (2010) Melatonin protects periventricular white matter from damage due to hypoxia. *J. Pineal. Res.,* 48:185–193.

[81] Laothong U, Pinlaor P, Hiraku Y, Boonsiri P, Prakobwong S, Khoontawad J, Pinlaor S (2010) Protective effect of melatonin against Opisthorchis viverrini-induced oxidative and nitrosative DNA damage and liver injury in hamsters. *J. Pineal. Res.,* 49:271–282.

[82] Nopparat C, Porter JE, Ebadi M, Govitrapong P (2010) The mechanism for the neuroprotective effect of melatonin against methamphetamine-induced autophagy. *J. Pineal. Res.,* 49:382–389.

[83] Ursing C, Wikner J, Brismar K, Rojdmark S (2003) Caffeine raises the serum melatonin level in healthy subjects: an indication of melatonin metabolism by cytochrome P450 (CYP)1A2. *J. Endo Inves,* 26:403–406.

[84] Xi SC, Siu SW, Fong SW, Shiu SY (2001) Inhibition of androgen sensitive LNCaP prostate cancer growth in vivo by melatonin: association of antiproliferative action of the pineal hormone with mt1 receptor protein expression. *Prostate,* 46:52–61.

[85] Shiu SY, Law IC, Lau KW, Tam PC, Yip AW, Ng WT (2003) Melatonin slowed the early biochemical progression of hormone-refractory prostate cancer in a patient whose prostate tumor tissue expressed MT1 receptor subtype. *J. Pineal. Res.,* 35:177–182.

[86] Joo SS, Yoo YM (2009) Melatonin induces apoptotic death in LNCaP cells via p38 and JNK pathways: therapeutic implications for prostate cancer. *J. Pineal. Res.,* 47: 8–14.

[87] Jung-Hynes B, Schmit TL, Reagan-Shaw SR, Siddiqui IA, Mukhtar H, Ahmad N (2011) Melatonin, a novel Sirtl inhibitor, imparts anti-proliferative effects against prostate cancer *in vitro* inculture and *in vivo* in TRAMP model. *J. Pineal. Res.,* 50:140–149.

[88] Oba S, Nakamura K, Sahashi Y, Hattori A, Nagata C (2008) Consumption of vegetables alters morning urinary 6-sulfatoxymelatonin concentration. *J. Pineal. Res.,* 45:17–23.

[89] Borah A, Mohanakumar K P (2009) Melatonin inhibits 6-hydroxydopamine production in the brain to protect against experimental parkinsonism in rodents. *J. Pineal. Res.,* 47:293–300.

[90] Guerrero, JM, Martinez-Cruz F, Elorza, FL (2008) Significant amounts of melatonin in red wine: Its consumption increases blood melatonin levels in humans. Food Chem doi:10.1016/j.*foodchem.,* 2008.02.007.

[91] Hardman WE, Ion G (2008) Suppression of implanted MDA-MB 231 human breast cancer growth in nude mice by dietary walnut. *Nutr. Cancer,* 60:666–674.

INDEX

#

A

B

C

D

G

H

I

J

K

L

M

N

O

P

Q

R

S

T

U

V

W

X

Y

Z